Laboratory Manual for

FUNDAMENTALS OF CHEMISTRY 3/E

Jo A. Beran
Texas A&I University
Kingsville, Texas

JOHN WILEY & SONS

New York Chichester Brisbane Toronto Singapore

Photo Credits

BRADY/HOLUM, *FUNDAMENTALS OF CHEMISTRY*, THIRD EDITION, JOHN WILEY & SONS, 1987

MF 18.1, page 187
MF 26.1, page 264

BRADY/HUMISTON, *PRINCIPLES AND STRUCTURE, GENERAL CHEMISTRY, FOURTH EDITION*, JOHN WILEY & SONS, 1986

Figure 12.1, page 127

FISHER SCIENTIFIC

"old lab", page ii
Figure T.13a, page 16
Figure T.13c, page 17
MF1.1, page 21
Hazard codes, page 32
Figure 2.3, page 38
Modern calorimeter, page 120
Figure 14.1, page 148
Vapor density bulb, page 170
Figure 27.1, page 273
Figure 27.2, page 273
Computer-aided titrimeter, page 286

METTLER INSTRUMENT CORPORATION

Figure T.13b, page 17

AMEND, *INTRODUCTORY CHEMISTRY: MODELS AND BASIC CONCEPTS*, JOHN WILEY & SONS, 1977

Figure 12.2, page 127

COURTESY THE DRACKETT COMPANY

Vanish Bowl Cleaner, page 292

ISBN 0-471-62798-4

Printed in the United States of America

10 9 8 7 6 5 4 3 2 1

PREFACE TO THE STUDENT

The chemistry laboratory will be one of the most interesting and rewarding academic experiences you will encounter as a college student. It will be one of those experiences that will enable you to apply the material presented in the textbook and in the lecture to observations made in the laboratory environment. The data that are collected in the laboratory can be observed, discussed, and interpreted in terms of these learned basic chemical principles. Predictions will be made as a result of the assimilation of the data. Solving for "unknowns" will be gratifying when you realize that your accumulated chemical knowledge is correctly used in your analysis of the observations.

The chemistry laboratory is also a social experience. Working with others while attempting to gain an understanding of your laboratory observations allows you to perceive the conceptual understanding of basic chemical principles from various viewpoints. More simply, sharing the rewards and frustrations of a laboratory assignment promotes lasting friendships. Recollections of the "lab" at class reunions or college social events vary from "the acid holes in the new pair of jeans" to "the percent antacid in Rolaids or the percent vitamin C in a vitamin tablet"–memorable events seem to happen in the laboratory. Its fun, challenging, rewarding, and memorable–enjoy it!

This manual covers a full year of general chemistry in which you may expect to spend an average of three hours per week in the laboratory. Advanced preparation for each experiment may decrease this time; no preparation may extend this time to four or five hours! Although the manual parallels the material in Brady and Holum's *Fundamentals of Chemistry, Third Edition,* the experiments are chosen and written so they may easily be adapted to any general chemistry text.

These experiments were chosen for you to gain an appreciation and an understanding of chemistry at a level that is necessary for your chosen major field of study, provided it is science–related. You need *not* be a chemistry major for a successful understanding of the material in this manual. The use of proper, but basic, laboratory techniques, the use and construction of the proper laboratory apparatus, and the strict adherence to proper laboratory safety procedures are points of emphasis throughout the manual.

Each experiment has six major divisions: **Objectives, Principles, Techniques, Procedure, Lab Preview,** and **Data Sheet**:

- OBJECTIVES. These, stated at the beginning of each experiment, indicate the purpose of the experiment and just what you are to learn and accomplish from the experiment.

- PRINCIPLES. Several paragraphs describe the nature of the material that is being investigated in the experiment. Appropriate equations, tests, colors, etc., are presented and/or illustrated.

- TECHNIQUES. Fifteen basic laboratory techniques are described at the front of the manual. Techniques that are appropriate for the successful completion of the experiment are listed. You will need to refer to these quite often.

- LAB PREVIEW. You will need to read and investigate *thoroughly* the Principles, Procedure, and Techniques sections of the experiment before the laboratory and answer corresponding questions *before* coming to the laboratory session. Its careful completion reduces the "waste of time and boredom" that the poorly–

prepared student experiences.

•DATA SHEET. This provides an outline for you to organize the collected data and to complete the calculations that are required. Questions reviewing the experiment and probing your interpretations appear on the Data Sheet. You will learn to depend on your data collection and analysis of the data for these answers.

Careful attention to safety is an essential component in the design of each experiment. While all potentially hazardous chemicals cannot be eliminated from the laboratory, their numbers are minimal in this manual. Those that may present a danger are flagged with a **Caution**, followed by a brief warning. Ultimately, it is your responsibility to practice safe laboratory guidelines, not only for your personal safety but for your friends and neighbors as well.

Preface to the Laboratory Instructor

Each experiment requires only basic chemicals and apparatus. Where appropriate, the apparatus required for the experiment or technique is shown. All of the experiments are considered "**safe**"; it is the joint responsibilities of you and the students to maintain a safe laboratory environment. Do not underestimate the value of safe laboratory!

Lists of the common and special laboratory equipment required in the use of this manual are at the front of the manual. Balances with ±0.001g sensitivity, visible spectrophotometers, and a slide projector are needed on occasion. A **special note** to adopters of the manual: the spectral slides in Experiment 12 are available from John Wiley & Sons, Publishers, Inc. upon request.

A **Teacher's Manual** is available upon request to the publisher. An effective teaching schedule is outlined for each experiment, including: a suggested prelab lecture outline, cautions, representative or expected data, common student questions about the experiment, answers to the Lab Preview and Data Sheet Questions, the chemicals and special equipment that are required, the preparation of reagent solutions, and the safety rules.

Several experiments were modified from those appearing elsewhere. The author thanks the following for their permission to include them, in principle, here.

James E. Brady, St. John's University, N.Y.
John R. Amend, Montana State University, Experiment 12
Jerry Mills, Texas Tech University, Experiment 22
John Holum, Augsburg College, Experiment 41

The valuable suggestions provided by the following reviewers were greatly appreciated:

Joan Reeder, Eastern Kentucky University
Robert Kowerski, College of San Mateo

Special recognition is given to Dennis Sawicki, Chemistry Editor with Wiley, for his guidance, patience, and valuable suggestions during the course of the project; the general chemistry faculty and staff at the University of Colorado who offered suggestions for this revision; and, most importantly, the general chemistry students, laboratory instructors, and my colleagues at Texas A&I University.

Finally the author appreciates the continued support of Judi, Kyle, and Greg. Without their patience, love, and understanding this manual would not have been satisfactorily completed.

The author invites corrections and suggestions for improvements of this manual from colleagues and students.

Jo A. Beran
Department of Chemistry, Box 161
Texas A&I University
Kingsville, Texas 78363

CONTENTS

EQUILIBRIA (PRINCIPLES AND ANALYSIS)

ELECTROCHEMISTRY

KINETICS

A SUMMARY OF INORGANIC CHEMICAL PRINCIPLES

ORGANIC CHEMISTRY AND BIOCHEMISTRY

Appendices

Common Laboratory Desk Equipment

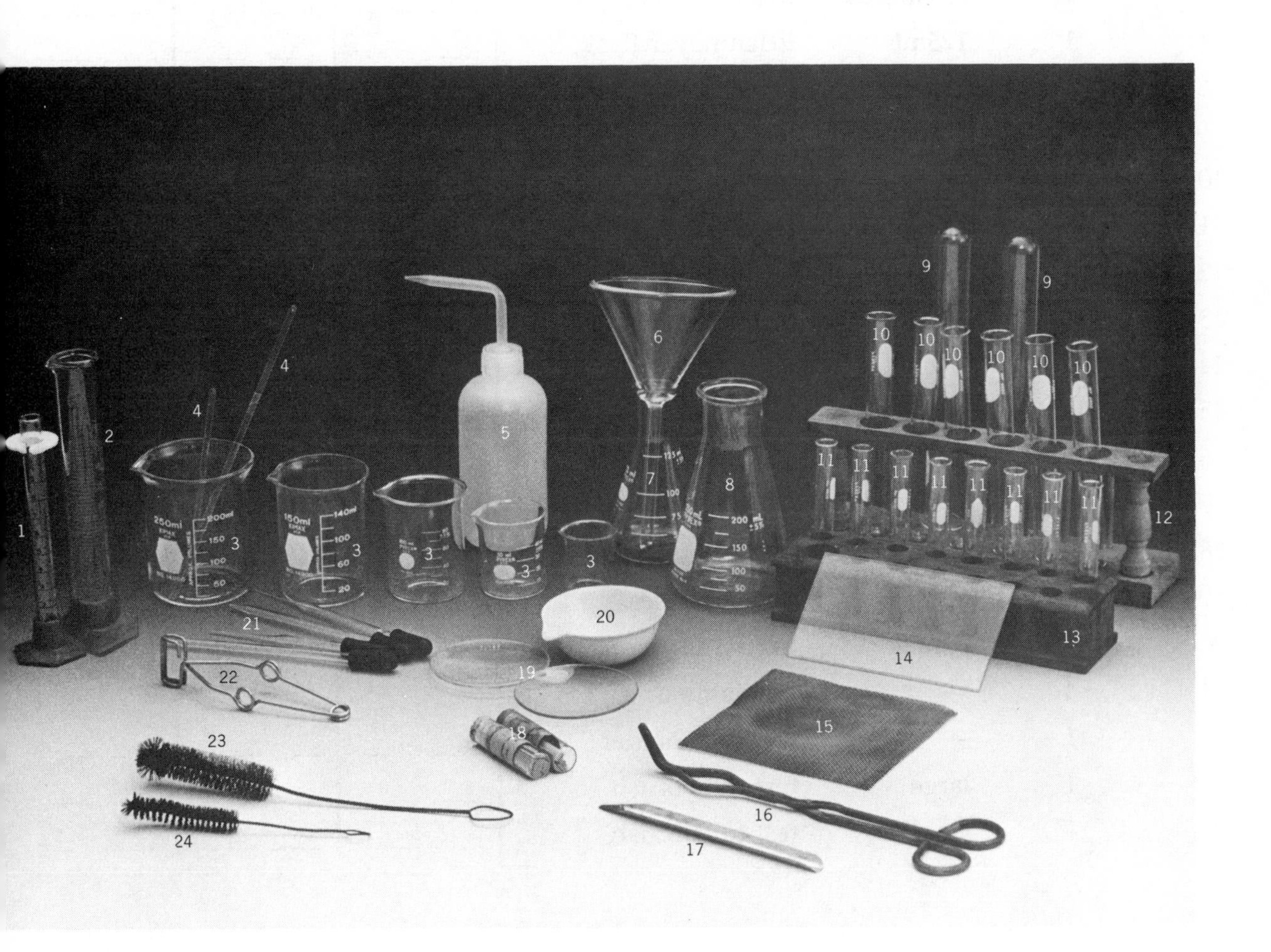

Common Laboratory Desk Equipment

No.	Quantity	Size	Item	First Term		Second Term		Third Term	
				In	Out	In	Out	In	Out
1	1	10mL	graduated cylinder	___	___	___	___	___	___
2	1	50mL	graduated cylinder	___	___	___	___	___	___
3	5	—	beakers	___	___	___	___	___	___
4	2	—	stirring rods	___	___	___	___	___	___
5	1	500mL	wash bottle	___	___	___	___	___	___
6	1	75mm, 60°	funnel	___	___	___	___	___	___
7	1	125mL	Erlenmeyer flask	___	___	___	___	___	___
8	1	250mL	Erlenmeyer flask	___	___	___	___	___	___
9	2	25x200mm	test tubes	___	___	___	___	___	___
10	6	18x150mm	test tubes	___	___	___	___	___	___
11	8	10x75mm	test tubes	___	___	___	___	___	___
12	1	large	test tube rack	___	___	___	___	___	___
13	1	small	test tube rack	___	___	___	___	___	___
14	1	—	glass plate	___	___	___	___	___	___
15	1	—	wire gauze	___	___	___	___	___	___
16	1	—	crucible tongs	___	___	___	___	___	___
17	1	—	spatula	___	___	___	___	___	___
18	2	—	litmus, red and blue	___	___	___	___	___	___
19	2	90mm	watch glasses	___	___	___	___	___	___
20	1	75mm	evaporating dish	___	___	___	___	___	___
21	4	—	medicine droppers	___	___	___	___	___	___
22	1	—	test tube holder	___	___	___	___	___	___
23	1	large	test tube brush	___	___	___	___	___	___
24	1	small	test tube brush	___	___	___	___	___	___

SPECIAL LABORATORY EQUIPMENT

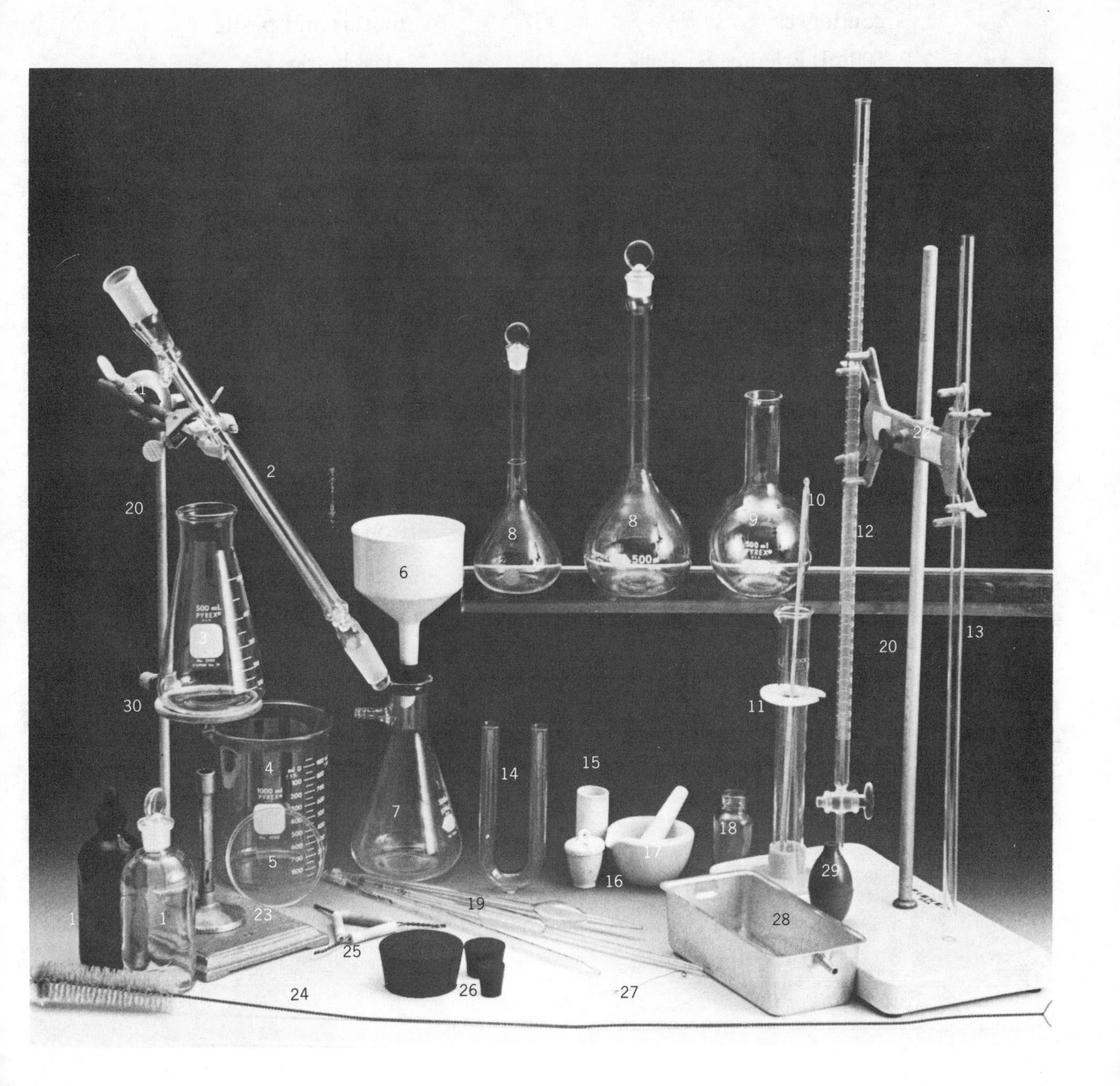

Special Laboratory Equipment

Number	Item	Number	Item
1	reagent bottles	16	crucible and cover
2	condenser	17	mortar and pestle
3	500mL Erlenmeyer flask	18	glass bottle
4	1000mL beaker	19	pipets
5	Petri dish	20	ring stands
6	Buchner funnel	21	buret clamp
7	Buchner flask	22	double buret clamp
8	volumetrick flasks	23	Bunsen burner
9	500mL Florence flask	24	buret brush
10	110°C thermometer	25	clay triangle
11	100mL graduated cylinder	26	rubber stoppers
12	50mL buret	27	wire loop for flame test
13	glass tubing	28	pneumatic trough
14	U–tube	29	rubber pipet bulb
15	porous cup	30	iron ring

LABORATORY TECHNIQUES

1. Inserting Glass Tubing (Thermometer or Glass Funnel) Through a Rubber Stopper

This procedure, when performed incorrectly, causes more serious injuries than any other single operation in the general chemistry laboratory. Please read this technique *CAREFULLY*!

Moisten the glass tubing, that has been previously firepolished, *and* the hole in the stopper with water or glycerol. Place your hand on the tubing 2 to 3 centimeters (1 inch) from the stopper. Protect your hands with a towel as shown in Figures T.1a, b, and c. Simultaneously *twist* and *push, slowly* and *carefully* the tubing through the hole. **Never** should there be more than 3 cm of glass tubing between your hand and the stopper. Wash off the excess glycerol.

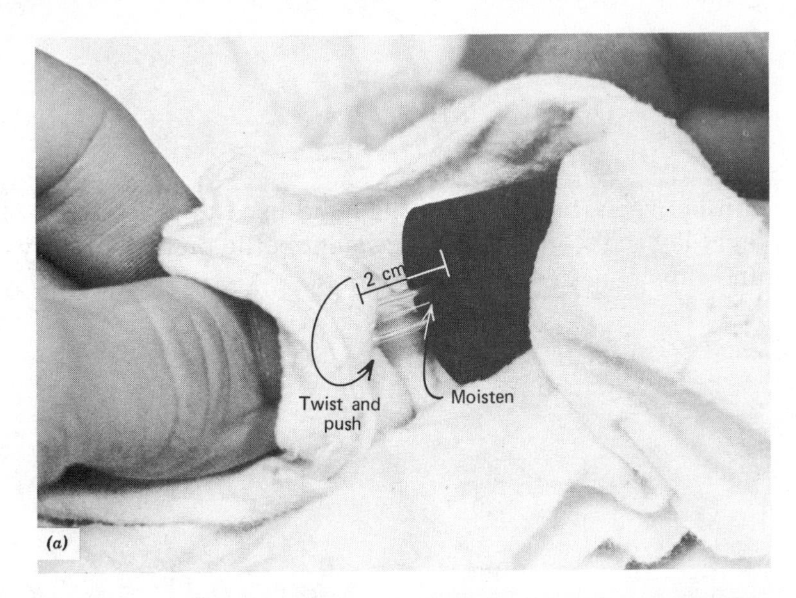

Figure T.1a
Inserting Glass Tubing into a Rubber Stopper

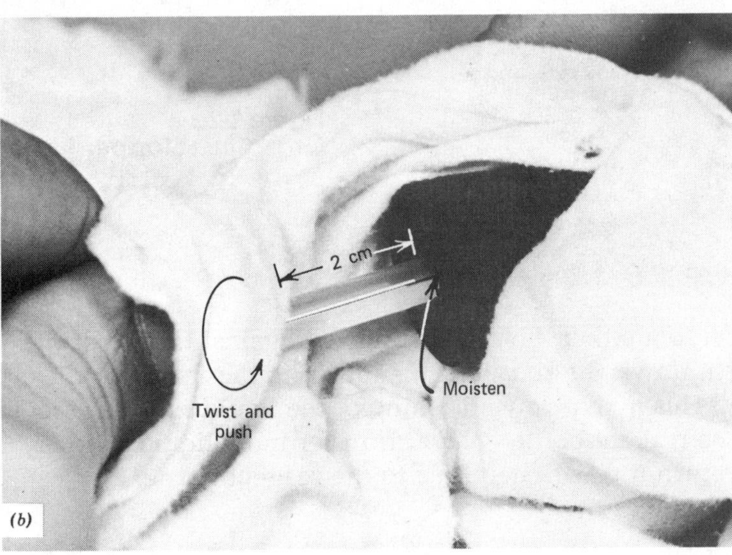

Figure T.1b
Inserting a Thermometer into a Rubber Stopper

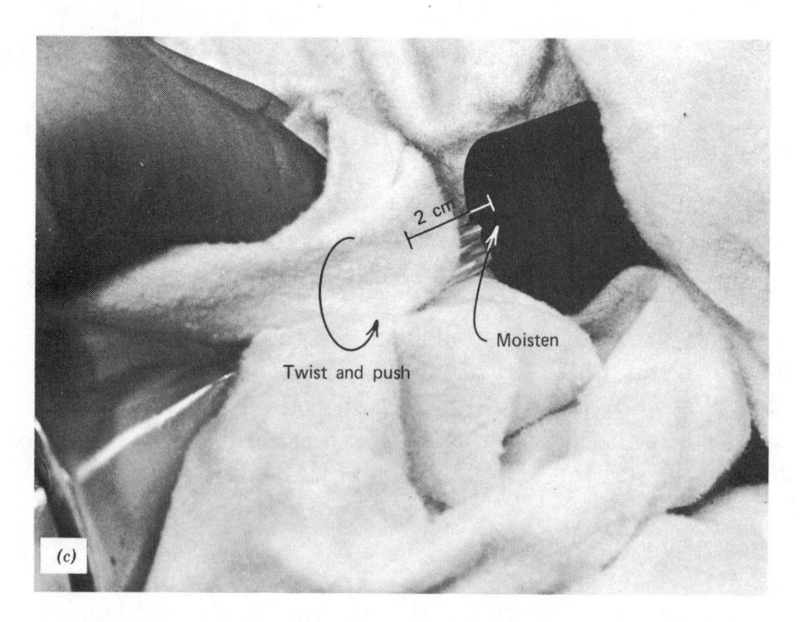

Figure T.1c
Inserting a Funnel into a Rubber Stopper

2. Transferring Liquid Reagents

Remove the glass stopper and hold it between the fingers of the hand used to grasp the reagent bottle (Figures T.2a and b). **Never** lay the stopper from a reagent bottle on a laboratory bench; impurities may be picked up and thus contaminate the solution.

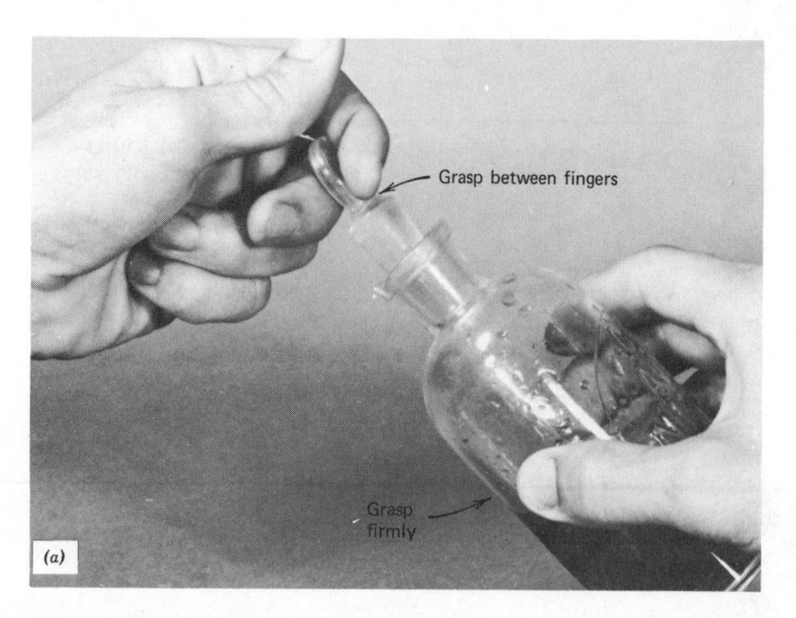

Figure T.2a
Grasp the Stopper

To transfer a liquid from one vessel to another, hold a stirring rod against the lip of the vessel containing the liquid and pour it down the rod which is touching the inside of the receiving vessel (Figures T.2b and c). This avoids any splashing of the liquid in the vessel and any loss of reagent down the side of the reagent bottle. **Never** transfer more liquid than is required for the experiment. **Never** return unused chemicals to the reagent bottle.

Figure T.2b
Pour the Liquid with the Aid of a Stirring Rod

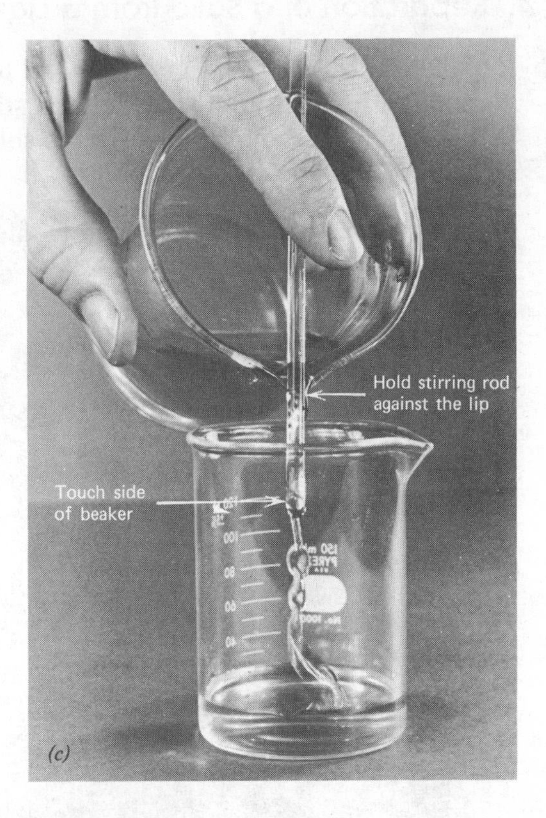

Figure T.2c
Transferring a Liquid with the Aid of a Stirring Rod

3. Transferring a Solid

First, read the label on the bottle to ensure the use of the correct reagent. Place the lid (glass stopper or screw cap) of the reagent bottle top-side-down. Hold the bottle with the label against your hand, tilt, and roll back and forth (Figure T.3) until the desired amount has been dispensed; try not to dispense more reagent than is needed. If too much reagent is removed, do not return the excess, but rather, share it with a friend. Do not insert a spatula or other object into the bottle. When finished return the lid.

Figure T.3
Dispensing a Solid from a Reagent Bottle

4. Separation of a Solid from a Liquid

a. DECANTATION. Allow the solid to settle in the beaker (Figure T.4a) or test tube. Then transfer the liquid with the aid of a stirring rod (Figure T.4b) as described in Technique 2. Do this slowly so as not to disturb the solid that has settled.

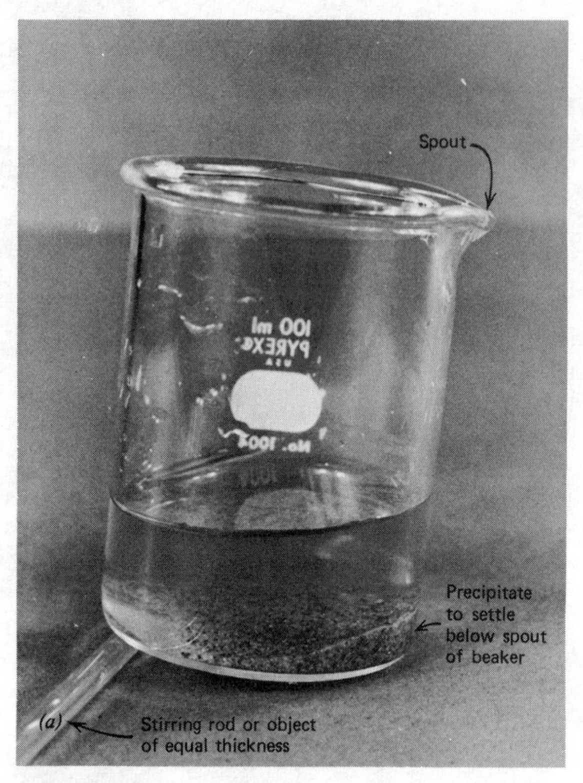

Figure T.4a
Settling of a Precipitate

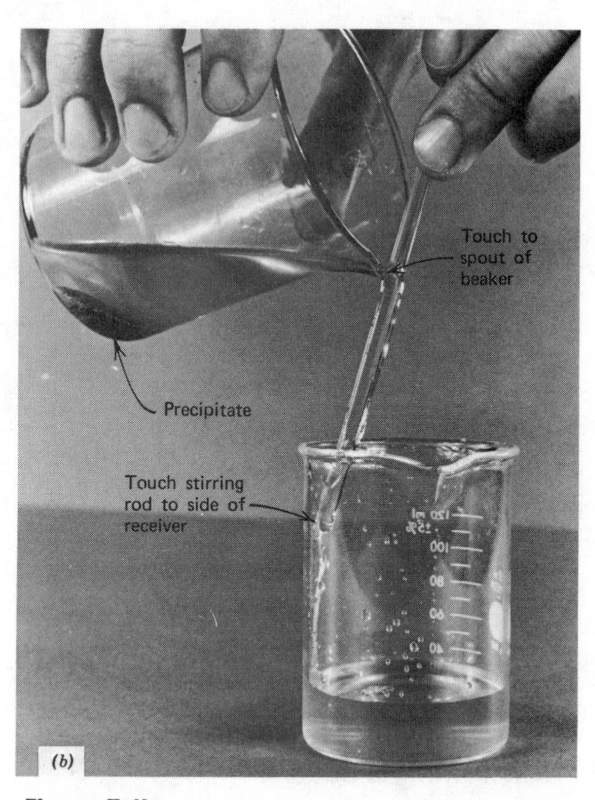

Figure T.4b
Transferring the Supernatant Liquid from a Precipitate

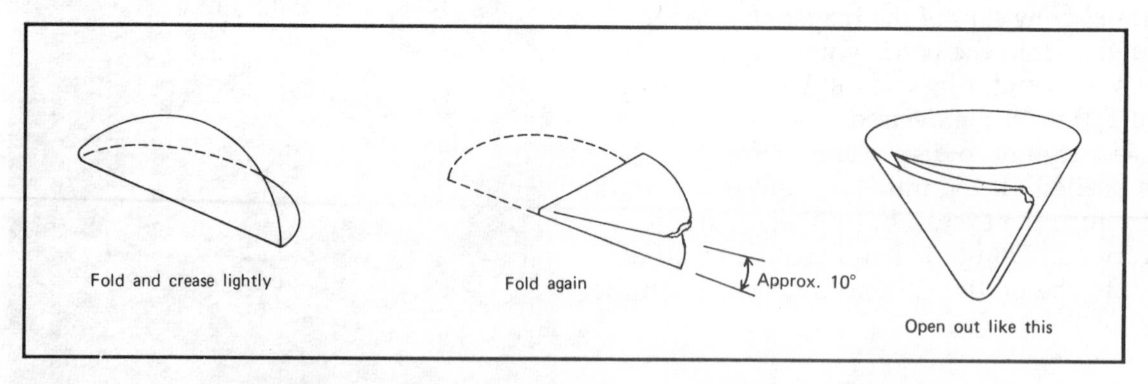

Figure T.4c
Sequence of Folding Filter Paper for a Gravity– Filtering Funnel

b. GRAVITY FILTRATION. Fold the filter paper in half (Figure T.4c); refold to within about 10° of a 90° fold; tear off the corner unequally; open. Place the folded filter paper snugly into the funnel. Moisten with the solvent being filtered and press the filter paper against the funnel's top wall to form a seal. Filter as shown in Figure T.4d. The funnel's tip should touch the beaker wall to reduce splashing. **Never** fill more than two-thirds full. Keep the funnel's stem filled with **filtrate** (the liquid passing through the funnel); the filtrate's weight creates a slight suction that hastens the filtration process. Steady the funnel with a support clamp.

Figure T.4e
Vacuum Filtering Apparatus

Figure T.4d
Gravity Filtering Apparatus

c. VACUUM FILTRATION. Figure T.4e shows a typical setup for a vacuum filtration. Although a gravity funnel/filter apparatus can be used, the apex of the filter paper is easily ruptured when the vacuum is applied. A Buchner funnel is normally used; a disc of filter paper fits over the flat, perforated bottom of this funnel. Applying a light suction to the filter paper, moistened with the solvent, creates a seal. When applying a vacuum with a water aspirator, open fully the faucet so that a maximum suction can be applied.

d. CENTRIFUGATION (USE OF A CENTRIFUGE). Precipitates that form in solution can be made more compact at the bottom of a test tube (or centrifuge tube) by using a centrifuge (Figure T.4f). The supernatant is then easily decanted without any loss of the precipitate (Figure T.4g). This quick and efficient separation requires 20 to 40 seconds.

Observe these precautions while operating a centrifuge:

- Never fill the centrifuge tubes to a height more than 1 cm from the top.
- Label the test (centrifuge) tubes to avoid confusion.
- Always operate with an even number of test (centrifuge) tubes, containing equal volumes of liquid, placed opposite one another; this *balances* the centrifuge and eliminates excessive vibration and wear. If only one tube needs to be centrifuged, balance it with a tube containing the same volume of solvent (Figure T.4h)

Figure T.4f
Centrifuge

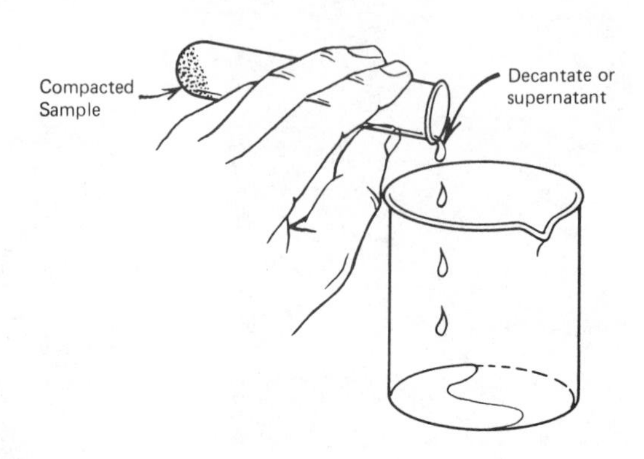

Figure T.4g
Decanting the Supernatant Liquid from a Compacted Precipitate in the Bottom of a Centrifuge Tube

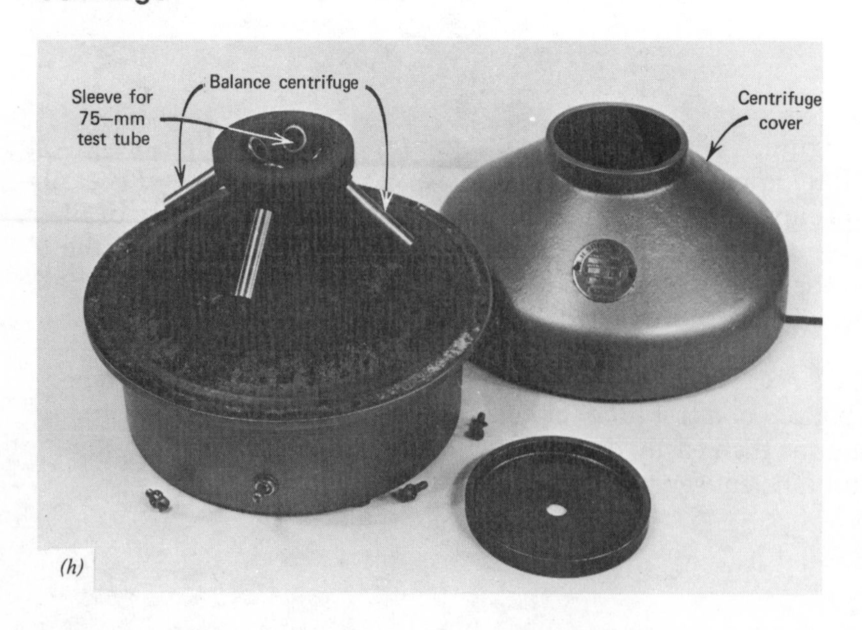

Figure T.4h
Balance the Centrifuge. Place Tubes with Equal Volumes of Solution Opposite Each Other Inside the Rotor's Metal Sleeves.

5. Flushing a Compacted Precipitate from a Beaker or Test Tube

Flush the precipitate with a wash bottle while holding the beaker or test tube over the receiving container (Figure T.5).

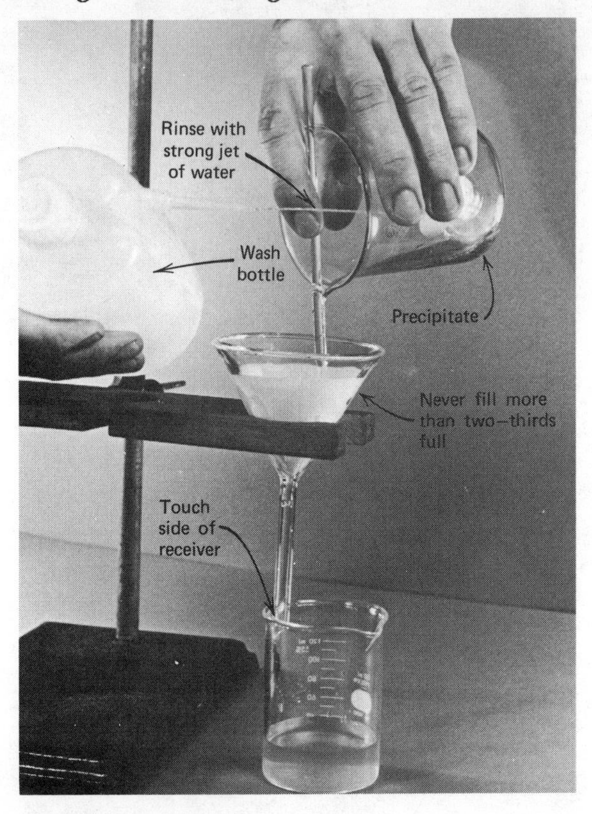

Figure T.5
Flushing a Precipitate from a Beaker

6. Heating Liquids

a. TEST TUBE. The test tube should be no more than one-third full. Place the flame at the same level as the top of the liquid, not at the base (Figure T.6a). Move the test tube in and out of the flame while swirling the contents. **Never** point the test tube toward anyone; the contents may be ejected violently if the test tube is not properly heated.

b. ERLENMEYER FLASK. An Erlenmeyer flask may be heated directly over a flame. Hold it with a piece of tightly folded paper or tongs and gently swirl (Figure T.6b). **Do not** place the hot flask on the laboratory bench; allow to cool by setting the flask on a wire gauze.

c. BEAKER (OR FLASK). Support the beaker (or flask) on a wire gauze. Place a glass stirring rod in the beaker; this avoids the problem of bumping (the sudden formation of superheated vapor near the flame). Position the flame directly under the tip of the stirring (Figure T.6c).

d. IN A HOT WATER BATH. A small quantity of solution in a test tube, that needs to be heated to a constant temperature, can be placed in a hot water bath (Figure T.6d). If the solution is in a beaker or Erlenmeyer flask instead of a test tube, place it in the next available-sized beaker, one–fourth filled with water, and heat to the desired temperature.

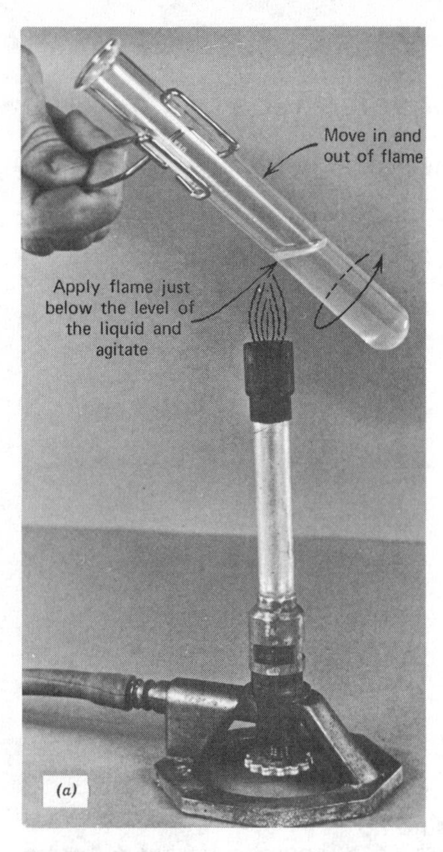

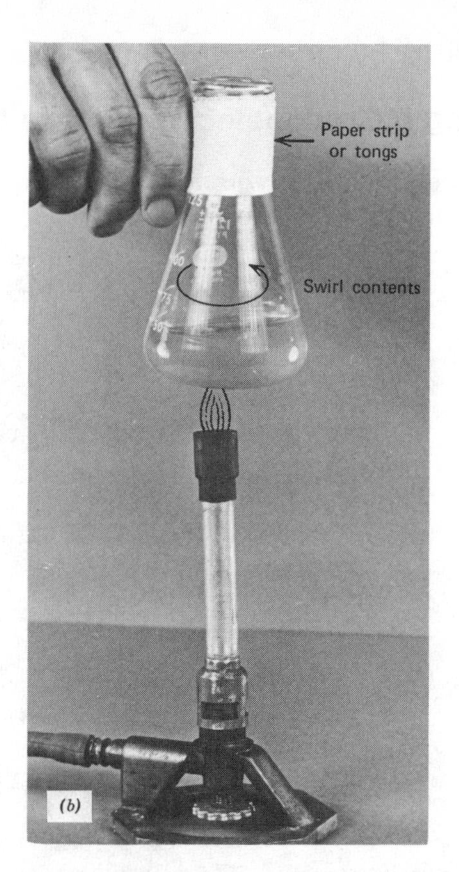

Figure T.6a
Heating a Liquid in a Test Tube with a Burner–Apply the Flame at the *Top* of the Liquid

Figure T.6b
Heating a Liquid in a Flask Directly Over the Flame while *Swirling*

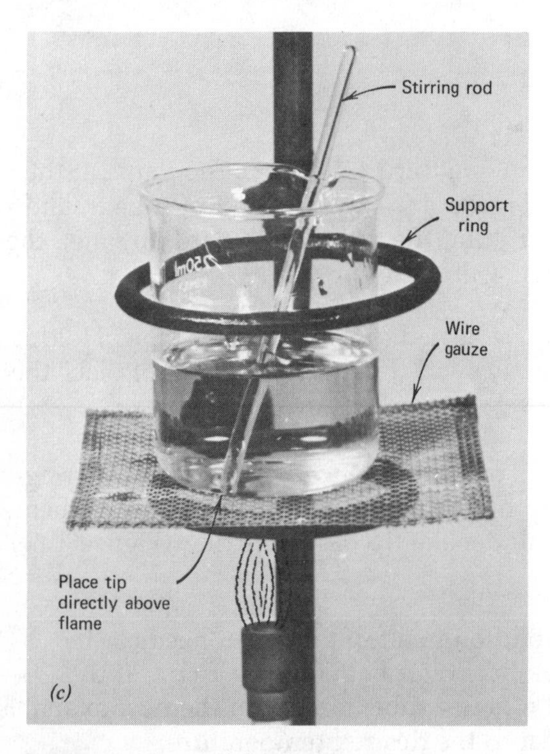

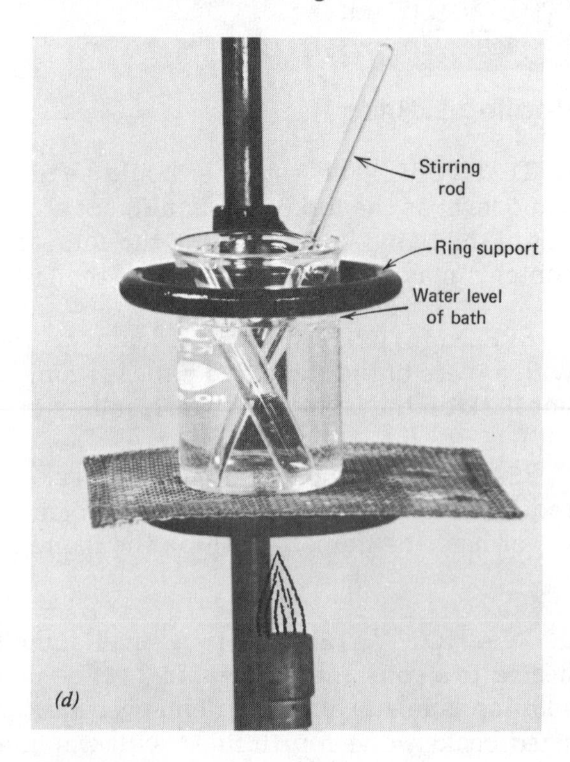

Figure T.6c
Place the Flame Directly Beneath the *Tip* of the Stirring Rod in the Beaker

Figure T.6d
A Hot Water Bath

7. Evaporation of Liquids

a. Nonflammable liquids may be evaporated from an evaporating dish with a *gentle* direct flame (Figure T.7a) or over a steam bath (Figure T.7b). Gentle boiling is more efficient that rapid boiling.

b. Flammable liquids may be similarly evaporated from an evaporating dish using a heating mantle (Figure T.7c) The use of a fume hood (Technique 11) is suggested if large amounts are evaporated in a laboratory with inadequate ventilation; consult with your laboratory instructor.

8. Ignition of a Crucible

a. DRYING AND/OR FIRING THE CRUCIBLE. Support the crucible on a clay triangle (Figure T.8a) and heat in a hot flame until it glows red. Rotate the crucible with tongs to ensure complete ignition. Allow the crucible and lid to cool to room temperature. Use crucible tongs for any transfer of the crucible and lid. If the crucible still remains dirty, add 2 mL of 6 M HNO_3 (**Caution**: *Avoid skin contact*) and evaporate to dryness.

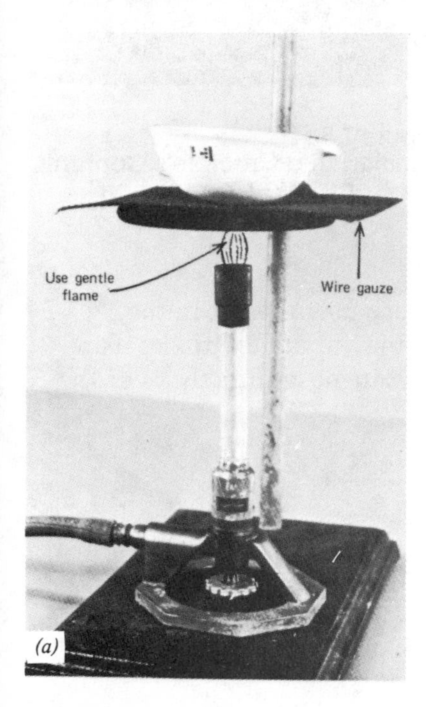

Figure T.7a
Evaporation of a *Non*flammable Liquid Over a Low, Direct Flame

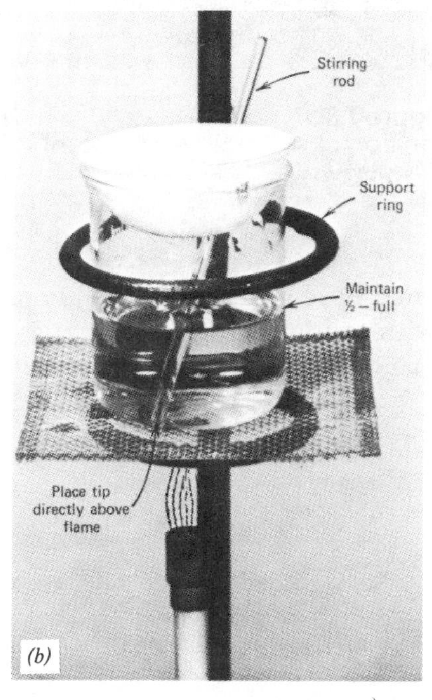

Figure T.7b
Evaporation of *Non*flammable Liquid Over a Steam Bath

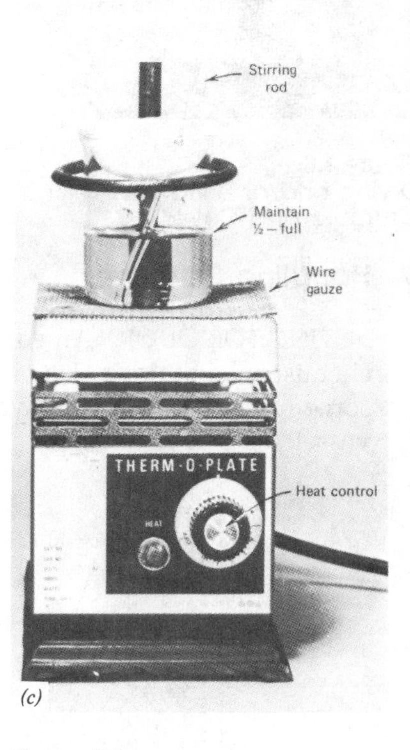

Figure T.7c
Evaporation of a *Flammable* Liquid Over a Steam Bath

b. IGNITION OF CONTENTS IN THE ABSENCE OF AIR. Set the crucible upright in the clay triangle; adjust the lid slightly off the lip of the crucible (Figure T.8b). Use the tongs for adjustment.

c. IGNITION OF CONTENTS FOR COMPLETE COMBUSTION. Tilt the crucible and adjust the lid so that only about half of the crucible is covered (Figure T.8c).

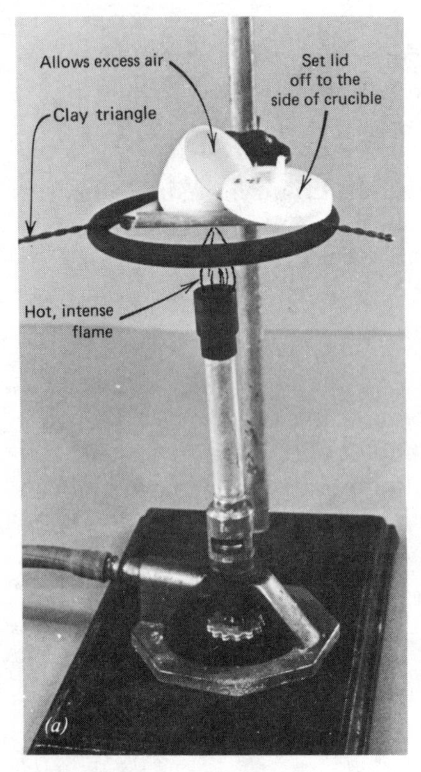

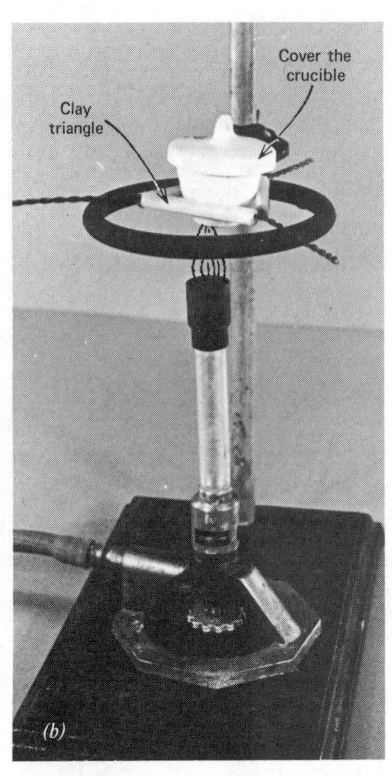

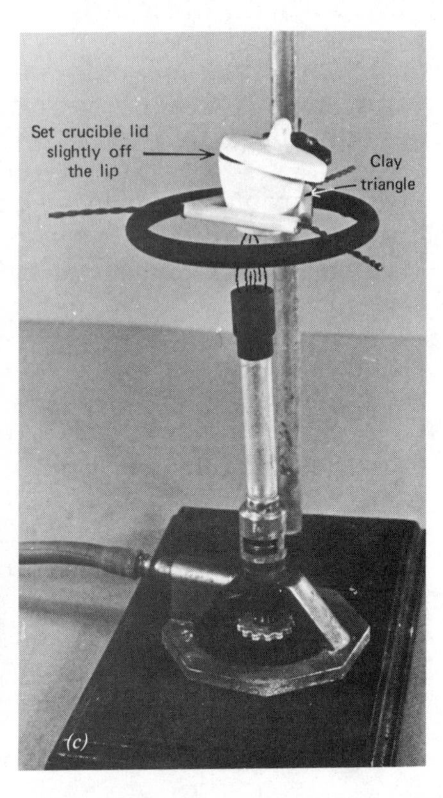

Figure T.8a
Drying and/or Firing of a Crucible and Cover

Figure T.8b
Ignition of a Crucible's Content *Without* Air

Figure T.8c
Ignition of a Crucible's Contents for *Complete* Combustion

9. Handling Gases

a. TESTING FOR ODOR. An educated nose is an important and a very useful asset in the chemistry laboratory. Use it with caution, however, because some vapors are toxic. Fan some of the vapor toward your nose (Figure T.9a); never hold your nose directly over the vessel.

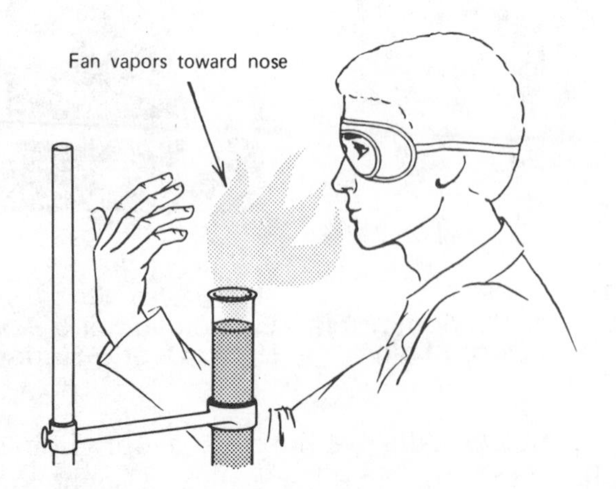

Figure T.9a
Testing Odors. Fan the Vapor Gently Toward the Nose

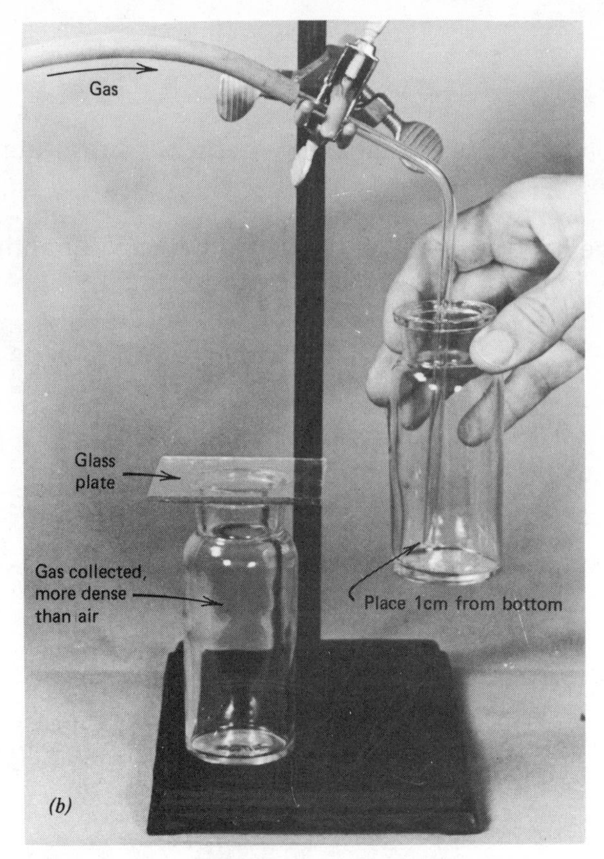

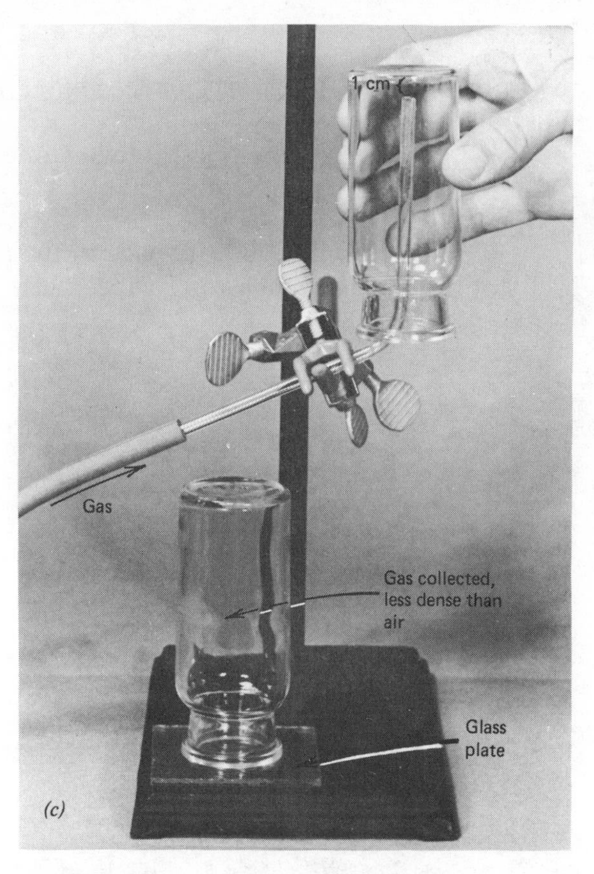

Figure T.9b
Collection of Gases *More* Dense than Air

Figure T.9c
Collection of Gases *Less* Dense than Air

b. COLLECTING GASES

i. By Air Displacement. Gases more dense that air may be collected by using the experimental setup shown in Figure T.9b. Gases less dense than air are collected using the apparatus in Figure T.9c.

ii. By Water Displacement. Gases that are relatively insoluble in water are collected using the apparatus shown in Figure T.9d.

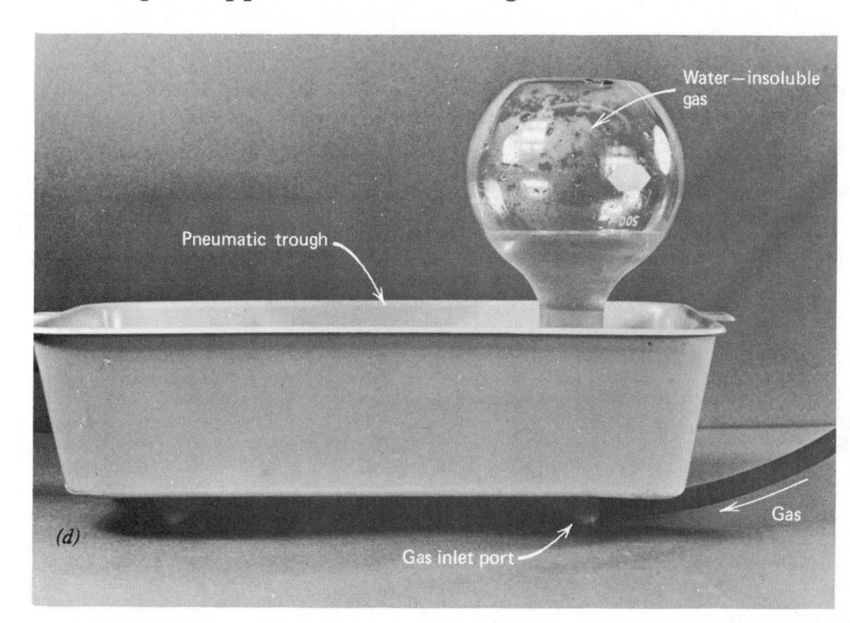

Figure T.9d
Collection of a *Water–Insoluble Gas* by Displacement

c. REMOVAL OF IRRITATING OR TOXIC GASES

i. Laboratory fume hood. When fume hood space is available, use it; do not substitute the improvised hood.

ii. Improvised Hood. If space in the fume hood is inadequate to remove *small* quantities of the toxic or nauseating vapors, use the improvised hood (Figures T.9e and T.9f)

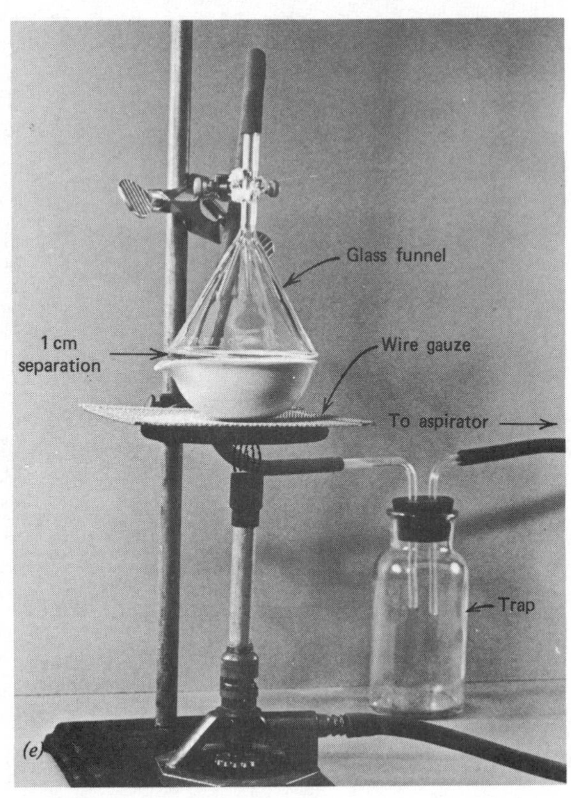

Figure T.9e
The Use of an Improvised Hood Over an Evaporating Dish

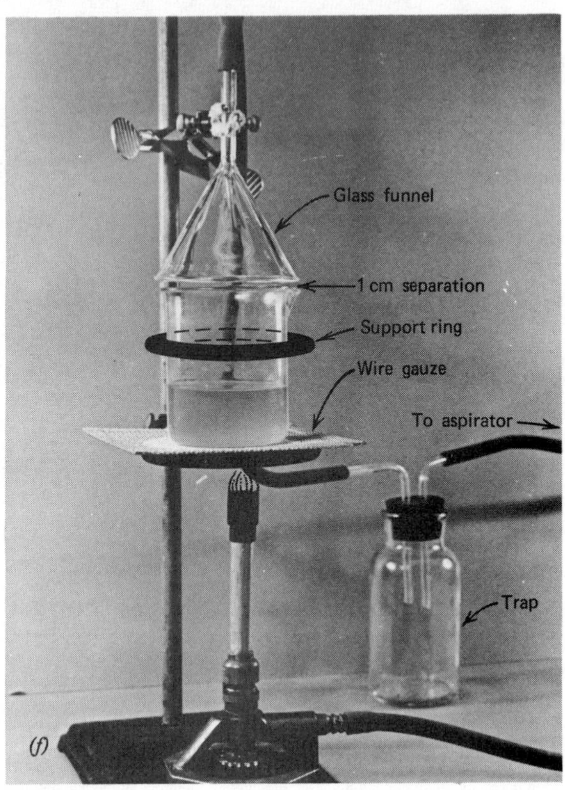

Figure T.9f
The Use of an Improvised Hood Over a Beaker

10. Reading a Meniscus

For exacting measurements of clear or transparent liquids in graduated cylinders, pipets, burets, and volumetric flasks, the solution's volume is read at the *bottom* of the meniscus. Steady the eye horizontal to the liquid's surface (Figure T.10a); position (horizontally) the top edge of a black mark (made on a white card) just below the level portion of the liquid. The black background reflects off the bottom of the liquid and better defines its lowest mark (Figure T.10b). Substituting a finger for the black mark on the white card is not as effective, but it does help.

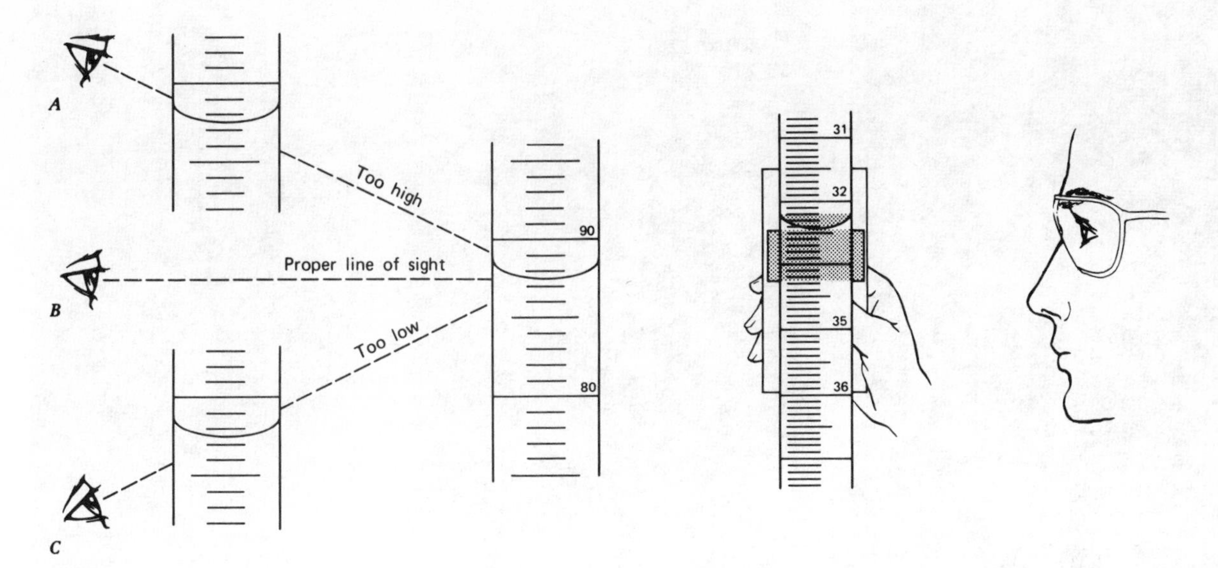

Figure T.10a
Reading the Meniscus with the Eye *Horizontal* to the *Bottom* of the Meniscus

Figure T.10b
Reading the Meniscus using a Black Mark on a White Card

11. Pipetting a Liquid or Solution

a. PIPET PREPARATION. Clean the pipet with a soap solution; rinse with tap water and then with distilled (or de-ionized) water. No water droplets should adhere to the pipet's inner wall. Transfer the solution that you intend to pipet from the reagent bottle to the beaker. **Do not** insert the pipet tip directly into the reagent bottle. Dry the pipet tip with a clean, dust free towel or tissue (e.g., a Kimwipe). Using suction from a collapsed rubber (pipet) bulb (**Never use your mouth**), draw several 2 to 3mL portions into the pipet as a rinse. Roll each rinse around in the pipet to make certain that the solution washes the entire surface of the inner wall. Deliver each rinse to a waste beaker.

b. PIPET DELIVERY. To fill the pipet, place the tip well below the solution's surface. Then using the collapsed rubber (pipet) bulb, draw the solution into the pipet until the level is 2 to 3cm above the "mark" (Figure T.11a). Remove the bulb and *quickly* cover the top of the pipet with your *forefinger, not* your thumb. Remove the tip from the solution, dry the tip with a dust-free towel, and holding in a vertical position over a waste beaker, control the delivery of the excess solution from the pipet until the solution level returns to the mark. Practice!! (Figure T.11b). See Technique 10 for reading the meniscus. Remove any drops suspended from the pipet tip by touching it to the wall of the waste beaker. Again wipe off the tip with a clean, dust-free towel or tissue, and deliver the solution to the receiving vessel (Figure T.11c); keep the tip above the level of the liquid and against the wall of the receiving vessel. **Do not** blow or shake out the last bit of solution that remains in the tip; this liquid has been included in the pipet's calibrated volume. See Experiment 2 for a discussion and procedure for the calibration of pipets.

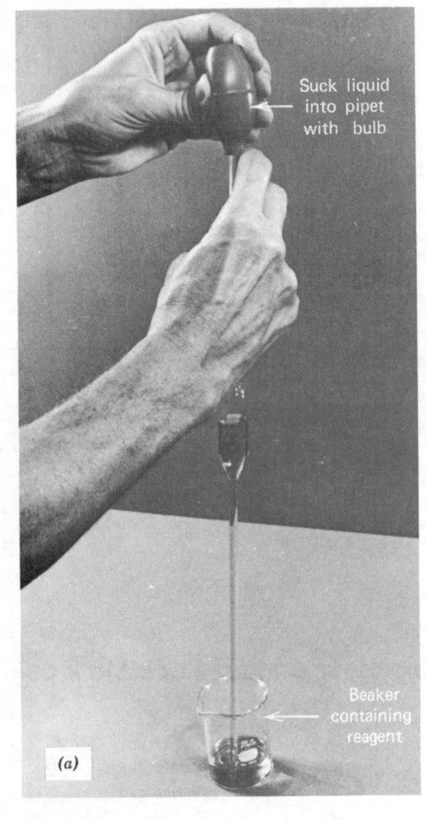

Figure T.11a
Drawing the Solution into the Pipet with a Rubber Bulb

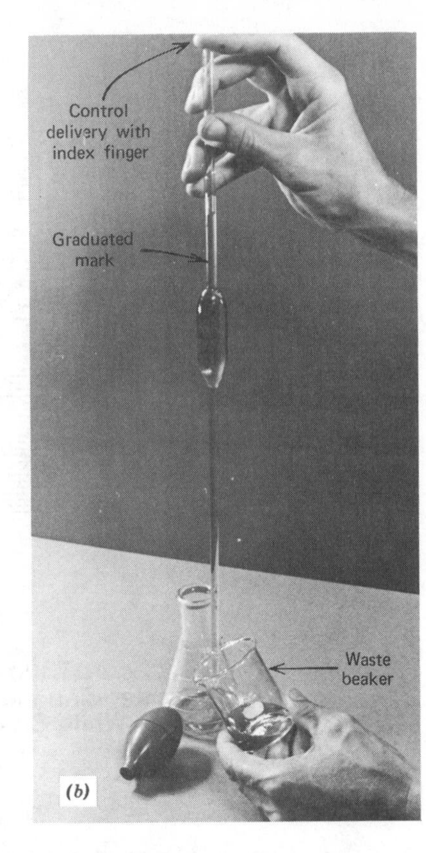

Figure T.11b
Controlling the Delivery with the *Forefinger*

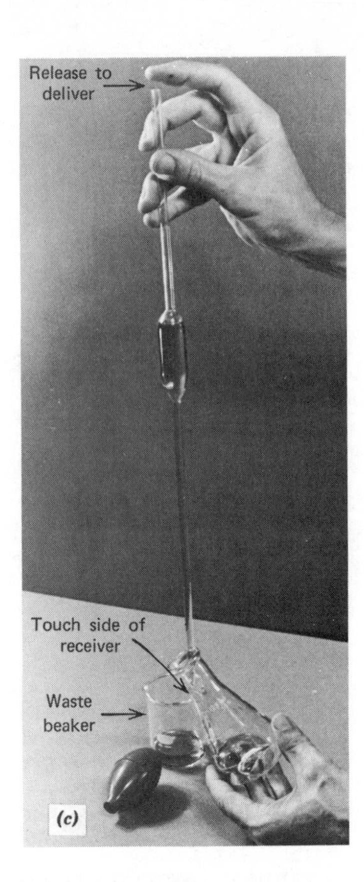

Figure T.11c
Delivering the Solution with the Pipet Touching the Side of the Receiving Flask

12. Titrating a Solution

a. CLEANING THE BURET. Clean the buret with a soap solution. If a buret brush is needed, be careful and prevent the wire handle from scratching its wall. Rinse the soap solution from the buret several times *through the stopcock,* first with tap water and then with distilled (or de-ionized) water. *No drops should adhere to the inner wall of a clean buret.* Close the stopcock. Rinse the buret with several 3 to 5mL portions of titrant. Tilt and roll so that the rinse comes into contact with the entire inner surface. Drain each rinse through the buret tip into a waste beaker.

b. FILLING AND OPERATING THE BURET. Support the buret with a buret clamp (Figure T.12a). Close the stopcock and, with the aid of a funnel, fill the buret to just above the zero mark. The solution in the buret is called the **titrant**. Open the stopcock briefly to remove any air bubbles in the tip. Allow 30 seconds for the titrant to drain from the wall; record the volume (±0.01mL). Note that the graduations increase in value from top to bottom on the buret. Operate the stopcock with your left hand (if right-handed, Figure T.12b) or right hand (if left-handed, Figure T.12c) during the titration to prevent the stopcock from sliding out of its barrel. Fill the buret after each titration.

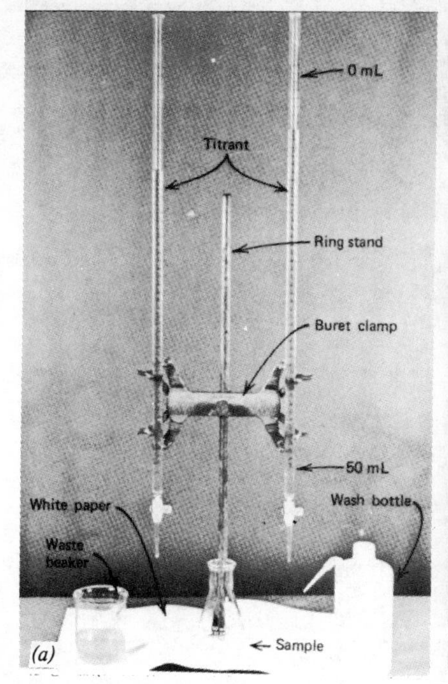

Figure T.12a
A Titration Apparatus

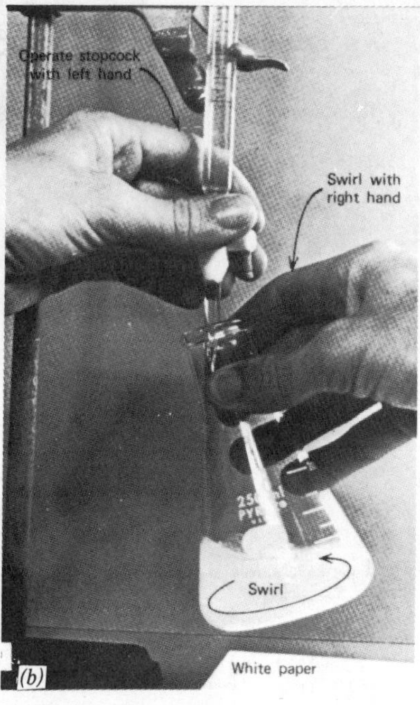

Figure T.12b
Titration Technique for *Right*-Handed Chemists

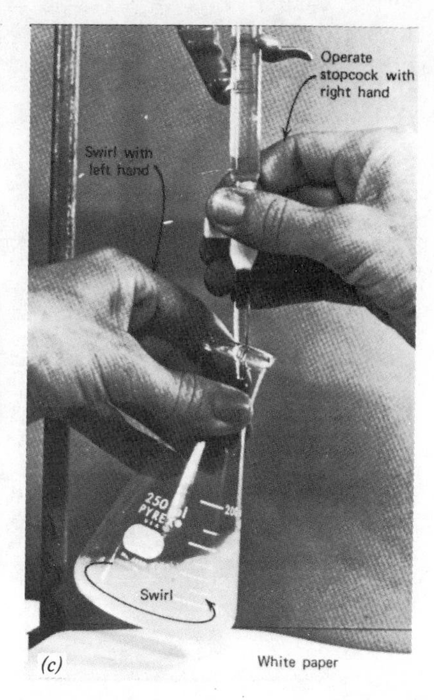

Figure T.12c
Titration Technique for *Left*-Handed Chemists

c. TRANSFERRING THE TITRANT TO THE RECEIVING VESSEL. Place a piece of white paper beneath the receiving vessel, generally an Erlenmeyer flask, so that the appearance of the endpoint (the point at which the indicator turns color) is more visual. If the endpoint results in the appearance of a light or white color, a black background is preferred. During the titration, operate the stopcock with your left hand (if right-handed) and constantly swirl the flask with the right hand. The buret tip should extend 2 to 3cm inside the mouth of the receiving vessel. Periodically stop adding the titrant to wash the wall of the flask (Figure T.12d). Near the endpoint (slower color fade, Figure T.12e), slow the rate of titrant addition until a single drop makes the color change persist for 30 seconds. **STOP**, allow 30 seconds for the titrant to drain from the wall, read and record the volume (±0.01mL). To add less than one drop of titrant, suspend the drop on the buret tip and wash it into the flask with distilled water.

d. CLEAN UP. After completing all of your titrations, drain the titrant from the buret, rinse with several portions of distilled water, and drain through the tip. Store the buret as directed by your instructor.

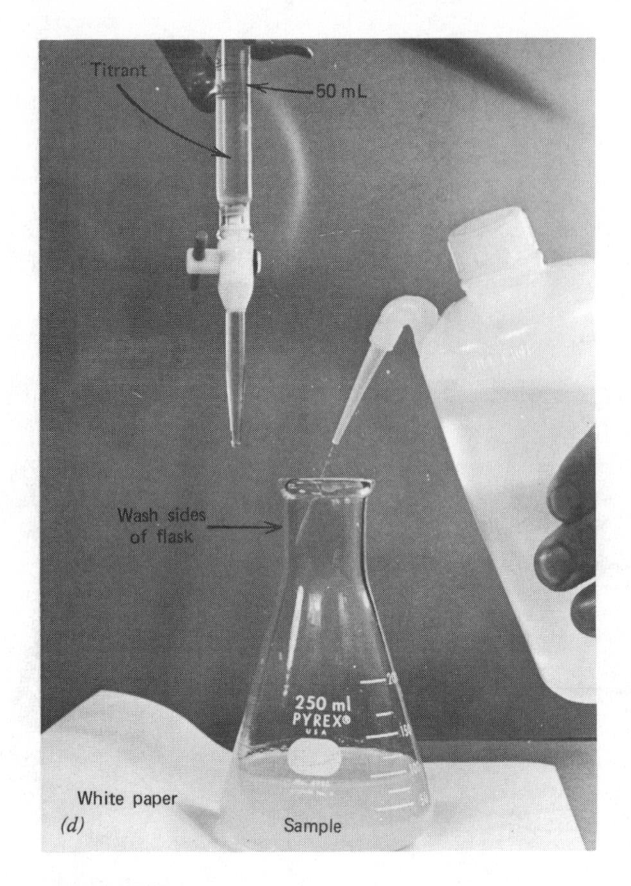

Figure T.12d
Wash the Wall of the Receiving Flask *Frequently* During the Titration

Figure T.12e
Note the Slow Color Fade Near the Endpoint in the Titration

13. Using the Laboratory Balance

Some various types and models of laboratory balances are shown. Select the balance that provides the sensitivity listed in the procedure.

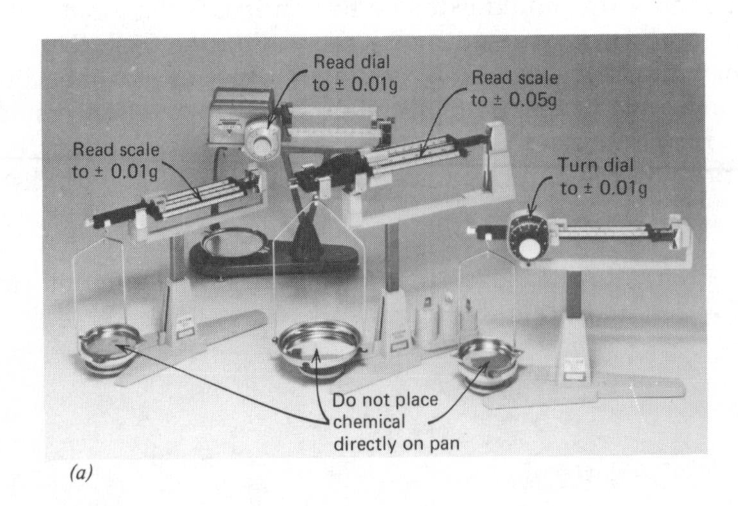

Figure T.13a
Triple Beam Balance, Sensitivity, ±0.01g

BALANCE/SENSITIVITY

- triple-beam balance (Figure T.13a)/±0.01g
- top-loader balance (Figures T.13b)/±0.01g or ±0.001g
- analytical balance (Figure T.13c)/±0.0001g

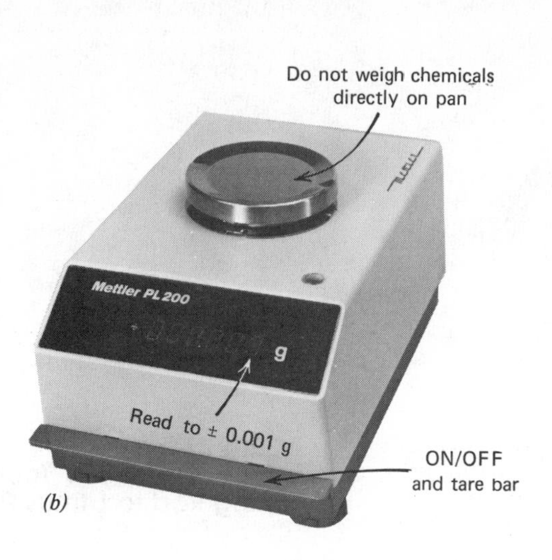

Figure T.13b
Top–Loader Balance, Sensitivity, ±0.01g or ±0.001g

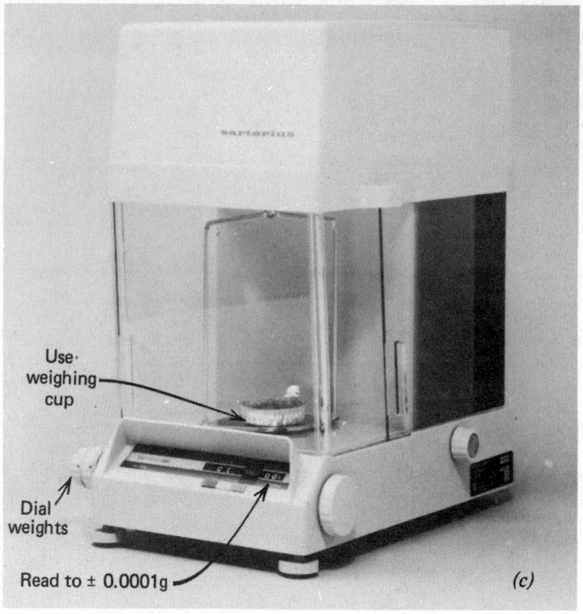

Use weighing cup
Dial weights
Read to ± 0.0001g
(c)

Figure T.13c
Analytical Balance, Sensitivity, ±0.001g

In using and caring for a balance, follow these guidelines:

- Handle with care; balances are expensive.
- Level the balance; see the instructor for assistance.
- Use a beaker, weighing paper, watch glass, or some other container to weigh laboratory chemicals. Do not weigh laboratory chemicals directly on the pan.
- Do not drop anything on the pan.
- Do not attempt to be a handyman. If the balance is not operating properly, see your instructor.
- After weighing, return the balance to the zero reading.
- Clean up any spillage of chemicals on the balance or in the balance area.

14. Testing with Litmus

To test the acidity/basicity of a solution with litmus paper, insert a stirring rod into the solution, withdraw it, and touch it to red or blue litmus (Figure T.14). Acidic solutions turn blue litmus red and basic solutions turn red litmus blue. **Never** place the litmus paper directly into the solution.

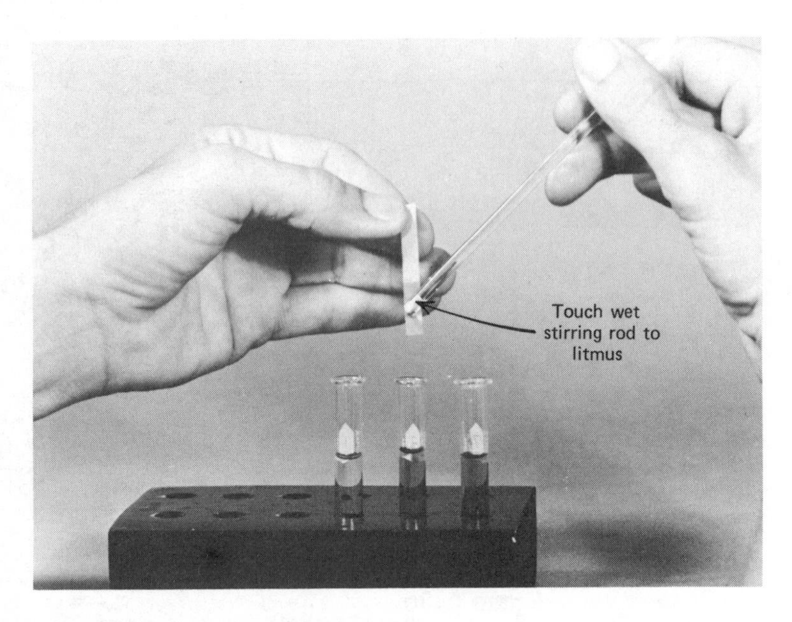

Figure T.14
Litmus Test for pH. Touch Wet Stirring Rod to Litmus Paper

15. Graphing Techniques

The most common graph used in the chemical laboratory is the line graph. The line graph has two lines drawn perpendicular to each other; these are called the axes: the x–axis is horizontal and the y–axis is vertical. The x–axis is called the **abscissa** and the y–axis is the **ordinate**. Each axis is scaled according to the units and range of the measurements; the scale on each axis need not be the same units or the same divisions–they only need to be consistent with the measurements. Each division or subdivision on the graph must have a constant integral value along each axis.

To illustrate the proper construction of a line graph, let's plot the following data for helium gas:

Volume, mL	20	24	30	35	37	41	46
Temperature, K	100	120	150	175	185	205	230

a. DRAW AND LABEL THE AXES. Generally the dependent variable (that which you measure) is plotted along the y–axis and the independent variable (that which you control in the experiment) is plotted along the x–axis. In our example, if we control the temperature of the system and observe the new volume, we are studying volume (V) as a function of temperature (T), or V *vs* T, and plot volume on the y–axis and temperature on the x–axis. The label on each axis should identify the parameter that is being plotted and its units (Figure T.15a).

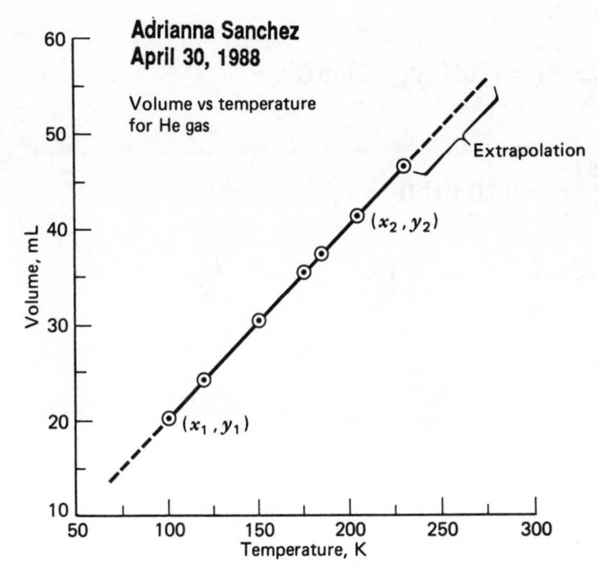

Figure T.15a
A Line Graph of Volume (V) vs Temperature (T) with Labeled and Scaled Axes; a Smooth Curve is Drawn Through the Data Points; the Data is also Extrapolated for Further Interpretation

b. SELECT THE SCALES SO THAT THE RANGE OF DATA POINTS FILL, OR NEARLY FILL, THE ENTIRE PAGE OF GRAPH PAPER. The subdivisions of the scale should be easy to interpret. The intersection of the two axes does not have to be the zero point for each. In our example, the volume units measure from 20 to 46mL; let's select an ordinate scale from 10 to 60mL with 5mL subdivisions. The temperature units range from 100 to 230K; let's select an abscissa scale from 50 to 250K with 25K subdivisions. This selection of ranges allows us to not only fill the page of graph paper with data but also allows us to easily interpret the value of each subdivision along each axis. Also our scale selection allows for some extrapolation of units beyond our current range of data. (Again refer to Figure T.15a)

c. PLACE EACH DATA POINT AT THE APPROPRIATE PLACE ON THE GRAPH. Draw a circle around each point. Ideally, the size of the circle should approximate the error in the measurement.

d. DRAW THE BEST SMOOTH LINE THROUGH THE DATA POINTS. The line does not have to pass through all of the points or, for that matter, any of the points–it merely represents the best averaging of all the data points. Notice that the line passes "undrawn" through the circled data points. The extrapolation of data (the dashed lines) extends the data for additional interpretations.

e. PLACE A TITLE ON THE GRAPH WELL AWAY FROM THE PLOT. If possible, the title, along with your name and date, should be placed in the upper portion of the graph.

f. FOR STRAIGHT–LINE GRAPHS, the drawn line is represented algebraically by the equation

$$\mathbf{y = mx + b}$$

m is the slope and **b** is the intersection of the line along the y–axis at x = 0. The slope is the ratio of the change in the ordinate (y–axis) values to that of the abscissa (x–axis) values, $\Delta y/\Delta x$ (Figure T.15b, page 20).

$$\textbf{slope, m} = \frac{(y_2-y_1)}{(x_2-x_1)} = \frac{\Delta y}{\Delta x}$$

In our example, the slope can be calculated from

$$x_2 = 200K, y_2 = 40mL; \; x_1 = 100K, y_1 = 20mL$$

$$m = \frac{(40 - 20)mL}{(200 - 100)K} = 0.20 \; mL/K$$

We can conclude that the volume changes 0.20mL for each 1K.

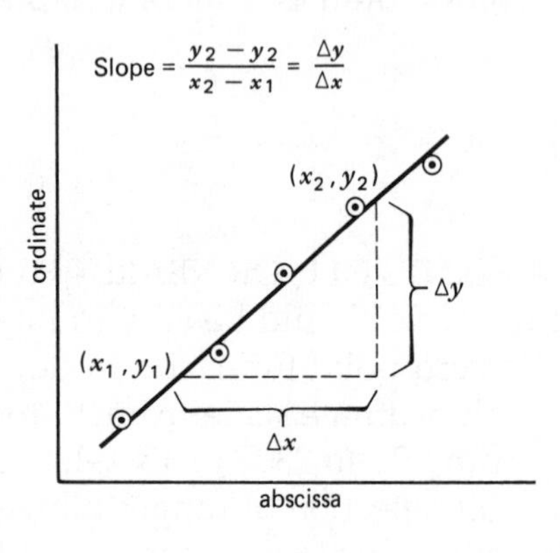

Figure T.15b
Determination of the Slope of a Straight Line

Experiment 1
Laboratory Safety and Glassworking

Objectives

- To check-in to the laboratory
- To learn various safety rules and procedures
- To identify and locate common laboratory equipment and safety devices
- To safely handle chemicals and glassware
- To light and properly adjust a Bunsen burner
- To shape glass rods and tubing

(Courtesy of Fisher Scientific, Inc.)

Principles

All of the chemical principles that you will study and use in this course are a result of careful laboratory experimentation. The techniques, procedures, and equipment for this laboratory were developed to provide a clearer understanding of those principles. No textbook or lecturer can substitute for the observations, the data, and the analyses that you will soon experience. Chemistry comes alive in the lab: observations are made, data are collected and analyzed, theories are stated, and principles are developed and/or understood. The understanding of these chemical principles can help us to better understand the real world of science and the science of everyday living.

You will use laboratory equipment in designing experiments aimed at gathering reliable data. As you work, record the data and use your knowledge of chemical principles to explain what is seen. A good scientist is a thinking scientist. Cultivate self-reliance and confidence in your work, even it it doesn't look right; this is how scientific breakthroughs occur.

In the first few laboratory periods, you are introduced to basic rules, tools, and techniques of chemistry and situations requiring their use. These include safety rules, the Bunsen burner, glassworking, the SI, and the analytical balance. The Techniques section (preceding Experiment 1) illustrates many laboratory techniques that will make your laboratory experience more meaningful. Other techniques are introduced as the course advances.

Techniques

- lighting and adjusting a Bunsen burner
- cutting and shaping glass rods and tubing

Procedure

A. Laboratory Check-In

In this first laboratory period you are assigned a drawer or locker containing your laboratory equipment. Place the equipment on the laboratory bench and, with the check-in list provided (page ix), check off your equipment. If you are unsure of the names of the items, refer to the list of chemical "kitchenware" on page viii. See your instructor about any item that is not in the drawer or locker. A good scientist is always neat and well-organized; keep your equipment clean and neatly arranged so that it is ready for instant use in the experiment. Always have a dish-washing soap or detergent available for cleaning glassware and paper towels on which to set cleaned items for drying.

B. Laboratory Safety

You and your laboratory partners must *always* practice laboratory safety. It is your responsibility, not the instructor's, to **play it safe**. On the inside front cover of this book, there is space to list the location of important safety equipment, telephone numbers, and reference information. Fill this out now.

Study each of the following safety rules and laboratory guidelines to answer the questions on the Lab Preview.

1. Self-Protection

a. Safety glasses, goggles, or eye shields must be worn at all times to guard against the laboratory accidents of others as well as your own. Contact lenses should not be worn; if they are, additional eye protection is necessary.

b. Laboratory aprons or coats (with snap fasteners only) should be worn to protect clothing. Above all, wear old clothing. The wearing of slacks (or jeans) and skirts that extend below the knee are permitted. Shorts and short skirts are not permitted.

c. Wear only shoes that shed liquids. Sandals or canvas shoes are not permitted.

d. Confine long hair. Hair will ignite near an open flame!

e. Wash your hands and arms before leaving the laboratory. Toxic chemicals may be transferred to the mouth.

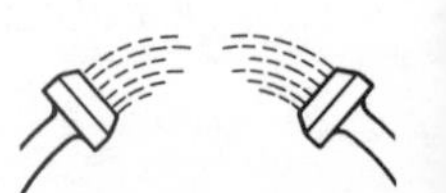

f. Whenever your skin (hands, arms, face,...) comes into contact with laboratory chemicals, wash it quickly and thoroughly with soap and warm water. Use the eye-wash fountain to flush chemicals from the eyes or face. **Do not** rub the affected area, especially the face or eyes, with your hands before washing.

g. If a chemical is spilled over a large part of the body, use the safety shower and flood the affected area for 5 minutes. Remove all

contaminated clothing. Get medical attention; you are not a physician...not yet!

h. Report any accident or injury to your instructor, even if it is apparently minor.

i. Learn how to use the fire extinguisher and other safety devices.

2. Laboratory Rules

a. Do not work alone in the laboratory. At least the instructor must be present.

b. Unauthorized experiments, including variations of those in the laboratory manual, are forbidden.

c. Do not waste time. Complete the Lab Preview and study the Objectives, Principles, Techniques, and Procedure for the experiment *before* lab. Always try to understand what you are doing and to *think*, whistle if you like, while working. Note beforehand the need for any extra equipment that is to be checked out from the stockroom.

What Else Do I Need?

d. No smoking, drinking, eating, or chewing is permitted. Chemicals may possibly enter through the mouth or lungs.

No smoking
No drinking
No eating
No tobacco

e. Horseplay or other careless acts are prohibited. Do not entertain guests.

f. Maintain an orderly, clean lab bench and drawer. Clean up all chemical and water spills with a damp towel (and discard); discard paper scraps in marked containers; and return the extra glassware to the drawer (Be sure that it is clean!). Clean the sinks of all debris. Keep drawers closed while working and the aisles free of any obstructions, such as book bags and athletic equipment.

g. Be aware of your neighbors' activities; you may be the victim of their mistakes. Advise them of improper techniques or unsafe practices. If necessary, report your concern to the instructor.

h. Carefully complete your Data Sheet for each experiment. Believe in your data. A scientist's most priceless possession is integrity. If the results "do not look right", repeat the experiment; don't alter the data to make it look better---be a scientist!

Record data *in ink* as you perform the experiment. Data on scraps of paper will be confiscated. Where calculations are required, be orderly for the first set of data analysis. Do not clutter the Data Sheet with arithmetic details.

i. **Think** while you perform the experiment. Questions at the end of the Data Sheet are designed to test your understanding of the experiment and the principles involved. Discussions with other scientists (your lecturer, laboratory instructor, or other students (although the latter source may be less reliable)), your text, or various reference books (such as the Chemical Rubber Company's *Handbook of Chemistry and Physics*) are often reliable sources of information.

3. Handling Chemicals and Glassware

a. Clean all glassware with soap and warm tap water. Rinse first with *tap* water and then once or twice with *small* amounts of distilled (or de–ionized) water. Never use distilled for washing glassware; it is too expensive.

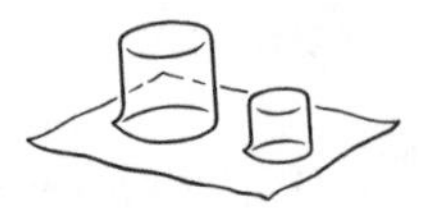

b. Invert clean glassware on a paper towel to dry; do not wipe or air–blow dry because of possible contamination. Do not dry heavy glassware (graduated cylinders, volumetric flasks, or bottles) over a direct flame.

c. Avoid direct contact with all laboratory chemicals. Avoid breathing chemical vapors. Never taste or smell a chemical unless specifically directed to do so. Any interaction with a chemical may cause a skin, eye, or mucous irritation. In other words, **play it safe!**

d. In transferring a chemical (solid or solution) read the label twice and transfer only the amount needed. The use of the wrong chemical (or wrong concentration of chemical) can lead to an "unexplainable" accident or result (and an argument with your laboratory instructor). **Never** use more reagent than the procedure calls for; **do not return the excess** to the reagent bottle——share it with a friend.

e. Add a reagent slowly; never dump it in. While stirring, slowly pour the ***more*** concentrated solution into water or into a less concentrated solution, never the reverse! This is especially true when diluting concentrated (conc) sulfuric acid, H_2SO_4.

Always Add The Concentrated To The Dilluted Solution Or Water
C → D Or W

f. Do not insert your own pipet, medicine dropper, or spatula into reagent bottles. Transfer the reagents as shown in Techniques 2 and 3.

Do Not Insert Glassware Into Reagent Bottles

g. Treat chemical spills as follows:

- Alert your neighbors and the laboratory instructor.
- Clean up the spill as directed by the laboratory instructor.

h. Dispose of waste chemicals as instructed or as follows:

- In the sink: non-flammable, non-toxic, water-soluble liquids followed by large amounts of water
- In waste jars (properly labeled): water-insoluble liquids solids, and toxic wastes
- In waste basket: paper products (such as litmus paper, filter paper, and matches) and broken glassware
- In covered containers (properly labeled): volatile liquids or very reactive chemicals

C. THE BUNSEN BURNER

Laboratory burners come in many shapes and sizes, but all accomplish one main purpose: to produce a combustible gas-air mixture that gives a hot, efficient flame. Because Robert Bunsen (1811-1899) was the first to build and perfect this type of burner, his name is commonly given to all burners of this type (Figure 1.1).

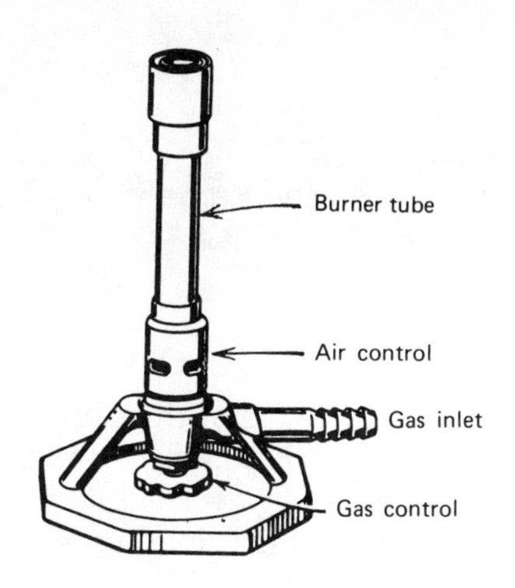

Figure 1.1
Bunsen–type Burner

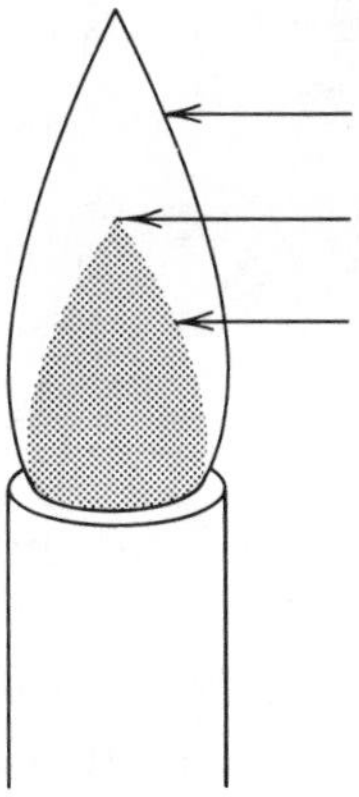

Violet oxidizing flame
Hottest part of flame
Bright blue reducing flame

Figure 1.2
Flame of a Properly Adjusted Bunsen Burner

The natural gas used in most laboratories is composed primarily of methane, CH_4. If supplied with sufficient oxygen, methane burns with a blue, *non*luminous flame yielding carbon dioxide and water. When there is insufficient oxygen, small particles of carbon are produced which when heated to incandescence produce a yellow, luminous flame.

1. Lighting the burner. Attach the burner's tubing to the gas outlet on the lab bench. Turn off the burner's gas control and fully turn on the gas valve at the outlet. Close the air holes at its base and open the gas control slightly. Bring a lighted match or striker up the outside of the burner tube until the escaping gas a the top ignites. After it is lighted, adjust the *gas control* until the flame is pale blue and has two or more distinct cones. Opening of the air control valve produces a slight buzzing sound characteristic of a burner's hottest flame. The addition of air may blow the flame out. When the best adjustment is reached, three distinct cones are visible (Figure 1.2) and the flame should "sing".

2. Flame Temperatures. Temperatures within a blue, *non*luminous flame approach 1500°C in certain regions. Test the temperature in different flame zones by holding with crucible tongs a wire gauze perpendicular to the burner tube (Figure 1.3) about 1 cm above the burner. Note the color and appearance of the gauze. Now move it up through the flame until it no longer gets hot. Close the holes of the *air* control and repeat the tests with a luminous flame. Now position the wire gauze in the flame parallel to the burner tube and centered in a nonluminous flame. This shows a vertical profile of the temperature regions of the flame.

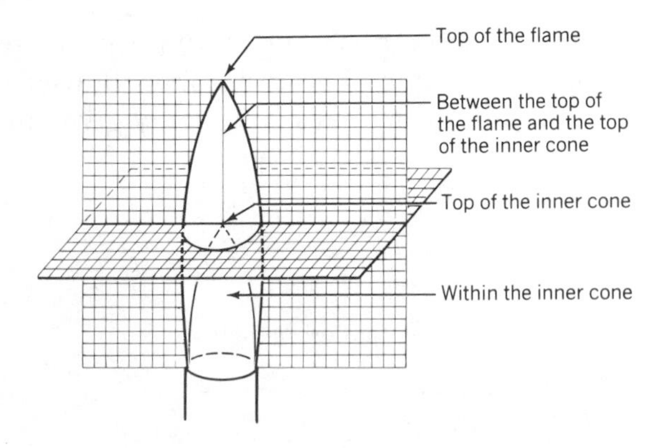

Figure 1.3
Regions of Temperature Measurement in a Bunsen Flame

D. Glassworking

Glass rods and tubing are so useful that the chemist must have knowledge of the simple manipulations involved in cutting, bending, and fire polishing glass, and in making capillary tubes. **Never** place hot glass tubing on anything combustible (**Caution**: *Hot glass and cold glass look the same.*). **Never** hand hot glass to anyone, especially the laboratory instructor, unless you are trying to end an friendship.

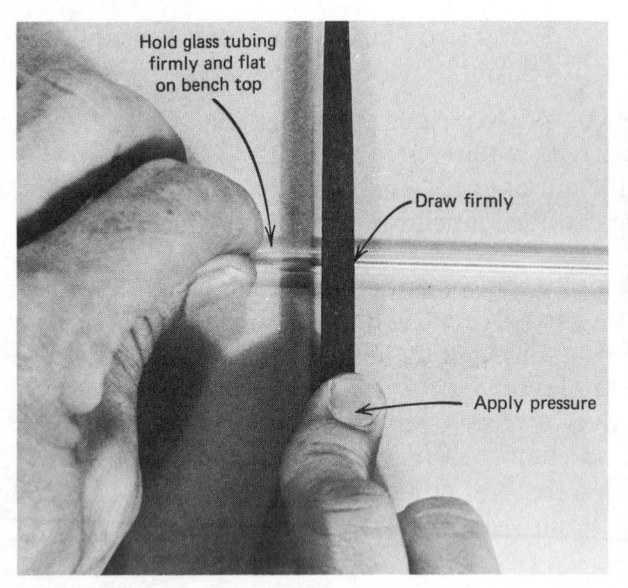

Figure 1.4
Scratch the Glass Tubing with a Triangular File

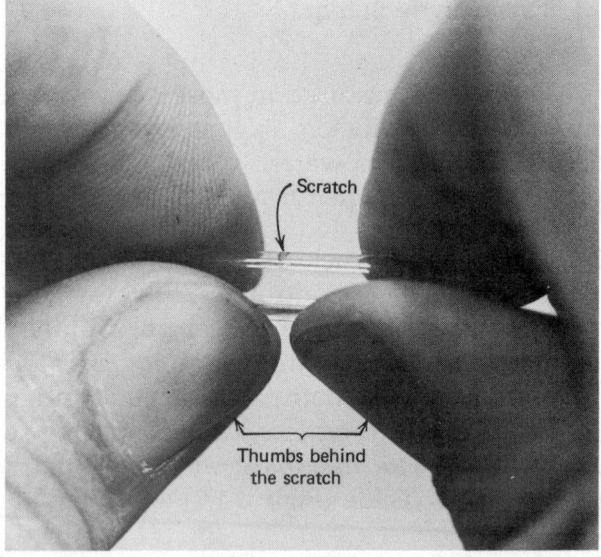

Figure 1.5
Position the Thumbs *Behind* the Scratch

1. Cutting Glass. Place a piece of 6mm glass tubing flat on the lab bench. Use a triangular file or glass scorer to make a scratch at the 20cm mark. Draw the file *once* across the glass; this requires some pressure and should not be a sawing action (Figure 1.4). Place a drop of water or saliva on the scratch to ensure a clean break. Place the thumbs, almost touching, on each side and *behind* the scratch; hold the tubing at the waist with the scratch facing *outward* from the body (Figure 1.5). Bend the tube backward as you pull it away from center (Figure 1.6). With practice, the tube breaks evenly. If not, try again making the scratch a bit deeper.

Cut a 15cm piece of 3mm glass rod in the same manner. This will be used as a stirring rod in later experiments.

2. Fire Polishing. Any cut piece of glass (tubing or rod) has a rough, sharp edge. Fire-polishing smooths the edge to guard against cuts and the scratching of glassware. Hold the cut glass in the flame and rotate it until the edge is fire-polished. Do not overheat; if the tubing is heated too much, the end closes (Figure 1.7). Fire-polish the glass tubing and rod that you

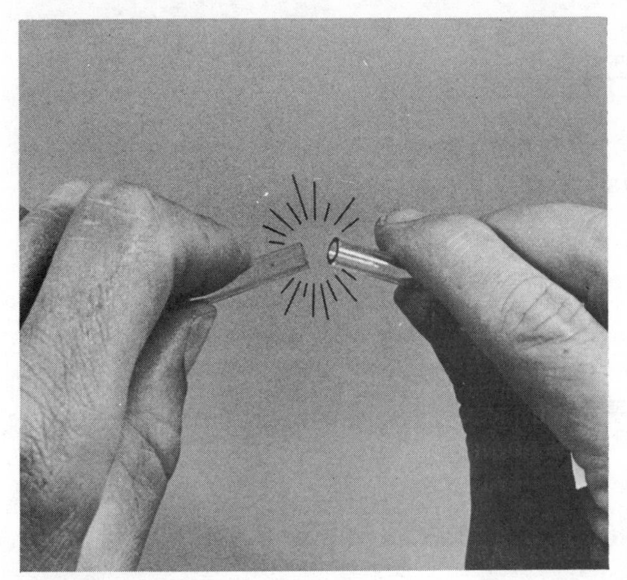

Figure 1.6
Pull and Bend–*Simultaneously*– to Break the Tubing

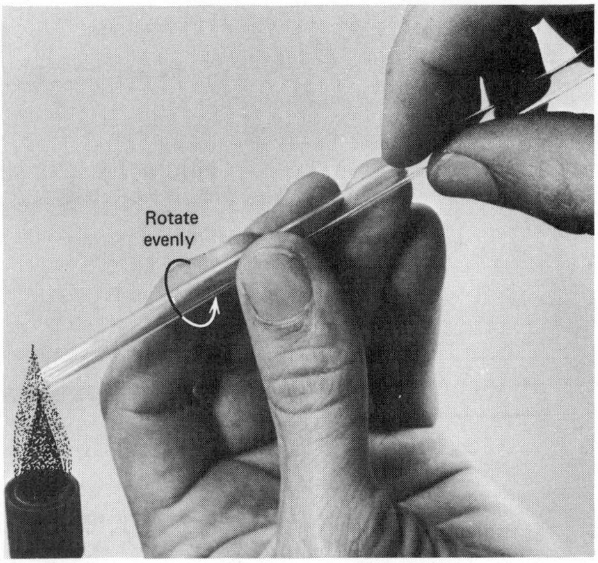

Figure 1.7
Fire Polishing to *Just* Round the Edges

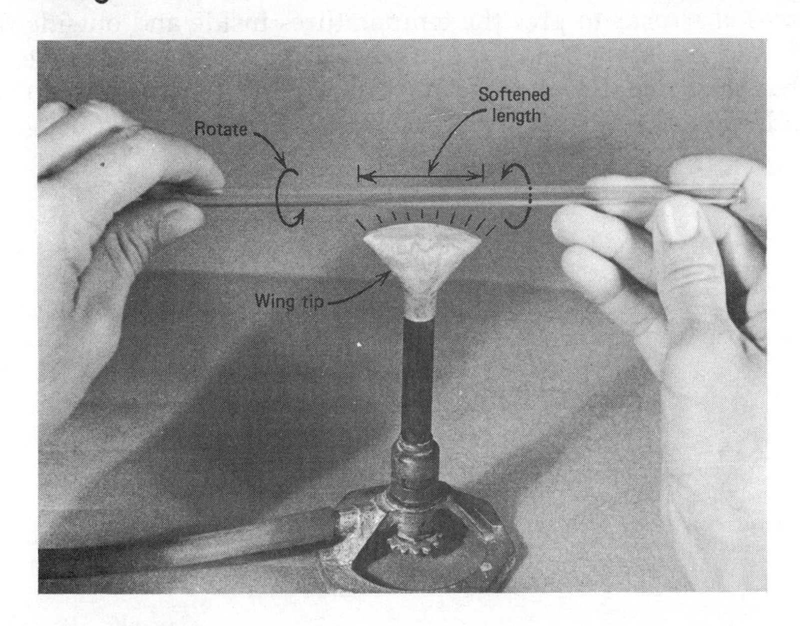

Figure 1.8
Slow Rotation in the Hottest Part of the Flame

3. Droppers and Capillaries. Attach a wing tip to the burner (Figure 1.8). Heat the middle of the 20cm piece of 6mm glass tubing, that was just cut and fire-polished, in the hottest part of the flame. Slowly rotate and heat the tubing until the glass becomes thoroughly softened and pliable. Remove the tubing and pull evenly at a moderate rate; it becomes longer and smaller in cross section. Do not attempt to reheat if "it doesn't look right"--start over. To

make a dropper, cut the glass and *carefully* fire-polish as shown in Figure 1.9. To prepare a capillary tube, cut and fire-polish as shown in Figure 1.10. Properly construct an 8cm dropper.

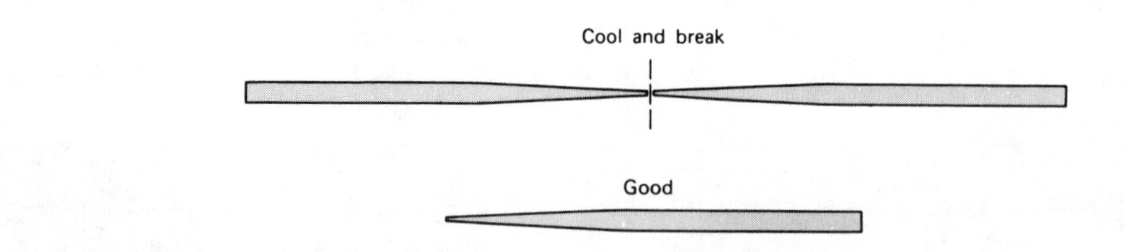

Figure 1.9 Construction of a Medicine Dropper

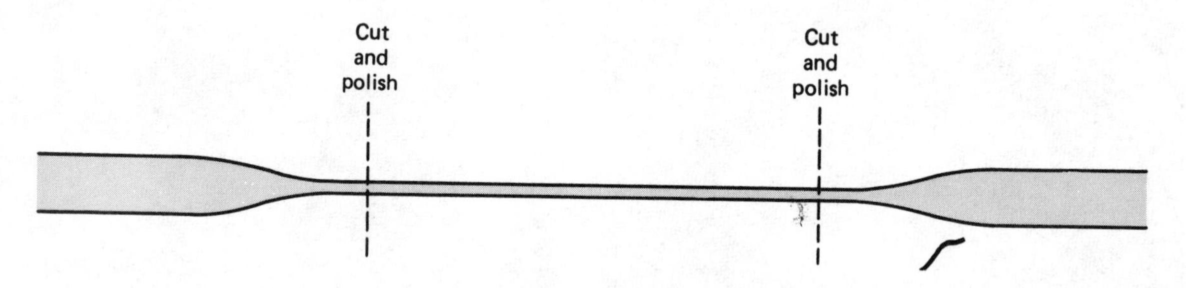

Figure 1.10 Construction of a Capillary Tube

4. Bending Glass. The secret to bending glass properly is to heat it sufficiently over the proper length. Heat and rotate a 10cm piece of 6mm glass tubing (cut and fire-polish as before) in the hottest part of the flame (using a wing tip on the burner) until you feel a slight sag. Remove the softened glass, hold it for *several seconds* to give the temperatures inside and outside the tube time to equilibrate. Slowly and smoothly bend the glass upward to the desired angle (Figure 1.11). A "good" bend has the same diameter throughout with no constriction at the bend; poor bends are shown in Figure 1.12. Practice! See the Data Sheet for the specific glass bends you are to construct.

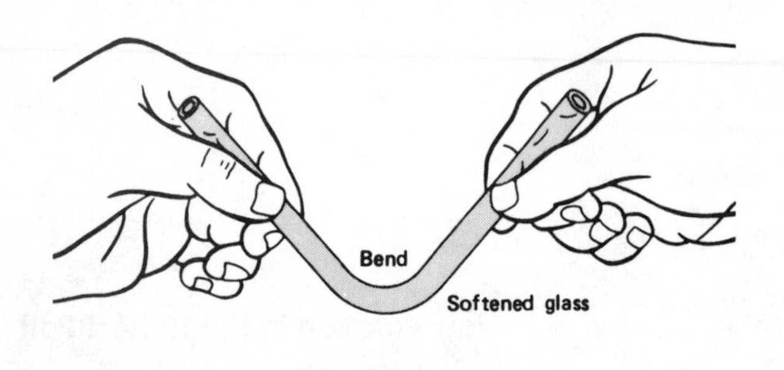

Figure 1.11
Bend the Softened Glass Tubing Slowly Upward

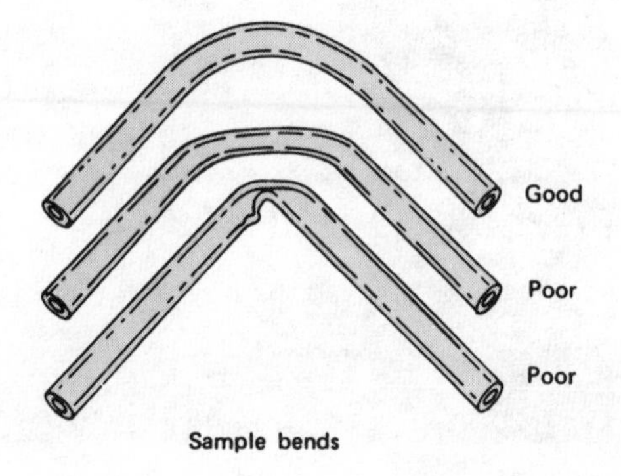

Figure 1.12
Good and Poor Glass Bends

LABORATORY SAFETY AND GLASSWORKING-LAB PREVIEW

Date________Name____Lab Sec. _____Desk No.____

True or False

____1. Paper towels are to be used for wiping dry your clean glassware.

____2. Sandals are *not* to be worn in the laboratory.

____3. Eye protection is worn only while you are conducting an experiment.

____4. Slacks, jeans, or skirts that extend below the knee are considered appropriate attire in the laboratory.

____5. You are encouraged to conduct experiments beyond those presented in the laboratory manual.

____6. Record data in pencil so that any errors in measurement can be corrected without presenting a messy Data Sheet.

____7. Do not discuss data, calculations, or the interpretations of the data with others.

____8. In diluting a chemical with water, always add the chemical to the water—never the reverse.

____9. To determine the safety of a chemical, read the label.

___10. Return all unused chemicals to the reagent bottle.

___11. Toluene, a water-insoluble liquid, should be discarded in the sink.

___12. Dry glassware quickly by using a direct flame.

___13. Hot and cold glass look the same.

___14. Bend glass tubing immediately after it is removed from the flame.

___15. A piece of glass tubing is scratched with a triangular in a *single* stroke.

___16. Fire-polishing makes the glass appear more transparent.

___17. Introduce yourself to the laboratory instructor by handing him a hot piece of tubing.

___18. Eating is permitted in the laboratory *after* lunch.

___19. Do *not* invite friends into the laboratory to show them your work.

___20. A properly adjusted Bunsen flame is a yellow, luminous flame.

21. What is the dominant color of the nonluminous flame in a Bunsen burner? Explain.

22. Arrange the following steps in the proper sequence when lighting a Bunsen burner:

a. Open the gas control valve slightly
b. Turn on the gas outlet valve
c. Close the air control valve
d. Close the gas control valve
e. Open the air control valve
f. Connect the burner's tubing to the gas outlet
g. Light the escaping gas

________, ________, ________, ________, ________, ________, ________

LABORATORY SAFETY AND GLASSWORKING-DATA SHEET

Date________Name____Lab Sec. ____Desk No.____

A. LABORATORY CHECK-IN

- Instructor's Approval________________________________

B. LABORATORY SAFETY

Complete the information on the inside front cover of the manual.

- Instructor's Approval________________________________

C. THE BUNSEN BURNER

At right, draw a picture of a nonluminous Bunsen flame and label the "cool" and "hot" regions. Use your observations from the horizontal and vertical gauze tests to draw your sketch.

D. GLASSWORKING

Construct the following:	Instructor's Approval (Good/Poor)
• 15cm stirring rod	____________________
• 8cm medicine dropper	____________________
• 90° bend with 5cm arms	____________________
• 120° bend with 10cm arms	____________________

Questions

1. A pipet, medicine dropper, or spatula should *not* be inserted into reagent bottles. Why?

2. Neither a stopper nor a medicine dropper that comes in contact with liquid or solid reagents should touch the reagent shelf or laboratory bench. Explain.

ChemAlert
The chemical label that makes safety instantly clear.
Fisher Scientific
Storage Codes
R STORAGE CODE RED
Red. Flammable. Store in area segregated for flammable reagents.
Y STORAGE CODE YELLOW
Yellow. Reactive and oxidizing reagents. May react violently with air, water or other substances. Store away from flammable and combustible materials.
B STORAGE CODE BLUE
Blue. Health hazard. Toxic if inhaled, ingested, or absorbed through skin. Store in secure area.
W STORAGE CODE WHITE
White. Corrosive. May harm skin, eyes, mucous membranes. Store away from red-, yellow-, and blue-coded reagents.
G STORAGE CODE GRAY
Gray. Presents no more than moderate hazard in any category. For general chemical storage.
STORAGE CODE STOP
Exception. When this symbol appears within the storage code bar, the reagent is incompatible with other reagents of same color bar. Store separately.
Fisher ChemAlert Guide
SAFETY CODE
FUME HOOD
EYE GUARD
PROPER GLOVES
SAFETY CLOTHING
NFPA CODE
FLAMMABILITY
HEALTH
REACTIVITY
3
1
2
NFPA Hazard Codes*
Health
Flammability
Reactivity
Special Warnings
Now, MSDS Information Is On The Label, Too.
Acetonitrile
HPLC GRADE
A998-4
NA1648
Fisher Scientific
Safety Codes
Eye guard
Proper gloves
Safety clothing
Fume hood

EXPERIMENT 2
SCIENTIFIC MEASUREMENTS

Objectives

- To develop skills in the use of the metric and SI systems
- To learn the skill needed for the correct operation of a laboratory balance
- To calibrate some glassware

Principles

A. THE METRIC AND SI SYSTEMS OF MEASUREMENT

Chemists and physicists throughout the world have used the metric system for standard units of weights and measurements for many years. A modern version of the metric system, adopted by the International Union of Pure and Applied Chemistry (IUPAC), provides a more logical and interconnected framework for all basic measurements. This is called the **SI** (Fr: **Le Systeme Internationale d'Unites**). In this system the **meter** is the basic unit of length, the **kilogram** is the basic unit of mass, and the **cubic meter** is the basic unit of volume. The cubic decimeter, dm^3, commonly known as the **liter,** is the more common unit of the SI used for volume measurements in the chemistry laboratory. The **gram** is the mass unit commonly used by chemists. Subdivisions and multiples of these are designated by powers of ten; prefixes for these powers of ten are listed in Table 2.1. Memorize these prefixes. Conversions within the SI are very simple if you know the definition for the prefixes and you can use the factor–label method (cancellation of units) in problem solving.

Table 2.1
The Meaning of Prefixes in SI

Prefix	Meaning (*Power of Ten*)	Abbreviation	Example using "grams"
femto-	10^{-15}	f	$10^{-15}g = fg$
pico-	10^{-12}	p	$10^{-12}g = pg$
nano-	10^{-9}	n	$10^{-9}g = ng$
micro-	10^{-6}	μ	$10^{-6}g = \mu g$
milli-	10^{-3}	m	$10^{-3}g = mg$
centi-	10^{-2}	c	$10^{-2}g = cg$
deci-	10^{-1}	d	$10^{-1}g = dg$
kilo-	10^{3}	k	$10^{3}g = kg$
mega-	10^{6}	M	$10^{6}g = Mg$
giga-	10^{9}	G	$10^{9}g = Gg$

In Table 2.1, the prefix means the power of ten. For Example, 8.3 *centi*meters means 8.3 $x\ 10^{-2}$ meter; *centi* has the same meaning as $x\ 10^{-2}$.

For example, to convert 6.6 kilograms to nanograms, first convert the given unit (kilograms, kg) to the basic unit of mass (the gram, g) by using the conversion factor, [10^3g/kg]:

$$6.6\ \text{kg} \bullet \left[\frac{10^3\text{g}}{\text{kg}}\right] = 6.6 \times 10^3\text{g}$$

Next convert from the basic unit to the desired unit (nanogram, ng) using the conversion factor, [ng/10^{-9}g]:

$$6.6 \times 10^3\text{g} \bullet \left[\frac{\text{ng}}{10^{-9}\text{g}}\right] = 6.6 \times 10^{12}\text{ng}$$

Or these two steps can be combined in consecutive operations:

$$6.6\ \cancel{\text{kg}} \bullet \left[\frac{10^3\cancel{\text{g}}}{\cancel{\text{kg}}}\right] \bullet \left[\frac{\text{ng}}{10^{-9}\cancel{\text{g}}}\right] = 6.6 \times 10^{12}\text{ng}$$

kg cancel g cancel

In the problem notice that the conversion factors, [10^3/kg] and [ng/10^{-9}g], do not affect the magnitude of the mass measurement since *10^3* means *kilo(k)* and *10^{-9}* means *nano(n)*--only the units for expressing the measurement are changed.

The English system of measurement is still in common use in the United States and a few other countries. Some comparisons are made in Table 2.2 and Appendix A.

Table 2.2
Comparison of the SI and English Systems of Measurement

Physical Quantity	SI unit	Conversion Factors
Length	meter (m)	1 km = 0.6214 mi 1 m = 39.37 in 1 in = 0.0254 m = 2.540 cm
Volume	cubic meter (m^3) [liter (L)]	1 L = $10^{-3}m^3$ = 1 dm^3 = 10^3mL 1 L = 1.057 qt 1 oz (fluid) = 29.57 mL
Mass	kilogram (kg) [gram (g)]	1 lb = 453.6 g 1 kg = 2.205 lb
Pressure	pascal (Pa) [atmosphere (atm)]	1 Pa = 1 N/m^2 1 atm = 101.325 kPa = 760 torr 1 atm = 14.70 lb/in^2 (psi)
Temperature	kelvin (K) [degrees Centigrade (°C)]	K = 273 + °C
Energy	joule (J) [calorie (cal)]	1 cal = 4.184 J 1 Btu = 1055 J

Conversions among the SI and English systems are quite valuable, especially since more and more international trade is based on SI (or metric) measurements. See Figure 2.1.

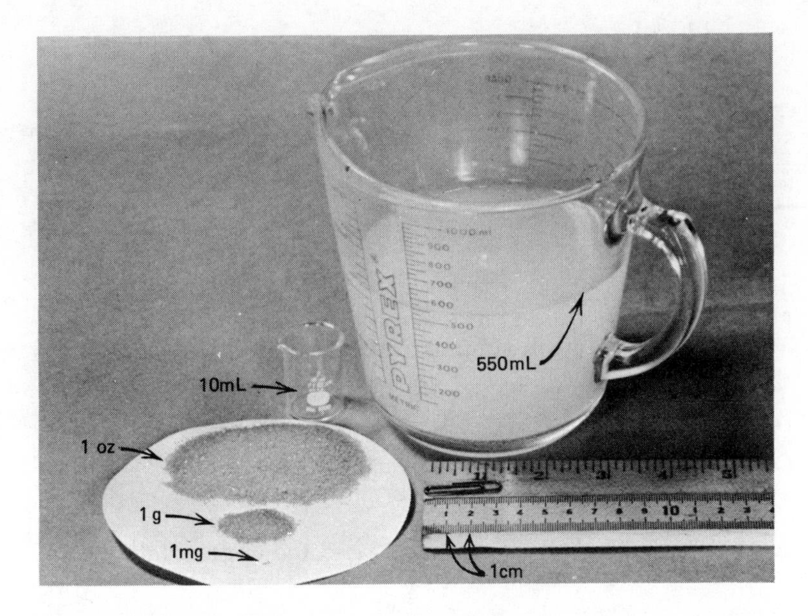

Figure 2.1
Comparisons of SI and English Quantities

Study carefully how each of the following conversions are made using Tables 2.1 and 2.2 and the factor–label method for problem solving.

• Express 3.84 millimeters (mm) in centimeters (cm).
Solution: Lets look at this problem in a series of steps.

Step 1. What do mm and cm mean?
mm = 10^{-3}m; cm = 10^{-2}m

Step 2. What conversion factors are possible?
[mm/10^{-3}m] and [10^{-3}m/mm]; [cm/10^{-2}m] and [10^{-2}m/cm]

Step 3. Set up the sequence of steps starting with the "end unit (cm)" equal to the "given unit (mm)" and then follow with conversion factors arranged so that units cancel from one factor to the next:

$$\text{cm} = 3.84\ \cancel{\text{mm}} \bullet \left[\frac{10^{-3}\cancel{\text{m}}}{\cancel{\text{mm}}}\right] \bullet \left[\frac{\text{cm}}{10^{-2}\cancel{\text{m}}}\right] = 3.84 \times 10^{-1}\text{cm} = 0.384\ \text{cm}$$

end unit   given unit   mm cancel   m cancel

end unit = ________________________________ solved unit

• Express 8.11 nanoliters (nL) in deciliters (dL).
Solution. Lets again look at this problem is a series of steps.

Step 1. What do nL and dL mean?
nL = 10^{-9}L; dL = 10^{-1}L

Step 2. Write the possible conversion factors.
[nL/10^{-9}L] and [10^{-9}L/nL]; [dL/10^{-1}L] and [10^{-1}L/dL]

Step 3. Set up the "end unit (dL)" and the "given unit (nL)" and then combine the appropriate conversion factors.

$$dL = 8.11\,nL \cdot \left[\frac{10^{-9}L}{nL}\right] \cdot \left[\frac{dL}{10^{-1}L}\right] = 8.11 \times 10^{-8} dL$$

end unit　given unit　nL cancel　L cancel

end unit = ________________________________ solved unit

- <u>What is the volume of 250 mL expressed in pints?</u>

Step 1. Write the meanings of all units that are needed to make the conversion.
mL = 10^{-3}L; 1 qt = 2 pt. We also need an SI to English volume conversion; from Table 2.2, 1 L = 1.057 qt.

Step 2. Write down the needed conversion factors.
[mL/10^{-3}L] and [10^{-3}L/mL]; [1 qt/2 pt] and [2 pt/1 qt];
[1 L/1.057 qt] and [1.057 qt/1 L]

Step 3. Write down the "end unit (pt)" and the "given unit (mL)" and the appropriate the sequence of conversion factors.

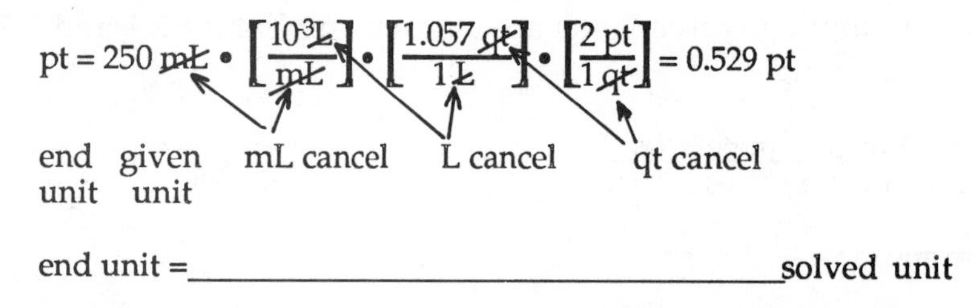

end unit = ________________________________ solved unit

Notice again that the conversion factors do not change the magnitude of the measurement, only its form of expression. Refer to the Data Sheet and complete the SI and SI-English conversions in the same manner as outlined in the previous three examples.

Techniques

- Technique 10, page 10 — Reading a Meniscus
- Technique 11, page 13 — Pipetting a Liquid or Solution
- Technique 13, page 16 — Using the Laboratory Balance

B. The Laboratory Balance

Several types of balances may be found in the chemistry laboratory. The triple–beam balance and the top–loading balance are the most common. Now read **Technique 13** carefully. As suggested on the Data Sheet, weigh several objects. Use the top–loader and the analytical balances only *after* the instructor explains its operation.

C. Volume Measurements

1. <u>Commercial Calibration of Graduated Cylinders.</u> Look closely at the 10mL and 50mL graduated cylinders. How accurately can each be read? Record this data on the Data Sheet as, for example, ±0.5mL, ±0.1mL, ±0.05mL, or ±0.01mL.

2. Volumes of Commonly Used Test Tubes. Fill a 50mL graduated cylinder with water; read and record the volume of water as accurately as the calibration allows (Technique 10). Transfer water from the 50mL graduated cylinder to a clean, dry 150mm test tube until the test tube is full. Read and record the volume of water remaining in the graduated cylinder. Repeat, using a clean, dry 75mm test tube. Determine the volume of each test tube.

3. Drops in 1.0mL of water. With a clean, dry medicine dropper (either one that you constructed in Experiment 1 or one from the stockroom) transfer drops of water from the 75mm test tube to a clean, dry 10mL graduated cylinder until 1.0mL has been transferred (Figure 2.2). Count and record the number of drops. What is the average volume per drop?

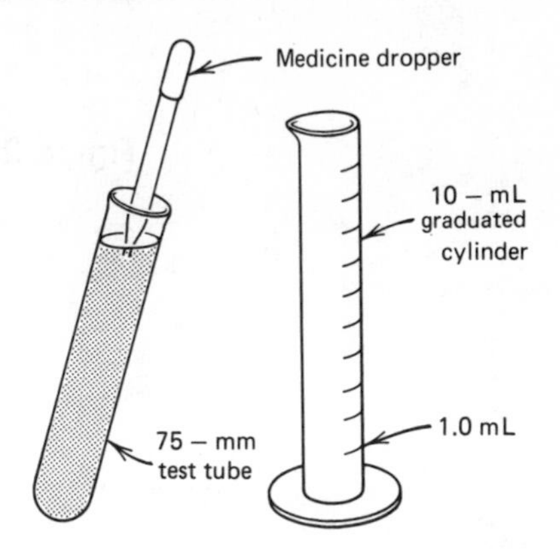

Figure 2.2
Determination of the Number of (Medicine) Drops in 1.0mL of Water →

4. Calibration of a Pipet. Obtain a pipet from the stockroom, its volume is not important. Look at its calibration. How accurately can its volume be recorded? If there is no indication of accuracy, it is called a Class B pipet having an accuracy range of ±0.012 to ±0.06mL for 1mL to 25mL pipets. If the pipet is labeled "A", for Class A pipet, the accuracy range is ±0.006 to ±0.03mL for 1mL to 25mL pipets.

   Weigh (±0.001 g) a clean, dry 100mL beaker. Fill the pipet to the etched mark with water (Technique 10). Transfer the water to the beaker (read Technique 11b carefully). Weigh the beaker and water on the same balance. Assuming the density of water to be 0.998203g/mL at 20°C, calculate the delivered volume of the pipet. Calculate the percent error in the calibration of the pipet.

   Note: Commercial pipets and some graduated cylinders are labeled as TD 20°C (Figure 2.3) or TC 20°C. TD 20°C (to deliver at 20°C) means that the pipet's volume is calibrated according to the volume it delivers from gravity flow only at 20°C. Some liquid is retained in the pipet tip and therefore is *not* to be blown out. Pipets labeled TD 20°C are used exclusively in this course. TC 20°C (to contain, or rinse out, at 20°C) means that the pipet must be rinsed out after all the liquid has drained and the rinsings added to the sample. Therefore a TC 20°C pipet is designed to contain the exact volume of the liquid whereas a TD 20°C pipet contains the calibrated volume plus the volume of liquid in the tip.

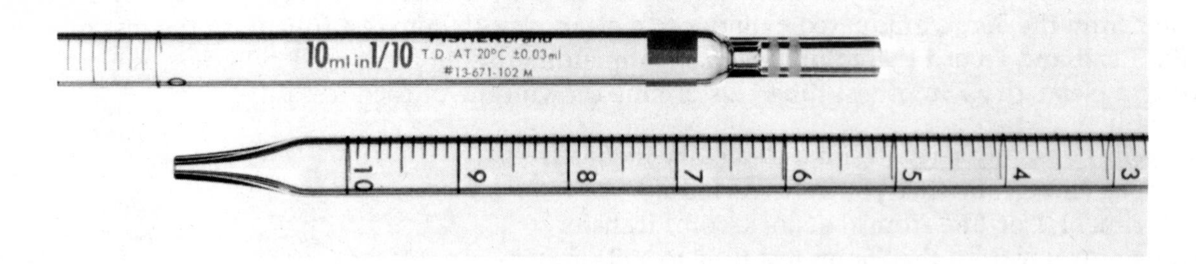

Figure 2.3 Pipet Calibrated to Deliver "TD"

SCIENTIFIC MEASUREMENTS-LAB PREVIEW

Date________Name____Lab Sec.____Desk No.____

1. Name the SI unit that chemists use for scientific measurements.

   a. length ________________________________

   b. volume ________________________________

   c. mass ________________________________

   d. temperature ________________________________

2. Diagram the cross section of a graduated cylinder, illustrating how a meniscus is read (Technique 10).

3. a. Explain haw to fill a pipet to the mark (Technique 11).

   b. How is the delivery rate of the liquid from the pipet controlled?

4. What does the label "TD 20°C" mean on a pipet?

5. A micropipet delivers 153 drops of alcohol for each milliliter. Calculate the volume, in mL, of each drop.

6. A 4g sample of zinc is weighed on a triple–beam balance. Which mass would be properly recorded (Technique 13)?

   a. 4.0g b. 3.99g c. 4g d. 3.994g

7. How should the mass in Question 6 be recorded if an analytical balance is used?

SCIENTIFIC MEASUREMENTS-DATA SHEET

Date________ Name____ Lab Sec. _____ Desk No._____

A. THE METRIC AND SI SYSTEMS OF MEASUREMENT

Convert each of the following using the definitions in Tables 2.1 and 2.2. Show the cancellation of units.

1. 16.2 cm • $\left[\frac{m}{cm}\right] \cdot \left[\frac{\mu m}{m}\right]$ = ____________________ μm

2. 4670 mL • [——] • [——] = ____________________ kL

3. 42 mg • [——] • [——] = ____________________ cg

4. 6.64×10^{-5} g • [——] • [——] = ____________________ μg

5. Express the size of a $^{9}/_{16}$-inch wrench in millimeters.

   $^{9}/_{16}$-inch • [——] • [——] = ____________________ mm

6. Indicate your or your friend's height in centimeters and mass in kilograms.

7. The metric mile is 1500 meters. Express 1500 meters in miles.

8. Determine the volume of a test tube (in m^3, mL, and dm^3) in your desk drawer by measuring its length and radius ($^{1}/_{2}$ of diameter) and substituting into the equation, $V = \pi r^2 l$

9. A beer can holds 12 fluid ounces. Express 12 fluid ounces in milliliters.

10. The amount of heat required to raise the temperature of 12 fluid ounces of water from room temperature to boiling is approximately 26.6 kilocalories. Express this quantity in joules, kilojoules, and British thermal units.

11. Some pickup trucks and vans are advertised as having a 6.2L diesel engine. Assuming an 8-cylinder engine, calculate the displacement volume, in mL, for *each* cylinder.

B. The Laboratory Balance

1. Weigh the following objects on the triple–beam balance. Express your answer to the nearest 0.01 g.

   a. 250mL beaker (g) ____________________

   b. 150mm test tube (g) ____________________

   c. 20cm piece of glass tubing (g) ____________________

   d. a quarter (g) ____________________

   e. a key (g) ____________________

2. Weigh the following objects on the top-loading balance. Express your answer according to the sensitivity of the balance.

a. a dime (g) ____________________

b. a dollar bill (g) ____________________

c. an empty 10mL graduated cylinder ____________________

d. a 10mL graduated cylinder containing 6mL of water (g) ____________________

e. 150mm test tube ____________________

C. VOLUME MEASUREMENT

1. Accuracy of 10mL graduated cylinder (mL) ________________

Accuracy of 50mL graduated cylinder (mL) ________________

2. a. Volume of water in graduated cylinder (mL) ________________

b. Volume of water after transfer to 150mm test tube (mL) ________________

c. Volume of water after transfer to 75mm test tube (mL) ________________

d. Volume of 150mm test tube (mL) ________________

e. Volume of 75mm test tube (mL) ________________

3. Drops in 1.0mL water ________________

Volume of each drop (mL) ________________

4. a. Accuracy of pipet (mL) ________________

b. Mass of 100mL beaker (g) ________________

c. Mass of beaker + water delivered from pipet (g)________________

d. Mass of water delivered (g) ________________

e. Calculated volume of pipet (mL) ________________

f. *Percent error in calibration (%) ________________

$$\text{*\% error in calibration} = \left[\frac{\text{absolute difference between calculated and labeled volume}}{\text{calculated volume}}\right] \times 100$$

Questions

1. In Experiment 1, you constructed an 8cm dropper. Explain how you could calibrate this to deliver a known volume of water.

2. Water drops will adhere to the inner wall of a dirty pipet. Does a dirty pipet deliver more or less that its calibrated volume? Explain

3. A TC 20°C pipet was assumed to be a TD 20°C pipet. Will a larger or smaller volume be delivered? Explain.

4. Suppose that in calibrating the volume of a pipet, sufficient time is not allowed for the liquid to drain from the pipet. Does this cause a higher or lower calculated volume for the pipet? Explain.

5. Suppose that in calibrating the volume of the pipet, the temperature of the water is not 20°C, but instead is 25°C. How does this affect the calibrated volume of the pipet? The density of water decreases with increasing temperature.

Experiment 3
Separation of a Mixture

Objective

- To separate a heterogeneous mixture based on the chemical and/or physical properties of each component.

Principles

A mixture is a physical combination of two or more pure substances in which each substance retains its own chemical identity. For example, in a salt (NaCl)–water mixture each component maintains the same chemical form as it does in the pure state; water is still H_2O molecules and salt is still Na^+ and Cl^- ions.

Mixtures are either homogeneous or heterogeneous. Homogeneous mixtures, also called solutions, consist of substances mixed at the molecular–size level to produce a mixture that has uniform properties and composition throughout. Air is an example of a homogeneous mixture of gases. Heterogeneous mixtures consist of distinct regions (or phases) having different properties and compositions; for example salt–sand and water–oil mixtures are heterogeneous.

Most substances, whether found in nature or prepared in the laboratory, are impure; that is, they are a part of a mixture. One goal of a chemist is to separate mixtures so that the "good stuff" can be used and the impurities can be discarded. The selection of a separation method depends on the differences in the chemical and/or physical properties of its components.

This experiment uses a combination of chemical and physical properties to separate a heterogeneous mixture of silicon dioxide, SiO_2, sodium sulfate decahydrate, $Na_2SO_4 \bullet 10H_2O$, and sodium chloride, NaCl.

Sodium sulfate decahydrate (also known as Glauber's salt) decomposes at 100°C

$$Na_2SO_4 \bullet 10H_2O\ (s) \xrightarrow{\Delta} Na_2SO_4\ (s) + 10\ H_2O\ (g)$$

The H_2O escapes as a gas creating a mass loss from the original mass of the sample. From this decomposition, 1.000g $Na_2SO_4 \bullet 10H_2O$ produces 0.559g of gaseous H_2O and 0.441g of solid Na_2SO_4; therefore the mass of $Na_2SO_4 \bullet 10H_2O$ in the sample can be calculated from this mass loss.

Sodium sulfate, Na_2SO_4, and NaCl are water soluble; SiO_2 is insoluble in water. After the mixture cools following the decomposition of $Na_2SO_4 \bullet 10H_2O$, the Na_2SO_4 and NaCl are dissolved in water and separated from the SiO_2. The wet SiO_2 is heated, driving off any water occluded to the SiO_2 particles. The mass of NaCl in the mixture is calculated by subtracting the sum of the $Na_2SO_4 \bullet 10H_2O$ and SiO_2 masses from the total mass of the sample.

Techniques

- Technique 4a, page 4 — Separation of a Solid from a Liquid
- Technique 7a, page 9 — Evaporation of Liquids
- Technique 13, page 16 — Using a Balance

Procedure

You are to complete two trials in determining the percent by mass of each substance in the mixture. Obtain no more than 5g of the sample mixture from your laboratory instructor.

1. Record the mass (±0.01g) of a clean, dry evaporating dish. Add 2g of the sample to the evaporating dish, weigh, and record. Remember to use the same balance for the remainder of the experiment.

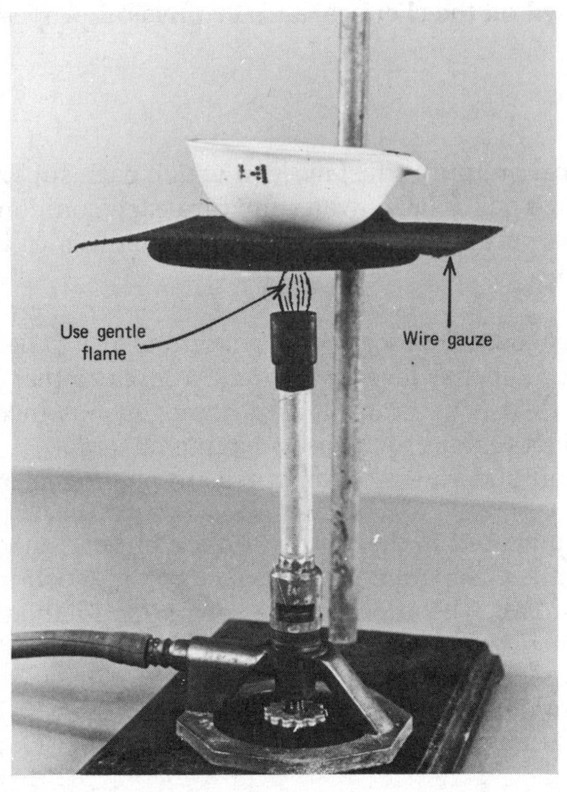

Figure 3.1
Heat to Decompose the $Na_2SO_4 \cdot 10H_2O$ in the Mixture

2. Place the evaporating dish and sample on a wire gauze (Figure 3.1). Gently heat the dish/sample for 3 minutes and then more strongly for 10 minutes. Allow the dish/sample to cool to room temperature—place in a desiccator to cool if available. Weigh and record.

3. Reheat the dish/sample for an additional 5 minutes. Cool, weigh, and record. This second weighing should be within 2% of the first; if not, continue the heat–cool-weigh cycle until ±2% reproducibility is achieved.

4. Add 5mL of distilled (or de-ionized) water to the solid mixture in the evaporating dish. Swirl and then discard the washing. Repeat four more times, each time being careful not to discard any SiO_2.

5. Return the evaporating dish containing the wet SiO_2 to the wire gauze. Heat slowly and then more strongly for 5 minutes. Cool, as in Part 2, weigh, and record. Repeat this heat-cool–weigh cycle until again ±2% reproducibility is achieved. Record the mass of the dish/SiO_2 after each cycle.

SEPARATION OF A MIXTURE-LAB PREVIEW

Date________Name____Lab Sec. _____Desk No.____

1. A 2.721g sample is known to contain 0.473g $Na_2SO_4 \cdot 10H_2O$, 1.497g SiO_2, and the remainder is NaCl. Determine the percent by mass of each substance.

2. In this experiment is the separation of NaCl from the original mixture a result of its chemical or physical properties? Explain.

3. a. What is used as an aid in decanting a liquid from a solid?_______________

   b. What sensitivity should the balance used in the experiment have?_______________

4. a. If in today's experiment a 0.53g mass loss occurs as a result of the initial heating, how many grams of $Na_2SO_4 \cdot 10H_2O$ are in the original sample?

   b. If the original sample weight is 1.83g, what is the percent $Na_2SO_4 \cdot 10H_2O$?

5. Suppose that in Part 4 only one 5mL rinse is used instead of five. How will this affect the reported percent SiO_2 in the sample?

SEPARATION OF A MIXTURE-DATA SHEET

Date________ Name____ Lab Sec. _____ Desk No._____

Unknown Number______________

	Trial 1	Trial 2
1. Mass of evaporating dish and sample (g)	___________	___________
2. Mass of evaporating dish (g)	___________	___________
3. Mass of sample (g)	___________	___________
$Na_2SO_4 \cdot 10H_2O$		
4. Mass of evaporating dish and dry mixture: 1st weighing (g)	___________	___________
2nd weighing(g)	___________	___________
3rd weighing (g)	___________	___________
5. Mass loss in sample from heating (g)	___________	___________
6. Mass of $Na_2SO_4 \cdot 10H_2O$ in sample (g)	___________	___________
7. Percent $Na_2SO_4 \cdot 10H_2O$ in the sample (%)	___________	___________
SiO_2		
8. Mass of evaporating dish and dry SiO_2: 1st weighing (g)	___________	___________
2nd weighing (g)	___________	___________
3rd weighing (g)	___________	___________
9. Mass of SiO_2 in sample (g)	___________	___________
10. Percent SiO_2 in sample (%)	___________	___________
NaCl		
11. Mass of NaCl in the original sample, #3-(#6 + #9) (g)	___________	___________
12. Percent NaCl in sample (%)	___________	___________

Questions

1. If only some of the $Na_2SO_4 \cdot 10H_2O$ decomposes in the heating in Parts 2 and 3, how will this inherent error affect the reported percent composition of

   a. $Na_2SO_4 \cdot 10H_2O$ in the sample?__________ Explain.

   b. SiO_2 in the sample? __________ Explain.

   c. NaCl in the sample? __________ Explain.

2. a. In Part 4 explain why it was necessary to add five 5mL portions of distilled water to the SiO_2 solid. What error(s) does it minimize?

b. Will additional 5mL washings improve the analysis?

Experiment 4
Water Analysis: Solids

Objective

- To determine the total, dissolved, and suspended solids in a water sample.

Principles

Surface water is used as the primary drinking water source for many municipalities. The water is piped into a water treatment facility where impurities are removed, bacteria are killed, and then placed into the water lines for distribution. The contents of the surface water must be known and predictable so that the treatment facility can properly and adequately remove these impurities. Tests are used to determine the contents of the surface water.

In this experiment we will analyze for total, dissolved, and suspended solids in a water sample. The water sample may be taken from the ocean, a lake, a stream, or from an underground water source (called an aquifer).

Water found in the environment has a large number of impurities with an extensive range of concentrations. **Dissolved solids** are water soluble substances, usually salts. Naturally–occurring dissolved solids generally result from the movement of water over or through mineral deposits of varying concentration. These dissolved solids, characteristic of the environment of the watershed, generally consist of the sodium, calcium, magnesium, and potassium cations and the chloride, sulfate, bicarbonate, carbonate, bromide, and fluoride anions. These dissolved solids are responsible for the "hard" water that exists in some locales. Anthropogenic (human–related) dissolved solids include nitrates from fertilizer runoff and human wastes, phosphates from detergents and fertilizers, and organic compounds from pesticides and industrial wastes. A measure of the dissolved solids in a water sample is **salinity**, defined as the grams of dissolved solids per kilogram of water.

Suspended solids are very finely divided particles that are water *insoluble* and are filterable. These particles are kept in suspension by the turbulent action of the moving water. Examples of suspended solids include decayed organic matter, sand, salt, and clay.

Total solids are, of course, the sum of the dissolved and suspended solids in the water sample. In this experiment the total solids and the dissolved solids are determined directly; the suspended solids are assumed to be the difference, since

total solids = dissolved solids + suspended solids

The U.S. Public Health Service recommends that drinking water not exceed 500mg total solids/kg water. However, in some localities, the total solids content may range up to 1000mg/kg of potable water; that's 1g/L!! An amount over 500mg/kg water does not mean the water is unfit for drinking; an excess of this amount is merely not recommended.

Techniques

- Technique 4b, page 5 — Separation of a Solid from a Liquid
- Technique 7a, page 9 — Evaporation of Liquids
- Technique 10, page 12 — Reading a Meniscus
- Technique 11, page 13 — Pipetting a Liquid or Solution
- Technique 13, page 16 — Using the Laboratory Balance

Procedure

Obtain 100mL of a water sample from your instructor. If permitted, bring your own water sample to the laboratory for analysis. Check out one additional evaporating dish from the stockroom.

Clean, dry, and weigh (±0.001g) two evaporating dishes. Be certain that you can identify each. Use the same balance for the remainder of the experiment.

A. Dissolved Solids

1. Gravity filter about 50mL of a thoroughly stirred or shaken sample into a clean, dry 100mL beaker. While waiting for the filtration to be completed, proceed to Part B.

2. Pipet a 25mL aliquot (portion) of the filtrate into the second clean, dry previously–weighed (0.001g) evaporating dish. Weigh the combined evaporating dish and sample. Place the dish/sample on the wire gauze (Figure T.7a) and *slowly* heat—do not boil— the mixture to dryness. As the mixture nears dryness, cover with a watch glass, and reduce the intensity of the flame—this reduces the spattering of the remaining solid and its subsequent loss in the analysis. If spattering does occur, allow the dish to cool to room temperature, rinse the adhered solids from the watch glass (Figure 4.1) and return the rinse to the dish. Again heat slowly, being careful to avoid further spattering. After all of the water has evaporated, maintain a "cool" flame beneath the dish for 3 minutes. Allow the dish to cool to room temperature, weigh, and record.

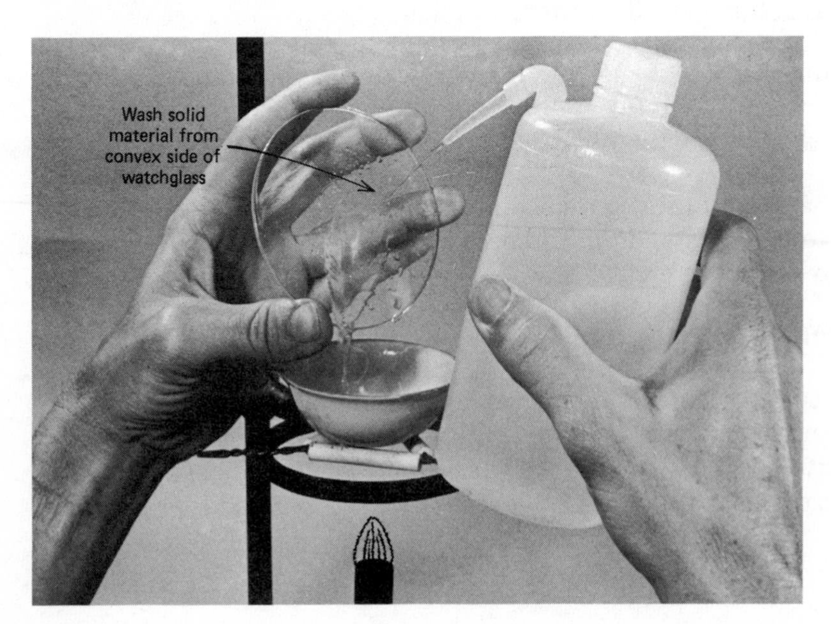

Figure 4.1
Wash the Spattered Material from the Convex Side of the Watch Glass

B. TOTAL SOLIDS AND SUSPENDED SOLIDS

1. Thoroughly stir or agitate 100mL of sample; pipet a 25mL aliquot of this sample into one of the evaporating dishes. Proceed to evaporate to dryness as in Part A.2. Calculate the amount of total and suspended solids in the sample, using data from Part A.

C. ANALYSIS OF DATA

1. Compare your data with three other chemists in your laboratory who analyzed the *same* water sample. Record their results on the Data Sheet.

2. The standard deviation, σ, for *n* sets of data is calculated using the equation

$$\sigma = \sqrt{\frac{d_1^2 + d_2^2 + d_3^2 + ...+ d_n^2}{(n-1)}}$$

d_1 is the difference between your suspended solids value (data set #1) and the average suspended solids value, *x*, obtained from the four other chemists. The meaning of the standard deviation value is that 68% of all subsequent solids determinations for that particular water sample will fall within the range of $x + \sigma$ to $x - \sigma$.

Portable water quality test kits can be used to perform many "quick" tests on a water sample. The following portable water quality test kits are commercially available from the Hach Company. The parameters for testing, along with their range of sensitivity, are

Tests	Concentration Range
aluminum	0-0.7mg/L
arsenic (EPA)	0-0.2mg/L
barium	0-250mg/L
benzotriazole	0-15mg/L
boron	0-15mg/L
cadmium	0-70µg/L
chloride	0-20mg/L
chlorine, free (EPA)	0-1.7mg/L
chlorine, total (EPA)	0-1.7mg/L
chlorine dioxide	0-0.055mg/L
chromium (EPA)	0-0.5mg/L
cobalt	0-1.2mg/L
color, true	0-70 units
copper (EPA)	0-150µg/L
cyanide(EPA)	0-0.12mg/L
cyanuric acid	0-50mg/L
N,N-diethyl hydroxylamine	0-300µg/L
detergents, anionic	0-0.2mg/L
erythorbic acid	0-1000µg/L
fluoride (EPA)	0-2mg/L
formaldehyde	0-650µg/L
hardness	0-2.5mg/L
hydrazine	0-0.015mg/L
iodine	0-7mg/L
iron, ferrous	0-2mg/L
iron, total (EPA)	0-2mg/L
lead (EPA)	0-150µg/L
manganese (EPA)	0-10mg/L
mercury	0-3µg/L
molybdenum	0-25mg/L
nickel (EPA)	0-1.5mg/L
nitrogen, ammonia (EPA)	0-2mg/L
nitrogen, nitrate	0-30mg/L
nitrogen, nitrite (EPA)	0-200mg/L
nitrogen, total	0-200mg/L
oil in water	0-60ppm
oxygen, dissolved	0-10mg/L
oxygen demand, chemical (EPA)	0-800mg/L
ozone	0-1.3mg/L
pH	pH 4-10
phenols (EPA)	0-0.2mg/L
phosphonates	0-20mg/L
phosphorus, reactive (EPA)	0-2mg/L
polyacrylic acid	0-20mg/L
potassium	0-6mg/L
residue, suspended solids	0-500mg/L
selenium	0-1.0mg/L
silica	0-15mg/L
silver	0-0.5mg/L
sodium chromate	0-1000mg/L
sulfate (EPA)	0-100mg/L
sulfide (EPA)	0-0.5mg/L
tannin and lignin	0-5mg/L
tolyltriazole	0-15mg/L
turbidity	0-400FTU
volatile acids	0-2500mg/L
zinc (EPA)	0-1mg/L

WATER ANALYSIS: SOLIDS-LAB PREVIEW

Date________ Name____ Lab Sec. _____ Desk No._____

1. a. Characterize a suspended solid.

   b. Identify three kinds of suspended solids.

   c. How do suspended solids stay suspended?

2. In evaporating a solution to dryness in an evaporating dish, why must the heating rate be decreased as the mixture nears dryness?

3. A 25mL aliquot of a well–shaken sample of river water was pipetted into a 27.211g evaporating dish. After the mixture was evaporated to dryness, the dish and remaining sample weighed 43.617g. Determine the total solids in the sample; express in units of g/kg sample. Assume the density of the sample to be 1.01 g/mL.

4. a. What is an aliquot of a sample?

   b. What is the filtrate in a filtration procedure?

   c. How full (the maximum level) should a funnel be filled with solution in a filtration procedure?

   d. What finger should be used in controlling the volume of solution in a pipet?

   e. How is a suspended drop removed from the tip of a pipet after delivery of the filtrate to the evaporating dish?

   f. What device or procedure is used to draw the filtrate into the pipet in Part A.2 of the procedure?

5. List several cations and anions that contribute to the salinity of a water sample.

WATER ANALYSIS: SOLIDS-DATA SHEET

Date________Name____Lab Sec. ____Desk No.____

Sample Number ________________Describe the nature of your water sample, i.e., its color, turbidity, etc.

A. DISSOLVED SOLIDS

	DATA SET #1
1. Mass of evaporating dish (g)	------------
2. Mass of evaporating dish + water sample (g)	------------
3. Mass of water sample (g)	------------
4. Mass of evaporating dish + *dried* sample (g)	------------
5. Mass of dissolved solids in 25mL aliquot (g)	------------
6. Mass of dissolved solids per total mass of sample (g solids/g sample)	------------
7. Dissolved solids or salinity (g solids/kg sample)	------------

B. TOTAL SOLIDS AND SUSPENDED SOLIDS

1. Mass of evaporating dish (g)	------------
2. Mass of evaporating dish + water sample (g)	------------
3. Mass of water sample (g)	------------
4. Mass of evaporating dish + *dried* sample (g)	------------
5. Mass of total solids in 25mL aliquot (g)	------------
6. Mass of total solids per total mass of sample (g solids/g sample)	------------
7. Suspended solids (g solids/kg sample) (g), ss_1	------------

C. Analysis of Data

	Chemist #1 (you)	Chemist #2	Chemist #3	Chemist #4
Dissolved Solids (g/kg)	____________	________	________	________
Total Solids (g/kg)	____________	________	________	________
Suspended Solids (g/kg)	ss_1=________	ss_2=_____	ss_3=_____	ss_4=_____

1. Average value of suspended solids from four chemists (x) = ___________

	$d_n = x - ss_n$	$d_n^2 = (x - ss_n)^2$
Chemist #1	_____________	_____________
Chemist #2	_____________	_____________
Chemist #3	_____________	_____________
Chemist #4	_____________	_____________

2. Standard deviation, σ = _________________________
   Show work here.

Questions

1. If the sample in the evaporating dish in Part B had not been heated to total dryness, how would this have affected the reported value for

   a. total solids? Explain.

   b. suspended solids? Explain.

2. Suppose that the water sample has a relatively high percent of volatile solid material. How would this affect the reported value for

   a. dissolved solids? Explain.

   b. total solids? Explain.

   c. suspended solids? Explain.

3. The average value for suspended solids for a water sample was 460g/kg. The standard deviation from six trials was 12g/kg.

   a. List the range in g/kg of precision for the suspended solid determinations.

   b. What percent of all subsequent determinations will lie within this range?

4. Suppose that in Part A of the procedure, the evaporating dish had not been properly cleaned of a volatile material before the sample was weighed. How would this error in technique affect the reported value for dissolved solids in the water sample? Explain

EXPERIMENT 5
CHEMICALS AND THEIR REACTIONS

Objectives

- To recognize and to write
    - symbols for elements
    - formulas for compounds
    - balanced equations for chemical reactions
- To characterize some physical properties of some compounds

Principles

The chemist communicates in a technical language that is unlike that of any other profession. Often the layman does not understand a lawyer's legal jargon, a physician's medical terms, or an announcer's descriptions of actions in a baseball game. However, if one is familiar with any of these professions, the terms are understood. A chemist's basic language is *symbols* for elements, *formulas* for compounds, and *equations* for *reactions*.

In this experiment we will observe a number of chemicals and chemical reactions, identify chemical change, and write balanced equations.

SYMBOLS

A **symbol** represents the name and also one atom of an element. The elements discovered by the early chemists are generally given Latin names that describe their appearance or chemical properties. The more recently discovered elements have English names. For example, **Au** represents aurum (gold), **Cu** is the symbol for cuprum (copper), but **Ni** represents nickel and **As** is arsenic. The periodic table (inside the back cover) lists the symbols for the elements, their atomic numbers (the whole number), and their relative atomic masses. You will eventually become *very* familiar with this table.

FORMULAS

A **formula** is a combination of the symbols of elements that represent the name and also one molecule (formula–unit) of a compound, a chemical combination of elements. Subscripts in the formula indicate the number of atoms of each element in the compound. For example, H_2O represents a chemical combination of 2 H atoms and 1 O atom; $C_{12}H_{22}O_{11}$ is the formula for sucrose (table sugar), a chemical combination of 12 C atoms, 22 H atoms and 11 O atoms. A change in subscript represents a change in the chemical composition of the compound.

EQUATIONS

A chemical equation is a shortened description of a chemical reaction in which a new compound is made from other compounds (or elements). The chemical equation has the general form,

$$\textbf{reactants} \rightarrow \textbf{products}$$

The "→" means "to produce" and a "+" sign means "reacts with" (it does *not* imply a summation). For example, the equation,

$$HCl(g) + NH_3(g) \rightarrow NH_4Cl(s)$$

indicates that the reactants, gaseous HCl and gaseous NH_3 react to produce the product, solid NH_4Cl. More specifically, the chemical equation states that one molecule of gaseous HCl reacts with one molecule of gaseous NH_3 to produce one ammonium chloride–that's a pretty wordy statement! Notice how the equation simplifies the presentation of the chemical reaction.

A chemical equation not only simplifies a statement, but it also indicates the number of atoms involved in the reaction and how they rearrange to form products. The above equation also states that 4 H atoms, 1 Cl atom, and 1 N atom rearrange to form a product that contains the 4 H atoms, 1 Cl atom, and 1 N atom–the atoms are conserved; if each atom has its own mass, then mass is also conserved.

A correct equation, then, must have the same number of atoms of each element appearing on both sides of the "→" . For example, solid sodium metal reacts with gaseous chlorine to produce solid sodium chloride.

$$Na(s) + Cl_2(g) \rightarrow NaCl(s) \quad \text{(unbalanced)}$$

These are the correct symbols for sodium and chlorine and the correct formula of sodium chloride. To indicate 2 atoms of chlorine on the product side of the equation, a coefficient of 2 is placed before NaCl (remember, we cannot change the formula of NaCl to balance the chlorine atoms).

$$Na(s) + Cl_2(g) \rightarrow 2\ NaCl(s) \quad \text{(unbalanced)}$$

Now that the equation shows 2 sodium atoms on the right a coefficient of 2 is placed before Na on the left.

$$2\ Na(s) + Cl_2(g) \rightarrow 2\ NaCl(s) \quad \text{(a balanced equation)}$$

Reactions

The product of a chemical reaction has chemical and physical properties unlike those of the reactants. For example, Na is a very reactive, silvery–white metal and Cl_2 is a very greenish–yellow, toxic gas; and yet, when they react, NaCl, a white, crystalline solid that is a part of our diet, is produced. Clearly, NaCl is unlike the elements from which it forms.

How does the chemist recognize chemical change? Generally, a change of color, the formation of a precipitate or gas, a detection of a different odor, and/or the absorption or evolution of heat may accompany a chemical reaction. Other evidence of chemical change may be more subtle.

Simple reactions can be easily categorized into several groups. We are going to observe several of these basic types of reactions.

- A single displacement reaction occurs when an element (usually a metal) displaces another element from a compound.

$$A + BY \rightarrow AY + B$$

• A double displacement reaction occurs when two elements substitute for each other in their respective compounds.

$$AX + BY \rightarrow AY + BX$$

• A direct combination reaction occurs when two elements combine to form a single compound.

$$X + Y \rightarrow XY$$

• A decomposition reaction occurs when heat or some other influence causes a compound to dissociate or decompose.

$$AQ \rightarrow A + Q$$

Techniques

• Technique 6a, page 7	Heating liquids
• Technique 9a, page 10	Handling gases
• Technique 14, page 18	Testing with Litmus

Procedure

A. Formulas and Compounds

A large display of compounds in their most stable state at room temperature is arranged in test tubes. Complete the table as suggested on the Data Sheet. Based upon your general observations, develop some generalizations about some compounds as suggested on the Data Sheet.

B. Single Displacement Reactions.

Set up the following experiments in 150mm test tubes.

1. Polish a Zn strip with steel wool and insert it into 6M HCl (**Caution**: *avoid skin contact; flush with water*). Record your observations on the Data Sheet.

2. Dissolve several crystals of $CuSO_4 \cdot 5H_2O$ in 3mL of water and insert a polished Zn strip. Allow the solution to set for the duration of the laboratory period. Wipe off and examine the reddish–black deposit and the zinc surface for pits or pock marks. Record.

3. Dissolve several crystals of $ZnSO_4$ in 3mL of water and insert a polished Cu strip. Allow the solution to set for the duration of the laboratory period. Do you see evidence of a chemical change?

C. Double Displacement Reactions

1. Set up 9 clean 75mm test tubes in your test tube rack. Label each and prepare accordingly.

   #1 Dissolve several crystals (or ≅10mg) of Na_2CO_3 in 1mL of water
   #2 Dissolve several crystals (or ≅10mg) of $BaCl_2$ in 1mL of water
   #3 Dissolve several crystals (or ≅10mg) of NH_4Cl in 1mL of water
   #4 Dissolve several crystals (or ≅10mg) of $CuSO_4$ in 1mL of water
   #5 Dissolve several crystals (or ≅10mg) of $NiCl_2$ in 1mL of water
   #6 Dissolve several crystals (or ≅10mg) of Na_3PO_4 in 1mL of water
   #7 1mL of 6M HCl (**Caution**: *handle with care and avoid skin contact*)
   #8 1mL of 6M NaOH (**Caution**: *handle with care and avoid skin contact*)
   #9 1mL of 6M NH_3 (**Caution**: *handle with care and avoid skin contact*)

2. Combine the contents of the test tubes as directed. Record your observations on the Data Sheet.

   a. Add $^1/_2$ of #8 to #3; heat over a cool flame. Smell *cautiously*. Test the vapor with red litmus paper.[1]
   b. Add #6 to #5.
   c. Add $^1/_2$ of #7 to $^1/_2$ of #1. Smell *cautiously*. Test the vapors with blue litmus paper.
   d. Add #9 to #4.
   e. Add #2 to the remainder of #1.
   f. Insert a thermometer into the remainder of #7 and add the remainder of #8.

3. Pour 5mL of 3M H_2SO_4 into a 150mm test tube. Insert a thermometer into the solution and add 5mL of 6M NaOH. This, an acid–base reaction, is called a **neutralization** reaction. Record. A more quantitative investigation of the heat evolved in this reaction is performed in Experiment 11.

4. Place a small (pinhead–size) portion of FeS in a 75mm test tube and add $^1/_2$mL of 3M HCl. Cautiously and with proper technique test the odor, the odor of rotten eggs. (**Caution**: *the odor is due to* H_2S, *a very poisonous gas and should be tested only long enough to detect the odor. Place the test tube in the fume hood*). Test the $H_2S(g)$ with moistened blue litmus paper.

D. Combination Reactions

1. Grip a 2cm piece of Mg ribbon with a pair of crucible tongs. Heat it directly in a Bunsen flame until it ignites. (**Caution**: *do not look directly at the burning Mg ribbon!*)

2. Place a very small piece (pin–head size) of sulfur in a 75mm test tube and heat directly with a Bunsen flame. Test the odor of the escaping gas (SO_2). Test the vapors with moistened blue litmus paper.

---

[1]To test the vapors with litmus paper, moisten it with water and hold it over the mouth of the test tube. If the blue litmus turns red, then the vapors are acidic; if the red litmus turns blue, the vapors are basic.

E. DECOMPOSITION REACTIONS

1. a. Place several crystals (pea-size sample) of $NaHSO_3$ in a 200mm test tube and heat cautiously with a Bunsen flame. Test and describe the odor (Use the proper technique for testing!). Test the vapors with blue litmus paper. Where, in this experiment, have you previously detected this odor?

   b. Now heat the contents more strongly until a reddish–brown color appears. Allow the contents to cool to room temperature. Add distilled (or de-ionized) water (the test tube should be 1/2 to 1/3 full) and agitate the test tube until the contents dissolve. Divide the solution into 2 equal volumes.

   #1. Add several drops of 0.1M $BaCl_2$.
   #2. Add several drops of 3M HCl. Test for odor–does it smell familiar?

   c. Dissolve several crystals of $NaHSO_3$ in water and repeat the tests in Part E.1b. How do the tests differ? Record.

2. **Demonstration only**. Place about 3g of $C_{12}H_{22}O_{11}$ into a porcelain evaporating dish. Place the dish in a fume hood. Add 3mL of conc H_2SO_4 (**Caution**: *conc* H_2SO_4 *causes severe skin burns and clothing to disappear! Immediately flush the skin with water*). Conc H_2SO_4 is a strong dehydrating agent, extracting H_2O molecules from the $C_{12}H_{22}O_{11}$ molecule. Record your observations and write a balanced equation.

Periodic Classification of the Elements

IA	IIA	IIIB	IVB	VB	VIB	VIIB	VIII			IB	IIB	IIIA	IVA	VA	VIA	VIIA	0
1 H 1.0079																1 H 1.0079	2 He 4.00260
3 Li 6.941	4 Be 9.01218											5 B 10.81	6 C 12.011	7 N 14.0067	8 O 15.9994	9 F 18.99840	10 Ne 20.179
11 Na 22.98977	12 Mg 24.305											13 Al 26.98154	14 Si 28.086	15 P 30.97376	16 S 32.06	17 Cl 35.453	18 Ar 39.948
19 K 39.098	20 Ca 40.08	21 Sc 44.9559	22 Ti 47.90	23 V 50.9414	24 Cr 51.996	25 Mn 54.9380	26 Fe 55.847	27 Co 58.9332	28 Ni 58.70	29 Cu 63.546	30 Zn 65.38	31 Ga 69.72	32 Ge 72.59	33 As 74.9216	34 Se 78.96	35 Br 79.904	36 Kr 83.80
37 Rb 85.4678	38 Sr 87.62	39 Y 88.9059	40 Zr 91.22	41 Nb 92.9064	42 Mo 95.94	43 Tc 98.9062	44 Ru 101.07	45 Rh 102.9055	46 Pd 106.4	47 Ag 107.868	48 Cd 112.40	49 In 114.82	50 Sn 118.69	51 Sb 121.75	52 Te 127.60	53 I 126.9045	54 Xe 131.30
55 Cs 132.9054	56 Ba 137.34	57 La* 138.9055	72 Hf 178.49	73 Ta 180.9479	74 W 183.85	75 Re 186.207	76 Os 190.2	77 Ir 192.22	78 Pt 195.09	79 Au 196.9665	80 Hg 200.59	81 Tl 204.37	82 Pb 207.2	83 Bi 208.9804	84 Po (210)	85 At (210)	86 Rn (222)
87 Fr (223)	88 Ra 226.0254	89 Ac** (227)	104 (260)	105 (260)													

*Lanthanum Series

58	59	60	61	62	63	64	65	66	67	68	69	70	71
Ce	Pr	Nd	Pm	Sm	Eu	Gd	Tb	Dy	Ho	Er	Tm	Yb	Lu
140.12	140.9077	144.24	(147)	150.4	151.96	157.25	158.9254	162.50	164.9304	167.26	168.9342	173.04	174.97

**Actinium Series

90	91	92	93	94	95	96	97	98	99	100	101	102	103
Th	Pa	U	Np	Pu	Am	Cm	Bk	Cf	Es	Fm	Md	No	Lr
232.0381	231.0359	238.029	237.0482	(244)	(243)	(247)	(247)	(251)	(254)	(257)	(258)	(255)	(256)

CHEMICALS AND THEIR REACTIONS-LAB PREVIEW

Date________Name______________________________________Lab Sec. ____Desk No.____

1. Write the English and Latin names for elements with the following symbols.

Symbol	Latin Name	English Name	Matching Latin Names
K	____________	____________	natrium
Pb	____________	____________	ferrum
Ag	____________	____________	aurum
Fe	____________	____________	stibium
Hg	____________	____________	kalium
Sb	____________	____________	plumbum
Sn	____________	____________	hydragyrum
Au	____________	____________	argentum
Na	____________	____________	stannum

2. At least 106 elements are known; however, only a few make a significant contribution to the earth's crust. Complete this table.

Name of Element	% of total atoms	Atomic Number	Symbol	Atomic Mass	Metal or Nonmetal
oxygen	46.6	_____	_____	_____	_______
silicon	27.2	_____	_____	_____	_______
aluminum	8.13	_____	_____	_____	_______
iron	5.00	_____	_____	_____	_______
calcium	3.63	_____	_____	_____	_______
sodium	2.83	_____	_____	_____	_______
potassium	2.59	_____	_____	_____	_______
magnesium	2.09	_____	_____	_____	_______

3. The predominant elements in the universe are the following. Complete this table.

Name of Element	% of total atoms	Atomic Number	Symbol	Atomic Mass	Metal or Nonmetal
hydrogen	93.3	_____	_____	_____	_______
helium	6.49	_____	_____	_____	_______
oxygen	0.063	_____	_____	_____	_______
carbon	0.035	_____	_____	_____	_______
nitrogen	0.011	_____	_____	_____	_______
neon	0.010	_____	_____	_____	_______
magnesium	0.003	_____	_____	_____	_______

4. All naturally occurring gases (except the noble gases), bromine, and iodine are diatomic molecules. List the elements that occur as diatomic molecules in their natural state.

________, ________, ________, ________, ________, ________, ________

5. Write the formula for the following compounds.

a. ammonia	1 N-atom, 3 H-atoms	_________
b. lye	1 Na-atom, 1 O-atom, 1 H-atom	_________
c. calcium carbonate	1 Ca-atom, 1 C-atom, 3 O-atoms	_________
d. glucose	6 C-atoms, 12 H-atoms, 6 O-atoms	_________
e. hydrogen peroxide	2 H-atoms, 2 O-atoms	_________
f. alcohol (ethanol)	2 C-atoms, 6 H-atoms, 1 O-atom	_________
g. octane	8 C-atoms, 18 H-atoms	_________

6. a. Describe the technique for testing a vapor with litmus paper.

   b. What is indicated when a vapor turns blue litmus red?

7. Describe the technique for testing the odor of a vapor.

CHEMICALS AND THEIR REACTIONS-DATA SHEET

Date________Name____Lab Sec. _____Desk No._____

A. FORMULAS AND COMPOUNDS

1. On a separate sheet of paper, construct a table with the headings shown below and complete it as best possible.

	Formula	Name	Physical State (gas, liquid, solid)	Color	Physical Appearance
a.	-------	----------------	------	------	---------
b.	-------	----------------	------	------	---------
•	-------	----------------	------	------	---------
•	-------	----------------	------	------	---------

2. Generalizations based on observations.

a. Blue-colored salts often contain the element __________.

b. Green-colored salts often contain the element __________.

c. Most salts exhibit a ____________________ color.

d. MnO_4^- salts exhibit a ____________________ color.

e. $Cr_2O_7^{2-}$ salts exhibit a ____________________ color.

f. Most CO_3^{2-} salts exhibit a ____________________ color. Exceptions are those that contain the elements __________ and __________.

B. SINGLE DISPLACEMENT REACTIONS.

Part	Chemical Reactants	Evidence of Reaction	Chemical Products	Balanced Equation
1.	------------	--------	$ZnCl_2 + H_2$	--------------------
2.	------------	--------	$ZnSO_4 + Cu$	--------------------
3.	------------	--------	no reaction	--------------------

C. DOUBLE DISPLACEMENT REACTIONS

Part	Chemical Reactants	Evidence of Reaction	Chemical Products	Balanced Equation
2a.	$NaOH + NH_4NO_3$	------	$NH_3 + H_2O + NaNO_3$	-----------------
Result of litmus test ------------------				
2b.	----------	------	$Ni_3(PO_4)_2 + NaCl$	-----------------
2c.	----------	------	$NaCl + CO_2 + H_2O$	-----------------
Result of litmus test ------------------				
2d.	----------	------	$[Cu(NH_3)_4{}^{2+}] + SO_4{}^{2-}$	-----------------
2e.	----------	------	$BaCO_3 + NaCl$	-----------------
2f.	----------	------	$NaCl + H_2O$	-----------------
3.	----------	------	$Na_2SO_4 + H_2O$	-----------------
4.	----------	------	$FeCl_2 + H_2S$	-----------------
Result of litmus test ------------------				

D. COMBINATION REACTIONS

Part	Chemical Reactants	Evidence of Reaction	Chemical Products	Balanced Equation
1.	------	-----------	MgO	-------------
2.	------	-----------	SO_2	-------------

E. Decomposition Reactions

1. a. Odor __________. Conclusion from litmus test__________.
Formula of gas __________.

b. Observation from $BaCl_2$ test ____________________. $BaSO_4$ is an insoluble salt; therefore what is one of the decomposition products of $NaHSO_3$?__________

Observation from 3M HCl test ____________________. From a detection of the odor, what is a second decomposition product of $NaHSO_3$?__________

Write a balanced equation for the thermal decomposition of $NaHSO_3$.

c. Observation from $BaCl_2$ test ____________________. How does the appearance of the system differ from that in Part E.1b (#1)?

Observation from 3M HCl test ____________________. How does the odor differ from that in Part E.1b (#2)?

2. What evidence of a chemical reaction is indicated?

The products of the reaction are carbon and water. Write a balanced equation for the decomposition of sugar.

Questions

1. In Part B.2, what happens with time to the color of the solution?____________________
Explain.

2. In Parts B.2 and B.3, the relative chemical reactivity of Zn and Cu is determined. Which metal is more reactive? __________ Explain.

3. What is the gas evolved in Part C.2a?__________ In Part C.2c? __________ In Part C.4?___________

4. In Part C.2b, a green precipitate forms. What element causes the green color? ____

5. When an antacid neutralizes excess stomach acid, is heat evolved or absorbed?____________________ What is *always* a product of the reaction?___________

6. What is the color of a cotton shirt or jeans after having spilled conc H_2SO_4 on them? (Note: cotton is chemically similar to sugar.)

Experiment 6
Identification of a Salt

Objectives

- To observe the chemical properties of the SO_4^{2-}, CO_3^{2-}, PO_4^{3-}, Cl^-, and S^{2-} anions
- To identify the presence of SO_4^{2-}, CO_3^{2-}, PO_4^{3-}, Cl^-, and S^{2-} anions in various salts

Principles

Evidence that a chemical reaction has occurred can be observed in a number of ways. You may note that the formation of a precipitate, the evolution of a gas, the change in color of the solution, and the evolution of heat are all indications that "something" has happened when the substances are mixed.

In this experiment we are going to look at the behavior and formation of precipitates. An aqueous solution consists of the solute and water. The solute may be quite soluble or relatively insoluble. Quantitatively we can determine the solubility of a solute by the number of grams that dissolve per amount (generally 100g) of water at a given temperature (usually 25°C). For example, the solubility of NaCl at 25°C is 35.8g NaCl/100g H_2O, whereas the solubility of $BaSO_4$ is only 2.4×10^{-4}g $BaSO_4$/100g H_2O at 25°C.

The solubility of a salt is often measured not only in pure water but also in an acidic solution. For example, sulfate salts, salts that have SO_4^{2-} as the anion, that have a relatively low solubility remain so in an acidic solution whereas many carbonate salts (CO_3^{2-} is the anion), that also have a low solubility in water, dissolve in an acidic solution because of the reaction

$$CO_3^{2-}(aq) + 2\,H^+(aq) \rightarrow H_2CO_3\,(aq) \rightarrow H_2O + CO_2(g)$$

Therefore, it is easy to distinguish between SO_4^{2-} and CO_3^{2-} salts with limited solubility. Many of the phosphate salts (PO_4^{3-} is the anion) also have limited solubility but, like the carbonate salts, dissolve in an acidic solution. How then can we distinguish between a CO_3^{2-} and a PO_4^{3-} salt? One observance is that the PO_4^{3-} salts do not produce a gas as do the CO_3^{2-} salts.

$$PO_4^{3-}(aq) + H^+(aq) \rightarrow HPO_4^{2-}(aq)$$

Secondly, we occasionally need to use an additional test to confirm our observations. In the case of the PO_4^{3-} test, we add ammonium molybdate, $(NH_4)_2MoO_4$, which precipitates the HPO_4^{2-} as a yellow solid. Neither the SO_4^{2-} or the CO_3^{2-} salts behave in a similar manner.

$$HPO_4^{2-}(aq) + 12\,(NH_4)_2MoO_4(aq) + 23\,H^+(aq) \rightarrow \underset{\text{(yellow)}}{(NH_4)_3PO_4(MoO_3)_{12}(s)} + 21\,NH_4^+(aq) + 12\,H_2O$$

Therefore, we should be able to easily distinguish between salts that have SO_4^{2-}, CO_3^{2-}, or PO_4^{3-} as the anion. Obviously there are many other anions that form salts. One of the most common anions is the chloride ion, Cl^-. Most Cl^- salts are considered soluble, the silver salt, AgCl, being an exception.

$$Ag^+(aq) + Cl^-(aq) \rightarrow AgCl(s)$$

Another anion common to salts is the sulfide ion, S^{2-}. A quick test for presence of S^{2-} is to acidify a concentrated solution of the salt and note the odor–its smell is characteristic!!–you'll know. As a secondary test we can expose the vapors to lead ion, Pb^{2+}; this forms a black precipitate.

$$Pb^{2+}(aq) + \text{vapor} \rightarrow PbS(s)$$

In today's experiment you will determine which of the SO_4^{2-}, CO_3^{2-}, PO_4^{3-}, Cl^-, and S^{2-} salts are present in an unknown. You will accomplish this by testing for solubility in water and in acidic solutions. Most of the solids will be white.

Techniques

- Technique 2, page 2 — Transferring Liquid Reagents
- Technique 6a, page 7 — Heating Liquids
- Technique 9a, page 10 — Handling Gases
- Technique 13, Page 16 — Using the Laboratory Balance
- Technique 14, page 18 — Testing with Litmus

Procedure

1. Weigh about 0.10g (±0.01g) of Na_2SO_4 on weighing paper. Note closely the amount used; you will estimate this mass of salt for subsequent tests in the procedure–you will *not* need further use of a balance.

2. Transfer the Na_2SO_4 to a 150mm test tube. Add 3mL of distilled (or de-ionized) water and agitate the solution for about 30s. Does the Na_2SO_4 dissolve? Record on the Data Sheet.

3. Add drops of 6M NH_3 until the solution is basic to litmus; then add several drops of a 0.2M $Ba(NO_3)_2$ solution and agitate. Does a precipitate appear? Record.

4. Add drops of 3M HNO_3 (**Caution:** *Do not allow* HNO_3 *to contact the skin. If it does, wash immediately with plenty of water.*) until the solution is acidic to litmus; then add 5–6 drops more and agitate the solution. Look at the solution; is there any evidence of a reaction? Moisten a strip of filter paper with a few drops of 2M $Pb(C_2H_3O_2)_2$, and hold it at the mouth of the test tube; heat *gently.* Have any changes occurred? To test for odor, use your nose (according to Technique 9a). Record all of your observations.

5. Carefully divide the solution into two equal volumes in 75mm test tubes; do not transfer any precipitate.

   (a) To one, add several drops of 0.1M $AgNO_3$.
   (b) To a second, add 1mL of 1M $(NH_4)_2MoO_4$, agitate, and warm *gently.*
   Record evidence of any precipitate.

6. Repeat Parts 1 through 4 substituting for the Na_2SO_4 the following salts: Na_2CO_3, Na_3PO_4, NaCl, and Na_2S.

7. Your instructor will issue two unknown samples. Sample A consists of a single salt and Sample B consists of two salts. In each sample you will identify the anion(s) present.

IDENTIFICATION OF A SALT-LAB PREVIEW

Date________Name_______________________________________Lab Sec. _____Desk No._____

1. Where should the flame be positioned when heating a liquid in a test tube?

2. Describe the procedure for testing the odor of a chemical.

3. Describe the procedure for using litmus paper in testing the acidity of a solution.

4. What chemical test(s) can you use to distinguish between calcium chloride, $CaCl_2$, and calcium carbonate, $CaCO_3$, both of which are white solids?

IDENTIFICATION OF A SALT-DATA SHEET

Date________ Name______________________________________ Lab Sec. ____ Desk No.____

Analyzing your unknown. The test salts and the unknown samples are *pure* samples. You must realize that these tests would not be as well defined if impurities were present in the salt. Sample A will contain just one salt, a salt that contains only one of the anions. However it may be a salt that is nearly insoluble. If so, you should begin your testing with Part 3 and continue through the procedure. Sample B will be a little more difficult. For example, if a precipitate in Part 2 dissolves in Part 3, you won't know immediately whether PO_4^{3-} or CO_3^{2-} is present; you will need to complete Part 4 to determine the presence of PO_4^{3-}. Also the $CO_2(g)$ formation may be difficult to observe—you must watch closely. It would be advisable for you to check Sample B two or three times to ensure confidence in your data.

	Na_2SO_4	Na_2CO_3	Na_3PO_4	NaCl	Na_2S	Unknown A	Unknown B
Soluble in Water?							
Precipitate with $Ba(NO_3)_2$?							
Reaction with HNO_3?							
Precipitate with $Pb(C_2H_3O_2)_2$?							
Odor Test							
Precipitate with $AgNO_3$?							
Precipitate with $(NH_4)_2MoO_4$?							

Sample A: Identification Number_____ ; Anion present__________

Sample B: Identification Number _____ ; Anions present__________

Questions

1. What single test reagent will distinguish a soluble CO_3^{2-} salt from a soluble PO_4^{3-} salt. Assume an acidic solution.

2. What single test reagent will distinguish a soluble Cl^- salt from a soluble SO_4^{2-} salt?

3. What single test reagent will distinguish $BaSO_4$ from $BaCO_3$?

4. Describe how you could identify the presence of *both* Na_2CO_3 and $BaCO_3$ in a salt mixture?

5. Describe how you could identify the presence of *both* Na_3PO_4 and $Ba_3(PO_4)_2$ in a salt mixture.

Experiment 7
Inorganic Nomenclature

Objectives

- To name common inorganic compounds
- To write the formulas for common inorganic compounds

Principles

You have probably noticed that members of virtually all professions, even chemists, have a specialized language. In order to communicate with chemists internationally, some standardization of the technical language is necessary. Historically, common names for many compounds were used and are still universally understood–for example, water, sugar, and ammonia–but with new compounds being prepared daily, a random system is no longer viable. In this assignment you will learn a few systematic rules, established by the International Union of Pure and Applied Chemistry (IUPAC), for naming and writing compounds. You are undoubtedly already familiar with some symbols for the elements and the names for several common compounds (see Experiments 5 and 6); for example, NaCl is sodium chloride. Continued practice and work in writing formulas and naming compounds will make you even more knowledgeable of the chemist's vocabulary.

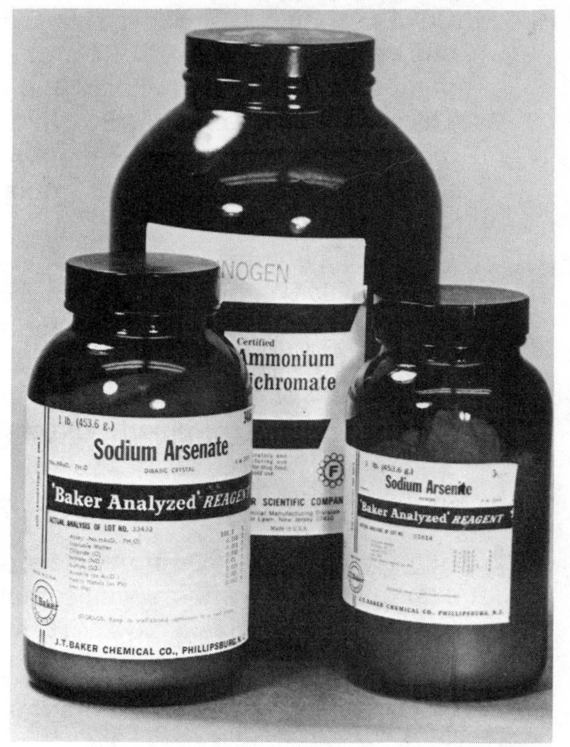

Inorganic compounds can be classified into several groups:

Binary Compounds. These compounds consist of only two elements. Three major subclassifications of binary compounds are:

Salts which consist of a metal and nonmetal
Compounds which consist of **two nonmetals**
Binary acids which consist of hydrogen and a more electronegative nonmetal *and* dissolved in water.

Compounds with Polyatomic Ions. These compounds *generally* consist of a metal cation and a polyatomic anion. One element in the polyatomic anion is almost always oxygen and the other is *usually* a nonmetal; the entire group of atoms has a negative charge.

NOMENCLATURE OF BINARY COMPOUNDS

1. Metal and Nonmetal—a Salt. In naming a binary salt, it is customary to list the more metallic (more electropositive) element first. The root of the second element is then listed with the suffix *-ide* added to it. NaBr is sodium brom*ide* Al_2O_3 is aluminum ox***ide.*** The names of a few have the-***ide*** ending, but are not binary compounds; these compounds contain the NH_4^+, OH^-, and CN^- ions. NH_4Cl is ammonium chlor*ide;* KOH is potassium hydrox*ide,* and NaCN is sodium cyan*ide.*

   Some metals, especially transition and post–transition metals, can form more that one cation resulting in the formation of more than one compound with a given anion. For example, copper can form Cu^+ and Cu^{2+} ions and therefore form two salts with chlorine: CuCl and $CuCl_2$. Two systems are used to distinguish the names of these salts. The older one applies a different suffix to the Latin root name for the metal: an ***-ous*** ending designates the ion with the lower charge and an ***-ic*** ending designates the ion with the higher charge. The newer one, called the Stock system, uses the English name for the metal but then indicates the charge on the ion with a Roman numeral in parentheses. The CuCl and $CuCl_2$ salts, as well as examples for iron and tin, are:

   - CuCl is cupr***ous*** chloride and copper(I) chloride
   - $CuCl_2$ is cupr***ic*** chloride and copper(II) chloride
   - $FeBr_2$ is ferr***ous*** bromide and iron(II) bromide
   - $FeBr_3$ is ferr***ic*** bromide and iron(III) bromide
   - SnO is stann***ous*** oxide and tin(II) oxide
   - SnO_2 is stann***ic*** oxide and tin(IV) oxide

   In the naming of salts, the Stock system is preferred.

2. Two Nonmetals. Two nonmetals oftentimes combine to form more than one compound; for examples carbon and oxygen combine to form carbon monoxide, CO, and carbon dioxide, CO_2, and nitrogen and oxygen combine to form N_2O, NO, NO_2, N_2O_3, N_2O_4, and N_2O_5. To distinguish the nitrogen oxides, we use the same system as that for the carbon oxides—Greek *prefixes* are used to indicate the number of atoms of each element in the compound. The common prefixes are in Table 7.2.

Table 7.1. Greek Prefixes

mono-*	one	hexa-	six
di-	two	hepta-	seven
tri-	three	octa-	eight
tetra-	four	nona-	nine
penta-	five	deca-	ten

*mono- is seldom used, since "one" is generally implied

3. Binary Acids. To name the binary acid, the prefix ***hydro-*** and the suffix ***–ic*** is added to the root name for the nonmetal and the word "**acid**" is added.
   - HCl is ***hydro***chlor***ic*** ***acid*** when hydrogen chloride is dissolved in water
   - HBr is ***hydro***brom***ic*** ***acid*** when hydrogen bromide is dissolved in water
   - H_2S is ***hydro***sulfur***ic*** ***acid*** when hydrogen sulfide is dissolved in water

NOMENCLATURE OF COMPOUNDS CONTAINING POLYATOMIC IONS

1. Metal Cation and a Polyatomic Anion Containing Oxygen—a Salt. When a metal combines with a polyatomic anion, a salt forms. The metal ion is named as it was for the binary salts (either using the "old" system or the Stock system); the polyatomic anion is named using the root of the "other" element, *not* the oxygen, and the suffix ***-ate*** or ***-ite***, depending on the number of oxygens—the polyatomic anion with the greater number of oxygens receives the ***-ate*** suffix. The sulf***ate*** ion is SO_4^{2-}, the nitr***ate*** ion is NO_3^-, and arsen***ate*** ion is AsO_4^{3-}; the sulf***ite*** ion is SO_3^{2-}, the nitr***ite*** ion is NO_2^-, and the arsen***ite*** ion is AsO_3^{3-}.

   - Na_2SO_4 is sodium sulf***ate*** ; Na_2SO_3 is sodium sulf***ite***
   - KNO_3 is potassium nitr***ate*** ; KNO_2 is potassium nitr***ite***

2. Acids Containing Polyatomic Anions (Oxoacids). Hydrogen substitutes for the metal ion in the salts to form the oxoacids. To name these acids we, first of all, do **not** name the hydrogen. We merely look at the polyatomic anion and change its ***-ate*** name to an "***-ic acid***" or its ***-ite*** name to an "***-ous acid***".

   - Na_2SO_4 is sodium sulf***ate*** ; H_2SO_4 is sulfur***ic acid***
   - Na_2SO_3 is sodium sulf***ite***; H_2SO_3 is sulfur***ous acid***
   - KNO_3 is potassium nitr***ate*** ; HNO_3 is nitr***ic acid***
   - KNO_2 is potassium nitr***ite***; HNO_2 is nitr***ous acid***

   Notice that the ***-ate*** suffix for the salt becomes the ***-ic*** suffix for the acid, while the ***-ite*** suffix for the salt becomes the ***-ous*** suffix for an acid.

   Some polyatomic anions have more than just a higher or lower number of oxygens, especially those having a halogen as the "other" element. In these other prefixes are added. For example, the chloro oxoacids and an appropriate salt are

   - $HClO_4$ is *per*chlor***ic*** acid; $KClO_4$ is potassium *per*chlor***ate***
   - $HClO_3$ is chlor***ic*** acid; $KClO_3$ is potassium chlor***ate***
   - $HClO_2$ is chlor***ous*** acid; $KClO_2$ is potassium chlor***ite***
   - $HClO$ is ***hypo***chlor***ous*** acid; $KClO$ is potassium ***hypo***chlor***ite***

   Bromine and iodine form similar oxoacids and salts. Notice again the ***-ic*** , ***-ate*** and the ***-ous*** , ***-ite*** relationships between the acids and the salts; the prefixes remain unchanged.

3. Acid Salts. When both a metal and hydrogen serve as the cations to the anion in a salt, it is called an **acid salt**. The presence of the number of hydrogens is indicated by the Greek prefix and its name. The suffix for the anion is that used for salts (binary or those with a polyatomic anion).

   - $NaHS$ is sodium hydrogen sulfide
   - NaH_2PO_4 is sodium dihydrogen phosphate
   - $CaHPO_4$ is calcium hydrogen phosphate
   - $Ca(H_2PO_4)_2$ is calcium dihydrogen phosphate

   An older system of naming acid salts substitutes the prefix ***bi-*** for a single hydrogen before naming the polyatomic anion; for example $NaHSO_4$ is sodium ***bi*sulfate** and $NaHCO_3$ is sodium ***bi*carbonate**.

WRITING FORMULAS

In writing formulas it is mandatory that the combined charges for the cations and anions in the salts/acids are equal to zero. Table 7.2 lists a number of common cations and anions. You should learn most, if not all, of those listed—others you will learn with experience. Lets try writing formulas for a few compounds (we'll need to use Table 7.2):

- barium fluoride. Barium is Ba^{2+} and fluoride is F^-. In order for the sum of the charges to be zero, one Ba^{2+} ion must combine with two F^- ions; the formula must be BaF_2.
- calcium nitride. Calcium is Ca^{2+} and nitride is N^{3-}. Three Ca^{2+} provides a 6^+ charge; this would balance two N^{3-} (a 6^- charge). The formula is Ca_3N_2.
- potassium oxalate. Potassium is K^+ and oxalate is $C_2O_4{}^{2-}$. Two K^+ balances one $C_2O_4{}^{2-}$; the formula is $K_2C_2O_4$.
- ferric chromate. Ferric is Fe^{3+} and chromate is $CrO_4{}^{2-}$. Two Fe^{3+} balances three $CrO_4{}^{2-}$; the formula is $Fe_2(CrO_4)_3$.

Table 7.2. Name and Charge of Common Ions for the Elements and of Polyatomic Ions

A. METALLIC AND POLYATOMIC CATIONS

Charge of 1^+

Ion	Name	Ion	Name
$NH_4{}^+$	ammonium	Li^+	lithium
Cu^+	copper(I), cuprous	K^+	potassium
H^+	hydrogen	Ag^+	silver
		Na^+	sodium

Charge of 2^+

Ion	Name	Ion	Name
Ba^{2+}	barium	Mn^{2+}	manganese(II), manganous
Cd^{2+}	cadmium	Hg^{2+}	mercury(II), mercuric
Ca^{2+}	calcium	$Hg_2{}^{2+}$	mercury(I), mercurous
Cr^{2+}	chromium(II), chromous	Ni^{2+}	nickel(II), nickelous
Co^{2+}	cobalt(II), cobaltous	Sr^{2+}	strontium
Cu^{2+}	copper(II), cupric	Sn^{2+}	tin(II), stannous
Fe^{2+}	iron(II), ferrous	$UO_2{}^{2+}$	uranyl
Pb^{2+}	lead(II), plumbous	VO^{2+}	vanadyl
Mg^{2+}	magnesium	Zn^{2+}	zinc

Charge of 3^+

Ion	Name	Ion	Name
Al^{3+}	aluminum	Co^{3+}	cobalt(III), cobaltic
As^{3+}	arsenic(III), arsenious	Fe^{3+}	iron(III), ferric
Cr^{3+}	chromium(III), chromic	Mn^{3+}	manganese(III), manganic

Charge of 4^+

Ion	Name	Ion	Name
Pb^{4+}	lead(IV), plumbic	Sn^{4+}	tin(IV), stannic

Charge of 5+

V^{5+}	vanadium(V)	As^{5+}	arsenic(V), arsenic

B. NONMETALLIC AND POLYATOMIC ANIONS

Charge of 1⁻

CH_3COO^-	acetate or	H^-	hydride
$C_2H_3O_2^-$	acetate	ClO^-	hypochlorite
Br^-	bromide	I^-	iodide
ClO_3^-	chlorate	NO_3^-	nitrate
Cl^-	chloride	NO_2^-	nitrite
CN^-	cyanide	ClO_4^-	perchlorate
F^-	fluoride	IO_4^-	periodate
OH^-	hydroxide	MnO_4^-	permanganate

Charge of 2⁻

CO_3^{2-}	carbonate	$C_2O_4^{2-}$	oxalate
CrO_4^{2-}	chromate	O_2^{2-}	peroxide
$Cr_2O_7^{2-}$	dichromate	SO_4^{2-}	sulfate
SiO_3^{2-}	silicate	SO_3^{2-}	sulfite
O^{2-}	oxide	S^{2-}	sulfide
		$S_2O_3^{2-}$	thiosulfate

Charge of 3⁻

N^{3-}	nitride	P^{3-}	phosphide
PO_4^{3-}	phosphate	BO_3^{3-}	borate
PO_3^{3-}	phosphite	AsO_3^{3-}	arsenite
		AsO_4^{3-}	arsenate

Procedure

On the Data Sheet your instructor will indicate which of the following exercises (or parts thereof) you are to answer.

1. Name the binary compounds

a. NaCl	e. Li_2S	i. SrS	m. Li_3N	q. IF_7
b. CsOH	f. Al_2O_3	j. V_2O_5	n. Ag_2O	r. $TiCl_4$
c. $CaBr_2$	g. ZnH_2	k. CaC_2	o. KCN	s. N_2O
d. MgO	h. AgI	l. BaI_2	p. NH_4Cl	t. NI_3

2. Name the following salts.

a. Na_2SO_4	g. Ag_2CrO_4	m. K_3PO_4	s. $VOCl_2$
b. KNO_3	h. $KC_2H_3O_2$	n. $(NH_4)_3PO_4$	t. UO_2Cl_2
c. Li_2CO_3	i. $NaMnO_4$	o. KIO_4	u. NaClO
d. $Cd(OH)_2$	j. $Li_2S_2O_3$	p. $KClO_3$	v. Na_2O_2
e. $Ca_3(PO_4)_2$	k. $Ba(NO_2)_2$	q. $Ca(C_2H_3O_2)_2$	w. $(NH_4)_2S_2O_3$
f. $K_2Cr_2O_7$	l. $AgClO_4$	r. $MgSO_4$	x. $K_2C_2O_4$

3. Name the following binary salts using the "-ic, -ous" system and the Stock system.

   a. CrS
   b. Fe_2O_3
   c. CrI_3
   d. $CuCl$
   e. PbO
   f. Hg_2Cl_2
   g. FeS
   h. As_2O_5
   i. $CuBr_2$
   j. $CoBr_3$
   k. $SnCl_2$
   l. PbO_2
   m. As_2S_3
   n. SnF_4
   o. CoO
   p. HgO
   q. $FeCl_3$
   r. SnO_2
   s. SnO
   t. $PbCl_2$
   u. Mn_2O_3
   v. NiS

4. Name the following salts using the "-ic, -ous" system and the Stock system.

   a. $FeSO_4$
   b. $As(NO_3)_3$
   c. $CuCN$
   d. $Hg(NO_3)_2$
   e. $CuCO_3$
   f. $CoSO_4$
   g. $Fe(OH)_3$
   h. $Cr(CN)_2$
   i. $Sn(NO_2)_4$
   j. $Co_2(CO_3)_3$
   k. $Pb(CH_3COO)_4$
   l. $As_2(SO_3)_3$
   m. $FePO_4$
   n. $Ni(NO_3)_2$
   o. $CuClO_4$
   p. $MnSO_4$
   q. $Hg_2(ClO_2)_2$
   r. $Hg(ClO_2)_2$
   s. $Co_3(PO_4)_2$
   t. $PbSO_4$

5. Name the following acids.

   a. HBr
   b. $HClO$
   c. H_2S
   d. H_2SO_3
   e. H_3PO_4
   f. H_3AsO_3
   g. $HMnO_4$
   h. H_2CrO_4
   i. H_3BO_3
   j. HCN
   k. HNO_2
   l. HIO_4
   m. H_2SO_4
   n. HNO_3
   o. $H_2S_2O_3$
   p. $HClO_4$
   q. H_3AsO_4
   r. H_3PO_3
   s. $H_2C_2O_4$
   t. H_2CO_3

6. Name the following acid salts. Use the "bi-" method wherever possible.

   a. $KHCO_3$
   b. $NaHSO_4$
   c. $Ca(HCO_3)_2$
   d. $KHSO_3$
   e. $NaHS$
   f. $NaHCO_3$
   g. $NH_4H_2PO_4$
   h. KH_2AsO_4
   i. Li_2HPO_4
   j. KHC_2O_4

7. Write formulas for each of the following compounds.

   a. ferrous sulfate
   b. potassium permanganate
   c. calcium carbonate
   d. iron(III) oxide
   e. cupric hydroxide
   f. aluminum sulfide
   g. mercury(II) chloride
   h. cadmium sulfide
   i. copper(I) chloride
   j. ammonium cyanide
   k. sodium chromate
   l. nickel(II) nitrate
   m. manganese(II) oxide
   n. manganese (IV) oxide
   o. lead(II) carbonate
   p. stannous chloride
   q. sodium nitrite
   r. barium acetate
   s. silver thiosulfate
   t. nitrogen triiodide
   u. potassium iodide
   v. sodium silicate
   w. calcium hypochlorite
   x. potassium chlorate
   y. cuprous iodate
   z. sodium dihydrogen phosphite
   aa. ammonium oxalate
   bb. potassium dichromate
   cc. copper(I) sulfate
   dd. cobalt(II) phosphate
   ee. chromium(III) phosphate
   ff. iron(II) oxalate
   gg. copper(I) sulfide
   hh. potassium periodate

8. Write formulas for each of the following compounds.

a. hydrogen fluoride
b. sulfuric acid
c. iodine pentafluoride
d. hydrobromic acid
e. hypobromous acid
f. hydrogen sulfide
g. sulfur trioxide
h. phosphorus pentafluoride
i. potassium borate
j. perchloric acid
k. nitrous acid
l. silicon dioxide
m. acetic acid
n. vanadyl acetate
o. tetraphosphorus decaoxide
p. dichlorine heptaoxide
q. mercurous cyanide
r. gold(III) nitrate
s. nitric acid
t. hydrocyanic acid
u. carbon monoxide
v. chlorous acid
w. uranyl sulfate
x. potassium phosphide
y. oxalic acid
z. thiosulfuric acid
aa. sodium bicarbonate
bb. calcium hydride
cc. boric acid
dd. carbonic acid
ee. zinc hydroxide
ff. dinitrogen pentaoxide
gg. hypochlorous acid
hh. xenon hexafluoride

Periodic Classification of the Elements

IA	IIA	IIIB	IVB	VB	VIB	VIIB	VIII	VIII	VIII	IB	IIB	IIIA	IVA	VA	VIA	VIIA	0
1 **H** 1.0079																1 **H** 1.0079	2 **He** 4.00260
3 **Li** 6.941	4 **Be** 9.01218											5 **B** 10.81	6 **C** 12.011	7 **N** 14.0067	8 **O** 15.9994	9 **F** 18.99840	10 **Ne** 20.179
11 **Na** 22.98977	12 **Mg** 24.305											13 **Al** 26.98154	14 **Si** 28.086	15 **P** 30.97376	16 **S** 32.06	17 **Cl** 35.453	18 **Ar** 39.948
19 **K** 39.098	20 **Ca** 40.08	21 **Sc** 44.9559	22 **Ti** 47.90	23 **V** 50.9414	24 **Cr** 51.996	25 **Mn** 54.9380	26 **Fe** 55.847	27 **Co** 58.9332	28 **Ni** 58.70	29 **Cu** 63.546	30 **Zn** 65.38	31 **Ga** 69.72	32 **Ge** 72.59	33 **As** 74.9216	34 **Se** 78.96	35 **Br** 79.904	36 **Kr** 83.80
37 **Rb** 85.4678	38 **Sr** 87.62	39 **Y** 88.9059	40 **Zr** 91.22	41 **Nb** 92.9064	42 **Mo** 95.94	43 **Tc** 98.9062	44 **Ru** 101.07	45 **Rh** 102.9055	46 **Pd** 106.4	47 **Ag** 107.868	48 **Cd** 112.40	49 **In** 114.82	50 **Sn** 118.69	51 **Sb** 121.75	52 **Te** 127.60	53 **I** 126.9045	54 **Xe** 131.30
55 **Cs** 132.9054	56 **Ba** 137.34	57 **La*** 138.9055	72 **Hf** 178.49	73 **Ta** 180.9479	74 **W** 183.85	75 **Re** 186.207	76 **Os** 190.2	77 **Ir** 192.22	78 **Pt** 195.09	79 **Au** 196.9665	80 **Hg** 200.59	81 **Tl** 204.37	82 **Pb** 207.2	83 **Bi** 208.9804	84 **Po** (210)	85 **At** (210)	86 **Rn** (222)
87 **Fr** (223)	88 **Ra** 226.0254	89 **Ac**** (227)	104 (260)	105 (260)													

***Lanthanum Series**

58	59	60	61	62	63	64	65	66	67	68	69	70	71
Ce	**Pr**	**Nd**	**Pm**	**Sm**	**Eu**	**Gd**	**Tb**	**Dy**	**Ho**	**Er**	**Tm**	**Yb**	**Lu**
140.12	140.9077	144.24	(147)	150.4	151.96	157.25	158.9254	162.50	164.9304	167.26	168.9342	173.04	174.97

****Actinium Series**

90	91	92	93	94	95	96	97	98	99	100	101	102	103
Th	**Pa**	**U**	**Np**	**Pu**	**Am**	**Cm**	**Bk**	**Cf**	**Es**	**Fm**	**Md**	**No**	**Lr**
232.0381	231.0359	238.029	237.0482	(244)	(243)	(247)	(247)	(251)	(254)	(257)	(258)	(255)	(256)

INORGANIC NOMENCLATURE-DATA SHEET

Date________Name____Lab Sec. ____Desk No.____

Problems assigned by laboratory instructor: ______, ______, ______, ______. Use the following table to answer the problems that you were assigned.

Formula	Name	Formula	Name

Instructor's Signature.

EXPERIMENT 8
FORMULA OF A HYDRATE

Objectives

- To determine the percent by weight of water in a hydrate
- To establish the mole ratio of salt to water

Principles

Many salts occurring in nature or purchased from chemical suppliers are hydrated ; that is, a number of water molecules is bound to the ions in the crystalline structure of the salt. The number of moles of water per mole of a given salt is usually constant. For example, ferric chloride is normally found as $FeCl_3 \bullet 6H_2O$, not as $FeCl_3$; cupric sulfate as $CuSO_4 \bullet 5H_2O$ or $CuSO_4 \bullet H_2O$, not as $CuSO_4$. For some salts, heat removes these hydrated water molecules:

$$Na_2CO_3 \bullet 10H_2O(s) \xrightarrow{\Delta} Na_2CO_3(s) + 10\,H_2O(g)$$

In sodium carbonate decahydrate, $Na_2CO_3 \bullet 10H_2O$, 10 moles of H_2O are bound to each mole of Na_2CO_3 or 180g of H_2O are bound to 106g of Na_2CO_3. The percent H_2O in the hydrated salt

$$\frac{180\text{ g } H_2O}{(180\text{g } H_2O + 106\text{g} Na_2CO_3)} \times 100 = 62.9\ \%\ H_2O$$

In other salts the water molecules cannot be easily removed, no matter how the intense the heat, e.g., $FeCl_3 \bullet 6H_2O$.

Techniques

- Techniques 8a, c, page 9 — Ignition of a Crucible
- Technique 13, page 16 — Using the Laboratory Balance, ±0.001g

Procedure

Complete 3 trials in this experiment. Obtain 3 crucibles and lids, identify each crucible and lid as a matched pair, and simultaneously perform the experiment in triplicate. While one crucible is cooling, another sample can be heated.

1. Support a clean crucible and lid on a clay triangle and heat with an intense flame for 5 minutes. Allow cooling. If the crucible is dirty, add a few milliliters of 6M HNO_3 (**Caution:** *Avoid skin contact, flush immediately with water*) and evaporate to dryness.[1] Weigh (±0.001g) the fired crucible and lid. Handle the crucible and lid with the crucible tongs for the rest of the experiment; do not use your fingers.

[1]Place the crucible and lid in a desiccator (if available) for cooling. Cool the crucible and lid in the same manner for the remainder of the experiment.

2. Add at most 3g of an unknown hydrate to the crucible and weigh (±0.001g) it, the lid, and sample.

3. Return the crucible with the sample to the clay triangle and set the lid off the crucible's edge to allow evolved gases to escape.

4. At first, heat the sample slowly and then gradually intensify the heat. Do not allow the crucible to become red hot. This can cause the anhydrous salt to decompose. Heat the sample for 15 minutes. Cover the crucible with the lid, cool to room temperature, and weigh it, the lid, and sample.

5. Reheat the sample for 5 more minutes. Reweigh. If the second weighing disagrees (>±2%) with first, repeat the heating until a constant weight is achieved.

FORMULA OF A HYDRATE-LAB PREVIEW

Date_________Name____Lab Sec. _____Desk No._____

1. Naturally occurring gypsum is a hydrate of $CaSO_4$. A gypsum sample weighing 4.335g, is heated in a crucible until reaching a constant weight. The weight of the anhydrous salt (salt without water), $CaSO_4$, is 3.428g.

   a. Calculate the percent by weight of water in the gypsum sample.

   b. Calculate the moles of H_2O removed and moles of $CaSO_4$ remaining in the crucible.

   c. What is the formula of gypsum?

2. Anhydrous $CaCl_2$ is used as a desiccant in desiccators—it removes water from the atmosphere within the desiccator to form a hydrate. A 16.43g $CaCl_2$ sample weighed 21.75g after being left in a desiccator for several weeks. What is the formula of the hydrated $CaCl_2$ salt?

3. In today's experiment what error in the data is likely to occur if the hydrated salt is heated too strongly? Read the Procedure.

4. How long should the hydrated salt be heated in removing the water?

5. After the crucible is weighed for the first time in Part 1 of the Procedure, you are advised to handle the crucible and lid with the crucible tongs only, not your fingers. Explain why this is good technique.

FORMULA OF A HYDRATE–DATA SHEET

Date________Name______________________________________Lab Sec. _____Desk No._____

Name of salt____________	Trial 1	Trial 2	Trial 3
1. Mass of crucible and lid (g)	______	______	______
2. Mass of crucible, lid, and hydrated salt. (g)	______	______	______
3. Mass of crucible, lid, and anhydrous salt 1st weighing (g)	______	______	______
2nd weighing (g)	______	______	______
3rd weighing (g)	______	______	______
4. Mass of hydrated salt (g)	______	______	______
5. Mass of anhydrous salt (g)	______	______	______
6. Moles of anhydrous salt (mol)	______	______	______
7. Mass of water lost (g)	______	______	______
8. Moles of water lost (mol)	______	______	______
9. Percent by mass of H_2O lost from hydrated salt	______	______	______
10. Average % H_2O in hydrated salt		______	
11. Mole ratio of anhydrous salt to water	______*	______	______
12. Formula of hydrate		______	

*Calculation for Trial 1. Show your work.

Questions

1. If some volatile impurities are not burned off in Part 1, but are removed in Part 4, is the mass of anhydrous salt too high or too low? Explain.

2. If the hydrated salt is not heated to a high enough temperature for a long enough period of time in Part 4, will the reported moles of water in the hydrated salt be too high or too low? Explain.

3. What happens to the sample's reported percent water if the salt decomposes, yielding a volatile product?

4. Anhydrous $CaCl_2$ removes water vapor from the atmosphere in a desiccator. Explain how $CaCl_2$ removes the water vapor.

EXPERIMENT 9
EMPIRICAL FORMULA OF A COMPOUND

Objective

- To determine the chemical formula of a compound between magnesium and oxygen or magnesium and chlorine.

Principles

The empirical formula of a compound specifies the simplest, whole–number, mole ratio of elements in the compound. This formula can be determined in the laboratory either by its synthesis from the elements or by a decomposition into its respective elements.

An example of determining the empirical formula from a decomposition reaction is given for mercuric oxide on the Lab Preview. Sodium chloride can be synthesized from its elements: 2.75g of sodium reacts with 4.25g of chlorine. The mole ratio of the two elements can be quickly determined:

$$\textbf{2.75g Na} \times \frac{\textbf{1mol Na}}{\textbf{23.0g Na}} = \textbf{0.120 mol Na}$$

$$\textbf{4.25g Cl} \times \frac{\textbf{1mol Cl}}{\textbf{35.45g Cl}} = \textbf{0.120 mol Cl}$$

Therefore, the mole ratio of sodium to chlorine is 0.120 to 0.120. Since the empirical formula must be a whole–number mole ratio, the mole ratio of sodium to chlorine is 1 to 1, the empirical formula is Na_1Cl_1 or NaCl.

The empirical formula also provides the mass ratio of the elements in the compound. The formula NaCl implies that 23.00g (1 mole) of sodium combines with 35.45g (1 mole) of chlorine. In NaCl the percents of sodium and chlorine are

$$\frac{23.00\text{g}}{(23.00\text{g} + 35.45\text{g})} \times 100 = 39.35\%\ \text{Na}$$

$$\frac{35.45\text{g}}{(23.00\text{g} + 35.45\text{g})} \times 100 = 60.65\ \%\ \text{Cl}$$

This experiment describes the syntheses of compounds between magnesium and oxygen and magnesium and chlorine. In each compound the empirical formula is determined using a series of techniques, measurements, and calculations.

Techniques

- Techniques 8a,b, page 9 — Ignition of a Crucible
- Technique 13, page 16 — Using the Laboratory Balance

Procedure

Three trials for at least one of the synthesis reactions are to be completed. Check out and identify additional crucibles and lids from the stockroom and follow the procedure in triplicate––while one synthesis reaction is occurring, prepare the next.

A. MAGNESIUM–OXYGEN SYNTHESIS

1. Fire a crucible and lid. Allow them to reach room temperature.[1] Weigh (±0.001g) the crucible only until a constant weight is achieved. For the rest of the experiment handle the crucible and lid with crucible tongs only.

2. Cut 0.2 to 0.3g of polished (with steel wool) Mg ribbon into short lengths. Place them into the crucible, weigh, and record.

3. Return the crucible, lid, and Mg to the clay triangle. Heat slowly, occasionally lifting the lid to allow air to the Mg (Figure 9.1). If too much air comes in contact with the Mg, rapid oxidation of the Mg occurs and it burns brightly (Experiment 5, Part D.1) with some probable loss of product. If this happens, immediately return the lid to the crucible.

4. Continue heating until no change is apparent in the Mg ash. Remove the lid and heat the open crucible with a hot flame for several minutes. Remove the heat; add a few drops of water to decompose any magnesium nitride[2] formed during combustion. Dry the ash with a low flame, cool, weigh the open crucible, and record.

5. Repeat Part 4 until repeated weighings <±3%.

---

[1] Allow the crucible to cool in a desiccator if available. Continue to use the desiccator for the remainder of the experiment.

[2] $3\,Mg(s) + N_2(g) \xrightarrow{\Delta} Mg_3N_2(s)$

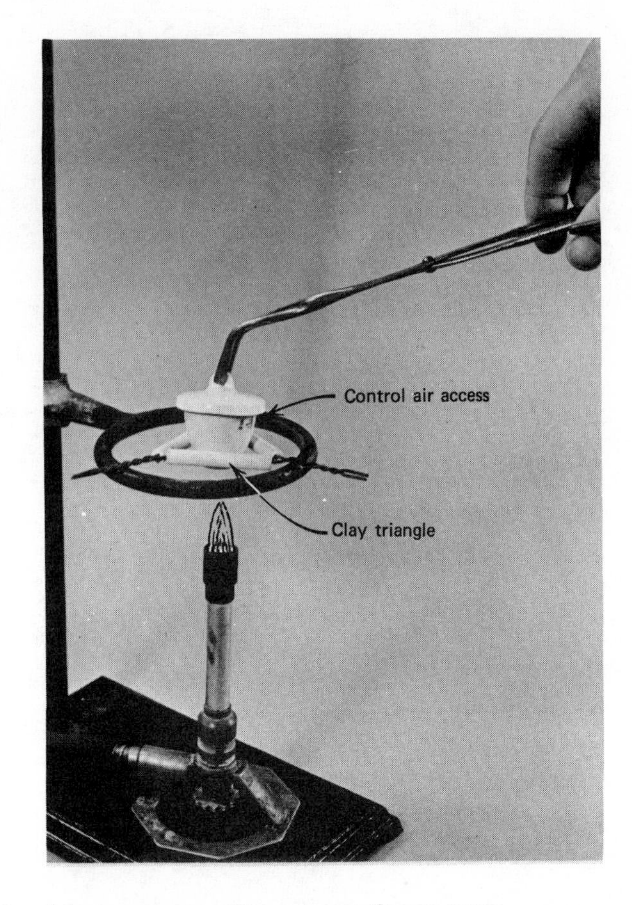

Figure 9.1
Controlling the Access of Air to the Mg Ribbon

B. MAGNESIUM-CHLORIDE SYNTHESIS

1. Prepare a crucible as described in Part A.1.

2. Cut 0.2 to 0.3g of polished (with steel wool) Mg ribbon into short lengths. Place them into the crucible, weigh, and record.

3. Place the crucible containing the Mg on a clay triangle. Slowly add drops of 6M HCl (**Caution**: *Avoid skin contact)* as the Mg reacts. After no reaction is apparent with the continued addition of the HCl, gently heat the reaction mixture and evaporate to dryness—do not boil. When the sample appears dry, continue heating for an additional minute; avoid excessive heating. Cover, cool (in a desiccator if available), weigh, and record.

4. Repeat Part 4 until successive weighings are $<\pm 3\%$

EMPIRICAL FORMULA OF A COMPOUND-LAB PREVIEW

Date________Name____Lab Sec. _____Desk No._____

1. Mercuric oxide decomposes with heat to form mercury metal and oxygen gas. When a 1.048g sample of mercuric oxide is heated, 0.971g of mercury remains.

   a. Calculate the moles of mercury and oxygen in the compound.

   b. What is the empirical formula of mercuric oxide?

2. A 2.60g sample of titanium metal reacts with 7.71g of chlorine gas to form a compound. Determine the empirical formula for the titanium chloride salt.

3. A 0.497g sample of chromium metal forms an oxide compound weighing 0.726g.
   a. Determine the empirical formula of the chromium oxide.

   b. Calculate the percent by mass of chromium and oxygen in the compound.

4. How many grams of oxygen gas combine with 0.843g vanadium to form V_2O_5?

5. Calculate the percent by mass of sulfur and bismuth in Bi_2S_3.

6. Explain how the mass of chlorine is determined in the synthesis of the magnesium-chlorine compound in Part B of the Procedure.

EMPIRICAL FORMULA OF A COMPOUND-DATA SHEET

Date________Name____Lab Sec. ____Desk No.____

SYNTHESIS______________	Trial 1	Trial 2	Trial 3
1. Mass of crucible and lid (g)	_______	_______	_______
2. Mass of crucible, lid, and Mg (g)	_______	_______	_______
3. Mass of crucible, lid, and sample after reaction			
1st weighing (g)	_______	_______	_______
2nd weighing (g)	_______	_______	_______
3rd weighing (g)	_______	_______	_______
4. Mass of Mg (g)	_______	_______	_______
5. Moles of Mg (mol)	______*	_______	_______
6. Mass of compound (g)	_______	_______	_______
7. Mass of oxygen/chlorine (g)	_______	_______	_______
8. Moles of oxygen/chlorine (mol)	_______	_______	_______
9. Mole ratio of Mg to oxygen/chlorine	_______	_______	_______
10. Empirical of compound	_______	_______	_______
11. Percent Mg by mass in compound	_______	_______	_______
12. Percent oxygen/chlorine by mass in compound	_______	_______	_______

* Show sample calculation for Trial 1

Questions

1. If in Part A.3 the magnesium oxidation is uncontrolled (burns brightly), how will this error in experimental technique affect the reported mass of

   a. magnesium in the sample? Explain.

   b. oxygen in the sample? Explain.

2. In this experiment the magnesium is weighed to the nearest milligram but not the oxygen or chlorine. Explain why this is unnecessary.

3. Suppose that in the synthesis reactions, the magnesium metal was not polished.

   a. How would this affect the calculated moles of magnesium as the starting material?

   b. How would this same error affect the calculated moles of oxygen/chlorine that combined with the magnesium?

4. In Part B, suppose all of the magnesium does not react with the HCl.

   a. How will this affect the report mass of chlorine in the compound?

   b. Does this affect the reported number of moles magnesium in the compound? Explain.

EXPERIMENT 10
LIMITING REACTANT

Objectives

- To determine the limiting reactant in the formation of a precipitate
- To determine the percent composition of a salt mixture

Principles

Two factors that limit the yield of products in a chemical reaction are (1) the amounts of starting materials (reactants) and (2) the percent yield of the reaction. Many experimental conditions (for example, temperature and pressure) can be adjusted to increase a reaction's yield, but because chemicals react according to fixed mole ratios (stoichiometrically), only a limited amount of product can formed from given amounts of starting material. The reactant restricting the amount produced is the limiting reactant in that chemical system.

To better understand the limiting reactant concept, let's look at the reaction studied in this experiment. The molecular form of the equation is

$$BaCl_2(aq) + Na_2SO_4(aq) \rightarrow BaSO_4(s) + 2\,NaCl(aq)$$

Since $BaCl_2$ (in the form of the dihydrate, $BaCl_2 \bullet 2H_2O$) and Na_2SO_4 are soluble salts and since $BaSO_4$ is insoluble, the ionic equation is

$$Ba^{2+}(aq) + 2Cl^-(aq) + 2\,Na^+(aq) + SO_4^{2-}(aq) \rightarrow BaSO_4(s) + 2\,Na^+(aq) + 2\,Cl^-(aq)$$

Cancelling spectator ions common to both sides of the equation, the net ionic equation is

$$Ba^{2+}(aq) + SO_4^{2-}(aq) \rightarrow BaSO_4(s)$$

One mole of Ba^{2+} [from 1 mole of $BaCl_2 \bullet 2H_2O$ (244.2g/mol)] reacts with one mole of SO_4^{2-} [from 1 mole of Na_2SO_4 (142.1g/mol)] to produce 1 mole of $BaSO_4$ precipitate (233.4g/mol), if the reaction proceeds to completion.

Suppose, however, only 2.00g of $BaCl_2 \bullet 2H_2O$ and 1.20g of Na_2SO_4 are in the reaction vessel. How many moles and grams of $BaSO_4$ are then produced? Since the equation reads in terms of moles, not grams, each reactant's number of moles (or millimoles, mmol) must be determined. Therefore,

$$2.00g\ BaCl_2 \bullet 2H_2O \times \frac{1 mol\ BaCl_2 \bullet 2H_2O}{244.2g\ BaCl_2 \bullet 2H_2O} = 0.00819\ mol\ BaCl_2 \bullet 2H_2O = 8.19\ mmol\ Ba^{2+}$$

$$1.20g\ Na_2SO_4 \times \frac{1\ mol\ Na_2SO_4}{142.1g\ Na_2SO_4} = 0.00844\ mol\ Na_2SO_4 = 8.44\ mmol\ SO_4^{2-}$$

Since 1.0 mol Ba^{2+} reacts with only 1.0 mol SO_4^{2-}, then the 8.19 mmol Ba^{2+} in the reaction vessel can only react with 8.19 mmol SO_4^{2-} (of the 8.44 mmol present) producing a maximum of 8.19 mmol $BaSO_4$. This consumes all of the Ba^{2+} in the vessel (Ba^{2+} is the *limiting reactant*) and leaves an excess of (8.44-8.19 =) 0.25 mmol of SO_4^{2-} (SO_4^{2-} is called the *excess reactant*).

Since the limiting reactant (Ba^{2+}) is now known, the theoretical yield of product ($BaSO_4$) is calculated from the balanced equation. Ba^{2+}, the limiting reactant, controls the moles and grams of $BaSO_4$ produced.

1.0 mol Ba^{2+} produces 1.0 mol $BaSO_4$

or 8.19 mmol Ba^{2+} produces 8.19 mmol $BaSO_4$ *or* 0.00819 mol $BaSO_4$

and $$0.00819 \text{ mol } BaSO_4 \times \frac{233.4\text{g } BaSO_4}{1\text{mol } BaSO_4} = 1.91\text{g } BaSO_4$$

Thus, 1.91g $BaSO_4$ form if the reaction is 100% complete.

In this experiment an unknown mixture of Na_2SO_4 and $BaCl_2 \bullet 2H_2O$ is added to water and $BaSO_4$ precipitates from the solution. We will measure the mass and calculate the moles of $BaSO_4$ that precipitate; from the balanced equation we can then calculate the moles and masses of $BaCl_2 \bullet 2H_2O$ and Na_2SO_4 in the original unknown mixture.

From a series of tests, the limiting and excess reactants are determined. The difference between the mass of the original salt mixture, $\mathbf{m_{SM}}$, and the mass of the limiting reactant, $\mathbf{m_{LR}}$, allows us to calculate the mass of excess reactant, $\mathbf{m_{XR}}$, in the salt mixture

$$\mathbf{m_{XR} = m_{SM} - m_{LR}}$$

The percent composition of the salt mixture can now be determined.

Techniques

- Technique 2, page 2 — Transferring Liquid Reagents
- Technique 4a, b, c, pages 4 & 5 — Separation of a Solid from a Liquid
- Technique 5, page 7 — Flushing a Precipitate from a Beaker
- Technique 13, page 16 — Using the Laboratory Balance

Procedure

Two trials are recommended for this experiment. To hasten the analyses, weigh duplicate unknown salt mixtures, and simultaneously follow the procedure for each. Label the beakers accordingly for Trial 1 and Trial 2 to avoid intermixing samples and solutions.

A. Precipitation of $BaSO_4$

1. Weigh (±0.001g) on weighing paper about 1.0g of the unknown salt mixture. Transfer it to a 400mL beaker and add 200mL (±0.2mL) of distilled (or de-ionized) water. Add, using a stirring rod, 1mL conc HCl (**Caution**: conc HCl *is a severe skin irritant. Flush the affected area with a large amount of water*). Stir the aqueous mixture with the stirring rod for about 1 minute and then allow the precipitate to settle.

2. Cover the beaker with a watchglass and maintain the solution at a temperature between 80°C and 90°C on a steam bath or with a low flame for 40-50 minutes (Figure 10.1). Allow the precipitate to settle.

3. Decant two 50mL (±0.2mL) volumes of the supernatant liquid into separate 100mL beakers, labeled Beaker I and Beaker II; save for Part B.

4. The $BaSO_4$ precipitate may either be gravity filtered or vacuum filtered. In either case, use fine porosity filter paper, such as Whatman No. 42 or Fisher*brand* Q2. Preweigh (±0.001g) the dry filter paper and seal it into the funnel with a few milliliters of distilled (or de-ionized) water. Discard this water from the receiving flask. Have your instructor approve your filtering apparatus.

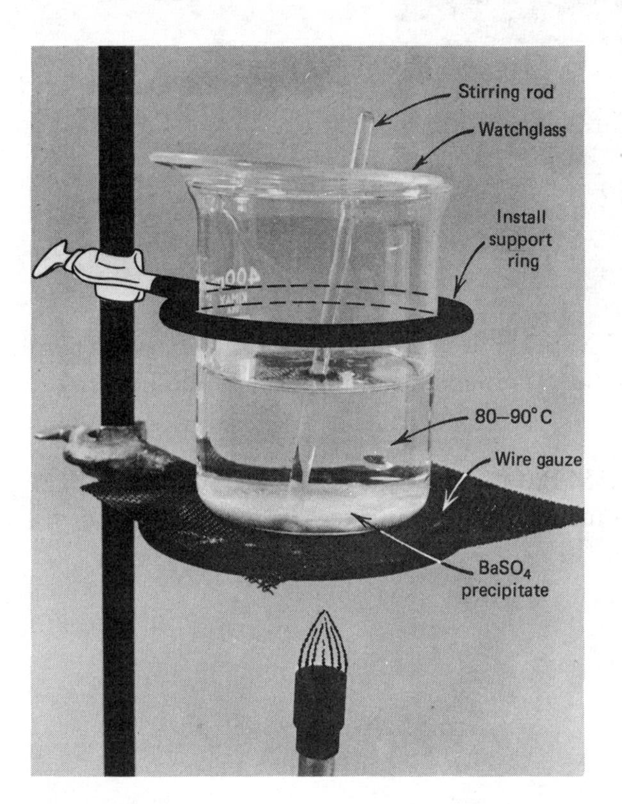

Figure 10.1
Warm the Precipitate to Digest the $BaSO_4$ Precipitate

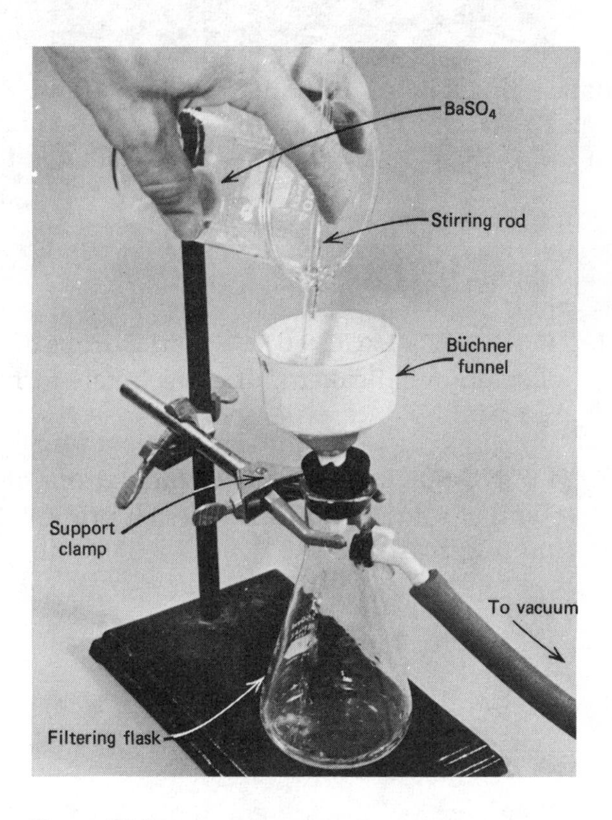

Figure 10.2
Quantitative Transfer of the Precipitate

5. While the solution is still warm, quantitatively transfer the precipitate to the funnel (Figure 10.2). Remove any precipitate from the beaker wall with a rubber policeman; use hot water to wash the precipitate onto the filter. Rinse the precipitate on the filter paper with two 5mL portions of hot water.

6. First, dry the precipitate on the filter paper; then dry overnight in a constant–temperature drying oven set at 110°C, or until the next laboratory period. Weigh the filter paper and precipitate. Record.

B. DETERMINATION OF THE EXCESS (AND THE LIMITING) REACTANT

The limiting reactant in the salt mixture is determined in the following tests. Use the two 50mL volumes collected in Part A.3. See Figure 10.3.

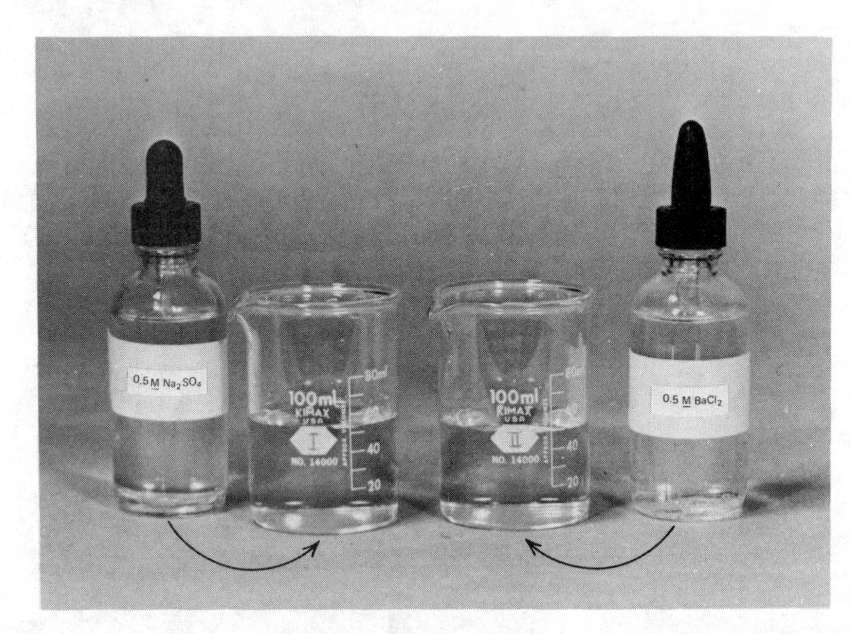

Figure 10.3
Testing for the Excess (and the Limiting) Reactant

1. TESTING FOR EXCESS BA^{2+}. Add 2 drops of 0.5 M SO_4^{2-} ion (from 0.5M Na_2SO_4) to the 50mL of solution in Beaker I. If a precipitate forms, Ba^{2+} is in excess and SO_4^{2-} is the limiting reactant.

2. TESTING FOR EXCESS SO_4^{2-}. Add 2 drops of 0.5M Ba^{2+} ion (from 0.5 M $BaCl_2$) to the other 50mL of solution in Beaker II. If a precipitate forms, SO_4^{2-} is in excess and Ba^{2+} is the limiting reactant.

LIMITING REACTANT-LAB PREVIEW

Date________Name____Lab Sec. _____Desk No._____

1. If 1.668g of $BaCl_2 \cdot 2H_2O$ is mixed with an excess of Na_2SO_4 to form 200mL of solution, what is the mass of $BaSO_4$ that can form?

2. A 1.668g sample of $BaCl_2 \cdot 2H_2O$ is mixed with 1.492g of Na_2SO_4 to form 200mL of solution.
   a. Calculate the moles of $BaCl_2 \cdot 2H_2O$ and the moles of Na_2SO_4 in the mixture.

   b. What is the limiting reactant?

   c. What is the mass of $BaSO_4$ that can form?

3. How is the test for the presence of excess $BaCl_2 \cdot 2H_2O$ in your unknown salt mixture conducted in today's experiment?

4. Describe the technique for transferring a precipitate from a beaker to a funnel.

5. A 1.582g sample of a $BaCl_2 \bullet 2H_2O/Na_2SO_4$ salt mixture, when mixed with water, filtered, and dried, produced 0.713g of $BaSO_4$. The Na_2SO_4 was determined to be the limiting reactant.
   a. Calculate the mass, in grams, of Na_2SO_4 in the mixture.

   b. Calculate the mass, in grams, of $BaCl_2 \bullet 2H_2O$ in the mixture.

   c. What is the percentage of Na_2SO_4 and $BaCl_2 \bullet 2H_2O$ in the salt mixture?

LIMITING REACTANT-DATA SHEET

Date________Name____Lab Sec. _____Desk No._____

A. PRECIPITATION OF $BaSO_4$

Unknown Number ______	Trial 1	Trial 2
1. Mass of salt mixture (g)	________	________
2. Mass of filter paper (g)	________	________
3. Instructor's approval of filtering apparatus ____________________		
4. Mass of filter paper and precipitate (g)	________	________
5. Mass of $BaSO_4$ precipitate (g)	________	________

B. DETERMINATION OF THE EXCESS (AND THE LIMITING) REACTANT

1. Limiting reactant in original salt mixture _______________

2. Excess reactant in original salt mixture _______________

CALCULATIONS

	Trial 1	Trial 2
1. Moles of $BaSO_4$ precipitate (mol)	________	________
2. Moles of $BaCl_2 \cdot 2H_2O$ reacted (mol)	________	________
3. Mass of $BaCl_2 \cdot 2H_2O$ reacted (g)	________	________
4. Moles of Na_2SO_4 reacted (mol)	________	________
5. Mass of Na_2SO_4 reacted (g)	________	________
6. Mass of salt mixture (from A.1), (g)	________	________
7. Mass of excess reactant (g)	________	________

8. Percent $BaCl_2 \cdot 2H_2O$ in mixture ____________________

9. Percent Na_2SO_4 in mixture ____________________

Questions

1. Since $BaSO_4$ is a very finely divided precipitate, some is lost in the filtering process. If a coarse filter paper is used instead of one with fine porosity, how will this affect the reported percent of limiting reactant in the original salt mixture?

2. The solubility of $BaSO_4$ at 25°C is 9.04 mg/L. How many grams and moles of $BaSO_4$ dissolved in the 200mL of solution?

3. How do excessive quantities of wash water affect the amount of $BaSO_4$ collected on the filter? Explain.

4. a. Na_2CO_3 is an unknown contaminant of Na_2SO_4. How does this affect the expected mass of $BaSO_4$ reported? $BaCO_3$ is also insoluble.

   b. Will this cause the reported percent of the limiting reactant to be high or low? Explain.

EXPERIMENT 11
CALORIMETRY

Objectives

- To determine the specific heat of a metal
- To measure the heat of neutralization for a strong acid–strong base reaction
- To measure the heat evolved or absorbed for the dissolution of a salt

Principles

In nearly all chemical and physical changes occurring in nature, heat (energy) is either evolved or absorbed. In the laboratory a calorimeter is used to measure this quantity of heat. Reactions which evolve heat are exothermic; their ΔH values (heat of reaction) are negative. Reactions that absorb heat are endothermic; their ΔH values are positive.

Three calorimetric measurements are made in this experiment. Each requires the measurement of temperature changes, masses of materials, and the plotting of the data.

SPECIFIC HEAT OF A METAL

The heat (in joules) required to change the temperature of 1g of a substance by 1 °C is the specific heat of that substance.

$$\textbf{specific heat (sp. ht.)} = \frac{\textbf{heat (joules)}}{\textbf{mass(grams)} \times \Delta T(°C)}$$

ΔT is the temperature change of the substance. The specific heat changes only slightly with temperature; we will assume that it remains constant over the temperature range used in this experiment. The specific heat of H_2O is 4.184J/g°C.

To determine the specific heat of a metal (non–reactive with water), it is first heated to a known temperature. The metal is then placed into a calorimeter containing a measured mass of (cooler) water at a measured temperature. The heat from the metal is transferred to the water until the metal and water reach the same temperature. This final equilibrium temperature is recorded. Expressed in equation form

$$\textbf{heat(J) lost by metal(M) = heat(J) gained by water(}H_2O\textbf{)}$$

Rearrange the (above) specific heat equation and solve for heat; substituting this gives

$$\text{sp. ht. (M)} \times \text{mass (M)} \times \Delta T(°C) = \text{sp. ht. }(H_2O) \times \text{mass }(H_2O) \times \Delta T(°C)$$

Rearranging to solve for the specific heat of the metal,

$$\text{sp. ht. (M)} = \frac{\text{sp. ht. }(H_2O) \times \text{mass }(H_2O) \times \Delta T\ (H_2O)}{\text{mass (M)} \times \Delta T\ \text{(M)}}$$

Each equation assumes no heat loss to the calorimeter.

HEAT OF NEUTRALIZATION FOR AN ACID-BASE REACTION

The reaction of a strong acid with a strong base produces water and heat.

$$H^+(aq) + OH^-(aq) \rightarrow H_2O + \text{heat} \quad (\Delta H_n \text{ is negative})$$

The heat of neutralization, ΔH_n, is calculated by (a) assuming the densities and specific heats of the acidic and basic solutions are the same as that for water and (b) measuring the temperature change when the two solutions are mixed.

$$\Delta H_n = \text{sp. ht. } (H_2O) \times \text{mass (acid + base)} \times \Delta T \text{ (solution)}$$

ΔH_n is generally expressed in units of kJ/mol of acid (or base) reacted.

HEAT OF SOLUTION FOR A SALT

The heat of solution of a salt may be endothermic or exothermic depending upon two factors often considered in the dissolution of the salt: the lattice energy of the salt and the hydration energy of the ions. The lattice energy is the energy required (endothermic process, a $+\Delta H_{LE}$) to vaporize one mole of the solid salt into its gaseous ions; the hydration energy is the energy released (exothermic quantity, a $-\Delta H_{hyd}$) when the gaseous ions are attracted to the water molecules. The heat of solution of a salt is the sum of these two factors (Figure 11.1 for NaCl).

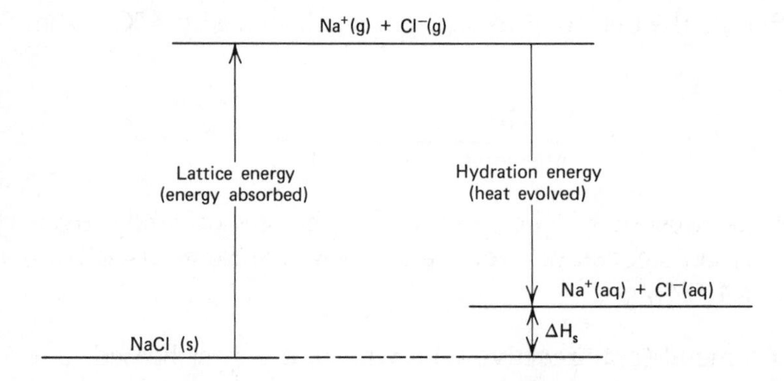

Figure 11.1
Energy Changes in the Dissolving of NaCl in Water

The heat of solution, ΔH_s, is determined by adding the heat changes for the salt and water.

$$\Delta H_s = \text{heat change } (H_2O) + \text{heat change (salt)}$$

$$\Delta H_s = \frac{\text{sp. ht.}(H_2O) \times \text{mass}(H_2O) \times \Delta T(H_2O) + \text{sp. ht.(salt)} \times \text{mass(salt)} \times \Delta T\text{(salt)}}{\text{mass(salt)}}$$

ΔH_s is generally expressed in units of kJ/g salt. The specific heats of some common salts are listed in Table 11.1.

Table 11.1 Specific Heat of Some Salts

Salt	Formula	Specific Heat (J/g°C)
ammonium chloride	NH_4Cl	0.506
ammonium nitrate	NH_4NO_3	1.28
ammonium sulfate	$(NH_4)_2SO_4$	1.18
sodium sulfate	Na_2SO_4	0.845
sodium thiosulfate pentahydrate	$Na_2S_2O_3 \cdot 5H_2O$	1.45
potassium bromide	KBr	0.435
potassium nitrate	KNO_3	0.895

Techniques

- Technique 1, page 1 — Inserting a Thermometer Through a Rubber Stopper
- Technique 6c, page 7 — Heating Liquids
- Technique 10, page 12 — Reading a Meniscus
- Technique 13, page 16 — Using the Laboratory Balance
- Techniques 15, page 18 — Graphing Techniques

Procedure

Ask your instructor which parts of the experiment you are to complete. You and a partner are to complete two trials for each part assigned. Two styrofoam cups, a lid, and a 110°C thermometer must be obtained from the stockroom–this will serve as your calorimeter. In this experiment do not use the thermometer as a stirring rod!!

A. Specific Heat of a Metal

1. Ask the instructor for 5–10g of an unknown metal sample. Weigh (±0.01g) it in a dry, previously weighed (±0.01g), 200mm test tube. Place the 200mm test tube in a 400mL beaker filled with water well above the level of the metal sample in the test tube (Figure 11.2). Heat to boiling and maintain this temperature for at least 5 minutes so that the metal reaches thermal equilibrium with the boiling water. Measure the temperature of the water (±0.1°C).

2. Set up the calorimeter (Figure 11.3). Using a graduated cylinder, add 25.0mL (±0.1mL) of water to the calorimeter. Be certain the thermometer bulb is below the water level. Read the temperature of the water several times. Once the temperature remains constant (at thermal equilibrium) record(±0.1°C) on the Data Sheet.

3. Now that thermal equilibrium has been reached in Parts A.1 and A.2, remove the test tube from the boiling water; quickly transfer the metal to the calorimeter. Be careful not to break the thermometer and not to splash water from the calorimeter. Replace the lid and swirl gently. Record the temperature (±0.1°C) as a function of time (10-20s intervals) on the table on the Data Sheet.

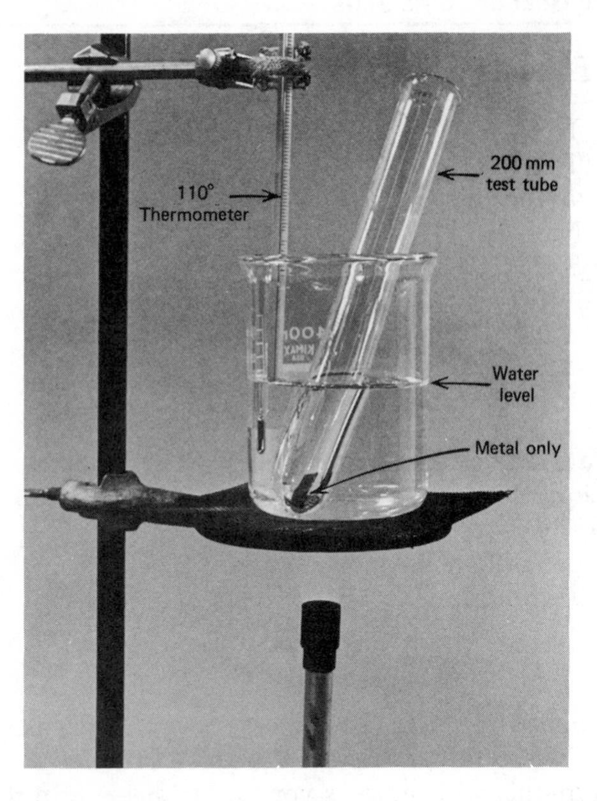

Figure 11.2
The Placement of the Metal *Below* the Level of the Water in the Beaker

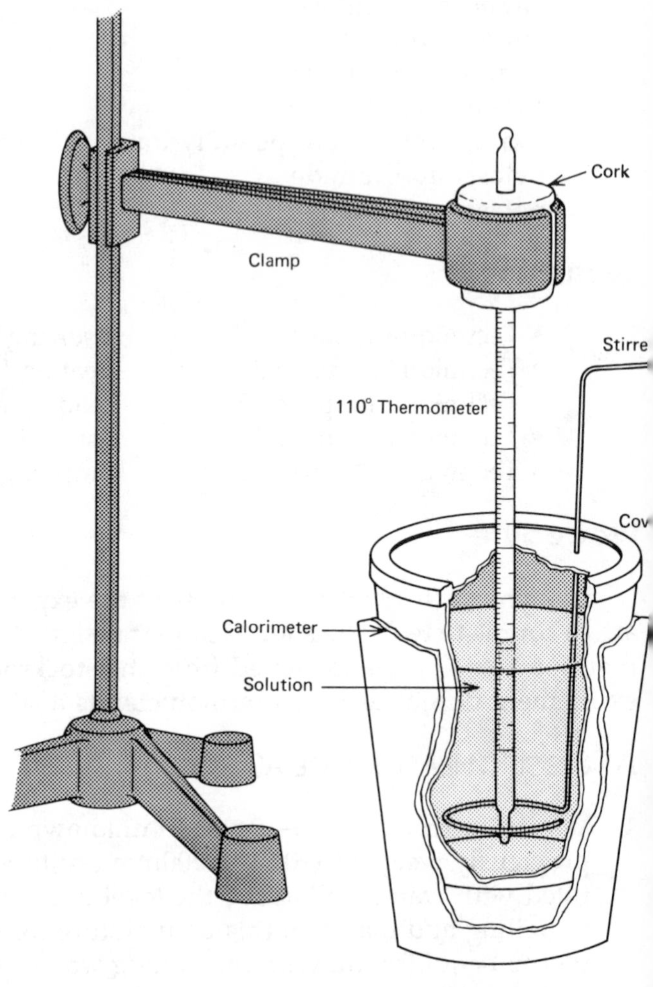

Figure 11.3
The Setup of a Calorimeter →

4. Plot on linear graph paper, temperature (ordinate) *vs* time (abscissa) (Figure 11.4) to determine the maximum temperature change. The maximum temperature occurs at the intersection of two drawn lines: a straight line drawn perpendicular to the initial temperature/time line at the point when the metal is added to the calorimeter and the best straight line drawn through the points after the maximum temperature is reached. The maximum temperature is never recorded because some heat is always being transferred to or from the wall of the calorimeter. Have your instructor approve your graph.

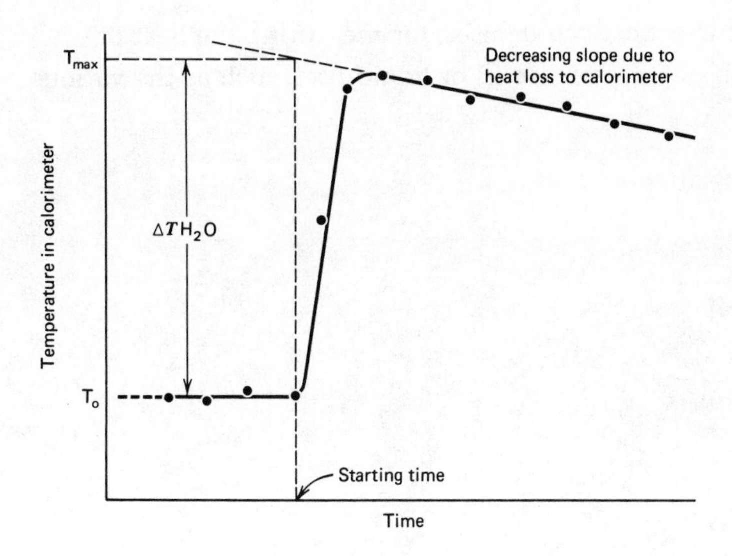

Figure 11.4
The Temperature of the Water in the Calorimeter as a Function of Time

B. Heat of Neutralization for an Acid-Base Reaction

1. Clean and dry the calorimeter. Using a graduated cylinder, pour 50.0mL (±0.1mL) of standard[1] 1.0M NaOH solution into the calorimeter, insert the thermometer through the lid down into the solution. Record the temperature (±0.1°C) of the NaOH solution. Record the exact concentration of the NaOH solution on the Data Sheet.

2. Measure 50.0mL (±0.2mL) of 1.1M HCl in a clean, graduated cylinder. Rinse the thermometer to remove any NaOH solution and then measure the temperature of the HCl solution. Carefully but quickly, add the acid to the base, replace the lid, and swirl gently. Record the solution temperature (±0.1°C) as a function of time on the Data Sheet. Plot on linear graph paper and interpret the maximum temperature obtained (Part A.4). Have your instructor approve your graph.

C. Heat of Solution for a Salt

1. Clean, rinse, and dry your calorimeter. Add 25.0mL (±0.1mL) of distilled (or de–ionized) water to the calorimeter and record the temperature (±0.1°C).

2. Weigh (±0.01g) on weighing paper about 5.0g of your assigned salt. Add the salt to the calorimeter, replace the lid, and swirl gently. Measure and record the temperature as a function of time on the Data Sheet.

3. Plot on linear graph paper and interpret the maximum temperature change (Part A.4). Have your instructor approve your graph.

---

[1]A standard solution is one whose concentration has been very accurately determined.

This is a plain jacket bomb calorimeter that can be used for measuring the heat of combustion (at constant volume), ΔH_{comb}, for any solid or liquid fuel, such as the various foodstuffs found in diet books.

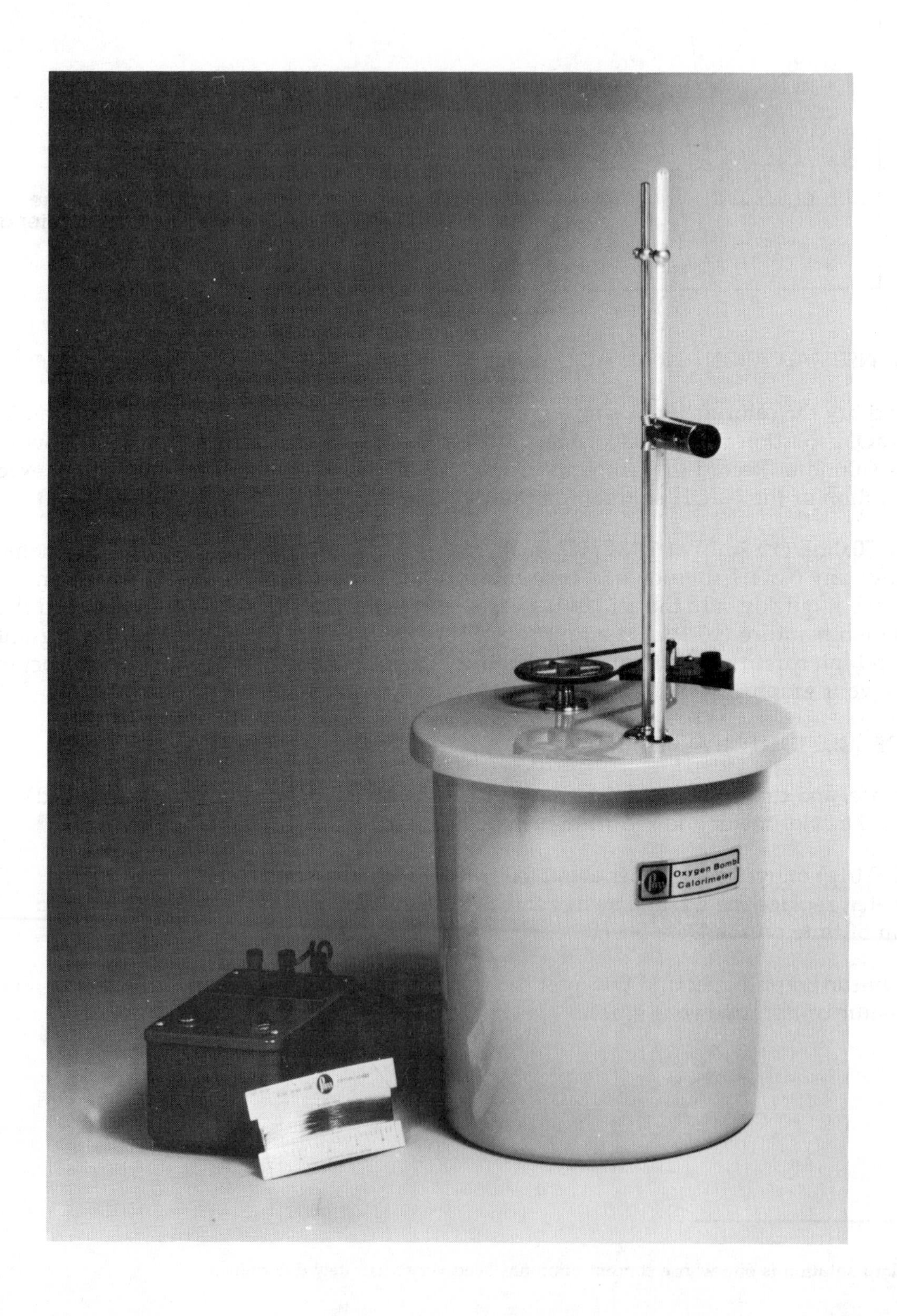

CALORIMETRY-LAB PREVIEW

Date________Name____Lab Sec. _____Desk No._____

1. A metal weighing 13.11g at 81.0°C is placed in a calorimeter containing 25.0mL of water at 25°C. The final equilibrium temperature is 30°C. What is the specific heat of the metal? The specific heat of H_2O is 4.184J/g°C. Assume the density of water to 1.00g/mL.

2. A 4.5g sample of a salt dissolves in 30.0mL of water initially at 25°C. The final equilibrium temperature is 18.0°C. What is the heat of solution per gram of salt? The specific heat of the salt is 0.692J/g°C.

3. In Part B, excess moles of HCl are added to the NaOH solution. Why is this procedure preferred rather that adding a 1:1 mole ratio according to the balanced equation?

4. Will the recorded temperature change for an exothermic reaction inside a glass calorimeter be greater or less than that for today's styrofoam "coffee cup" calorimeter? Assume glass to be a better conductor of heat than styrofoam.

5. (a) What lubricant should be used when inserting a thermometer through a rubber stopper?

(b) How can "bumping" be avoided when water is being heated in a beaker?

(c) When determining the volume of solution in a graduated cylinder, you should always read the ____________________of the meniscus.

(d) A balance with _____ g sensitivity is used in today's experiment.

(e) The margin of error in reading the thermometer in today's experiment is ± _____°C.

(f) What are you *not* to do with your thermometer in today's experiment?

CALORIMETRY-DATA SHEET

Date________Name____Lab Sec. _____Desk No._____

A. SPECIFIC HEAT OF A METAL

Unknown number ______________	Trial 1	Trial 2
1. a. Mass of test tube + metal (g)	__________	__________
b. Mass of test tube (g)	__________	__________
c. Mass of metal (g)	__________	__________
d. Temperature of metal (°C)	__________	__________
2. a. Volume of water in calorimeter (mL)	__________	__________
b. Mass of water (g); assume density of water is 1.0g/mL	__________	__________
c. Temperature of water (°C)	__________	__________
3. a. Instructor's Approval of graph	__________	__________
b. Maximum temperature of metal and water from graph (°C)	__________	__________

CALCULATIONS

	Trial 1	Trial 2
1. Temperature change of water, ΔT (°C)	__________	__________
2. Heat gained by water (J)	__________	__________
3. Temperature change of metal, ΔT (°C)	__________	__________
4. Specific heat of metal (J/g°C)	__________	__________

5. Average specific heat of metal (J/g°C) ______________

B. HEAT OF NEUTRALIZATION FOR AN ACID-BASE REACTION

	Trial 1	Trial 2
1. a. Concentration of NaOH solution (mol/L)	__________	__________
b. Initial temperature of NaOH solution (°C)	__________	__________
c. Volume of NaOH solution (mL)	__________	__________

2. a. Initial temperature of HCl solution (°C)	----------	----------
b. Volume of HCl solution (mL)	----------	----------
3. a. Instructor's approval of graph	----------	----------
b. Maximum final temperature of mixture from graph (°C)	----------	----------

CALCULATIONS

1. Average initial temperatures (NaOH + HCl)	----------	----------
2. Temperature change, ΔT (°C)	----------	----------
3. Mass of final mixture. Assume density of mixture is 1.0g/mL	----------	----------
4. Specific heat of solution	4.184J/g•°C	
5. Heat evolved (J)	----------	----------
6. Moles of OH^- reacted (limiting reactant) (mol)	----------	----------
7. Moles of H_2O formed (mol)	----------	----------
8. Heat evolved per mole of water (J/mol H_2O)	----------	----------
9. Average ΔH_n (kJ/mol H_2O)	______________	

C. HEAT OF SOLUTION FOR A SALT

Salt______________________	Trial 1	Trial 2
1. a. Volume of water (mL)	----------	----------
b. Mass of water (g). Assume density of H_2O is 1.0g/mL	----------	----------
c. Initial temperature of water (°C)	----------	----------
2. Mass of salt (g)	----------	----------
3. a. Instructor's approval of graph	----------	----------
b. Maximum (or minimum) temperature of solution from graph (°C)	----------	----------

CALCULATIONS

1. Change in temperature of solution, ΔT (°C) ________ ________
2. Heat change of water (J) ________ ________
3. Heat change of salt (J). See Table 11.1 ________ ________
4. Heat of solution, ΔH_S (J) ________ ________
5, ΔH_S per gram of salt (J/g) ________ ________
6. Average ΔH_S per gram salt (J/g) ________________

Specific Heat of a Metal				Heat of Neutralization for an Acid-Base Reaction				Heat of solution for a Salt			
Trial 1		Trial 2		Trial 1		Trial 2		Trial 1		Trial 2	
Temp	Time	Temp	Time	Temp	Time	Temp	Time	Temp	Time	Temp	Time

Questions

1. The coffee cup calorimeter, although a good insulator, absorbs some heat when the system is above room temperature.

   a. How does this affect the specific heat of the metal?

   b. How does this affect the reported ΔH_n value in the acid-base reaction?

2. The specific heat of the styrofoam calorimeter is 1.34 J/g°C. If we assume that the entire inner cup reaches thermal equilibrium with the solution, how much heat is lost to the calorimeter in Part B? Assume the inner cup weighs 2.35g.

3. If the maximum recorded temperature in Part B is used to determine ΔT rather than the maximum extrapolated temperature, will the reported ΔH_n be greater or less than the actual ΔH_n for this reaction? Explain.

4. Suppose that when the metal was transferred to the calorimeter in Part A, some of the water splashed from the calorimeter. How does this affect the reported specific heat of the metal? Explain.

EXPERIMENT 12
SPECTROSCOPY

Objectives

- To account for a line spectrum
- To observe the flame emission colors from excited metallic ions
- To study the hydrogen spectrum
- To identify an element from its spectrum

Principles

The current model of the atom includes a nucleus, comprised of protons and neutrons, surrounded by electrons. From experimental observations of atomic absorption and emission spectra, and from a quantum mechanical interpretation of spectra, the energy of an electron in an atom is assumed to be *quantized*; that is, an electron in an atom has only discrete energies.

When an atom absorbs energy from a flame or electric discharge, it absorbs the energy necessary to excite one of its electrons to a higher energy state—the atom is now in an excited state. When the electron returns to its original energy state, it emits the energy previously absorbed in the form on one or more photons. The energy of the excited atom decreases by ΔE_{atom} when the electron returns from a higher, E_h, to a lower, E_l, energy state. This energy change of the atom equals the energy of one photon.

$$\Delta E_{atom} = E_h - E_l = E_{photon}$$

E_h ——— Higher energy state

→ E_{photon}

E_l ——— Lower energy state

The photon energy, E_{photon}, equal to the energy difference between two energy states for the electron, is inversely proportional to its wavelength, λ, by the equation,

$$E_{photon} = h\frac{c}{\lambda} = h\nu$$

c equals the speed of light, 3.00×10^8 m/s, **h** is Planck's constant, 6.63×10^{-34} J s/photon, ν is the frequency of the photon, and λ is the wavelength of the photon in meters.

Many possible energy states exist for electrons in an atom. For example, suppose a large number of atoms of a particular element are excited–the electrons now occupy excited energy states. As the electrons return to their ground state (the lowest energy state) photons of different energies are emitted; the number of photons depends on the stability of the intermittent energy states of the atom. The photons are of characteristic energies, characteristic of these electronic energy states in the atom. When these photons pass through a prism, they produce a **line**

spectrum (Figures 12.1 and 12.2); each line corresponds to electron transitions which release photons of a characteristic energy and wavelength.

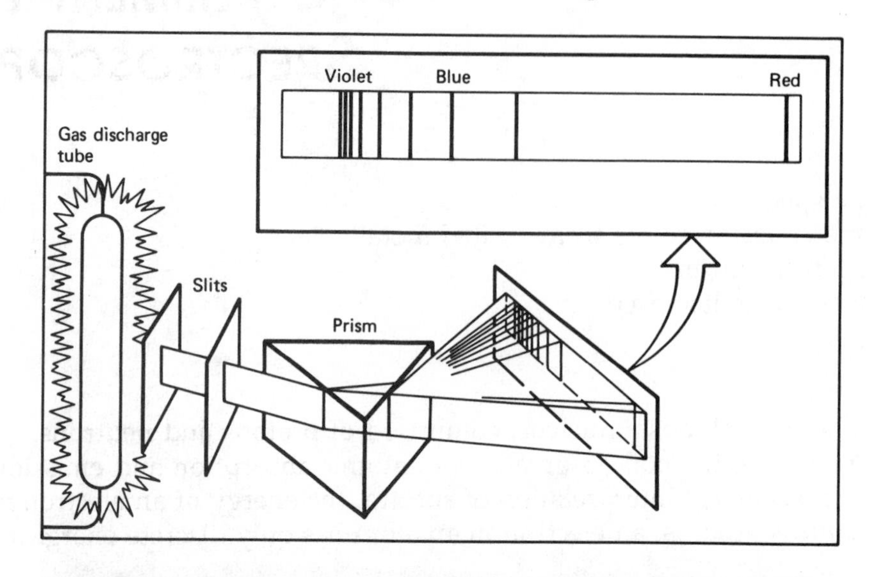

Figure 12.1
A Line Spectrum

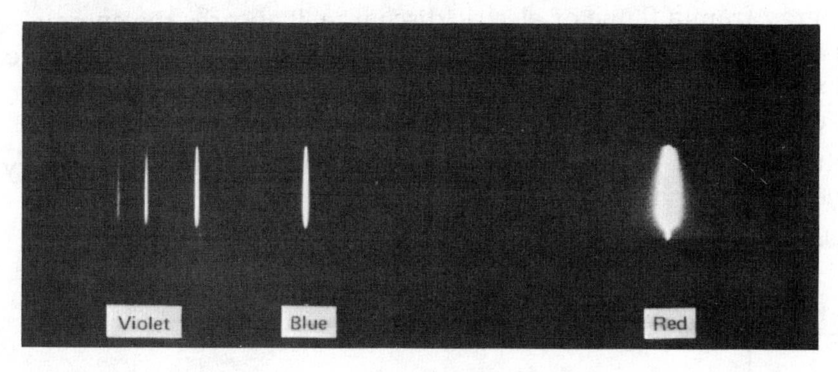

Figure 12.2
The Visible Line Spectrum of Hydrogen Gas

Each element has its own unique line spectrum because the energy states for electrons in the atoms of each element are unique. Sodium having 11 electrons has a different set of electronic energy states, and thus exhibits a different line spectrum, than does barium or mercury. The most prevalent mode of electron de-excitation produces the most intense line in the line spectrum, producing the color characteristic of that element. For example, many of the electrons in excited sodium atoms commonly de-excite to produce a photon having a wavelength in the orange part of the visible spectrum. We, therefore, associate excited sodium atoms as being orange.[1] By the same account, we associate blue light with mercury vapor lamps (a de-excitation of electrons in excited mercury atoms with an energy change equivalent to a photon in the blue region of the spectrum) and green with excited barium atoms. The characteristic colors associated with a number of excited atoms are observed in this experiment, Part A.

Hydrogen Spectrum

The value or designation of each electronic energy state in a hydrogen atoms is calculated in this experiment, using the equations,

---

[1](Orange) sodium vapor lamps are slowly replacing (blue) mercury vapor lamps for lighting streets and highways in public areas.

$$E = h\frac{c}{\lambda} = h\nu\ ; \quad \frac{1}{\lambda} = R\left[\frac{1}{n_l^2} - \frac{1}{n_h^2}\right]$$

The wavelength, λ, of photons emitted from excited hydrogen atoms is related to whole number integers, n, which identify the higher energy states for the electron before de-excitation, n_h, and the lower energy states after its de-excitation, n_l. For all electron de-excitations, $n_h > n_l$; for the visible part of the spectrum, n_l equals 2. R, the Rydberg constant, is approximately $1.1 \times 10^{-2}\ nm^{-1}$.

In Part B of this experiment, we will observe the hydrogen spectrum, estimate the wavelengths of the emitted photons, and calculate the energy change of the atom associated with each photon. Next we will calculate an n_h value for each line in the spectrum; if this value is not a whole number, then we will round it off to the *nearest* integer. A plot of the data will enable us to determine a more precise value of the Rydberg constant, R. This is accomplished by rearranging the equation $\frac{1}{\lambda} = R\left[\frac{1}{n_l^2} - \frac{1}{n_h^2}\right]$ into an equation for a straight line, $y = mx + b$.

$$\begin{array}{ccccc} y = & m & x & + & b \\ \frac{1}{\lambda} = & -R & \left[\frac{1}{n_h^2}\right] & + & \left[\frac{R}{n_l^2}\right] \end{array}$$

Since $n_l = 2$ in the *visible* spectrum for hydrogen, a plot of $1/\lambda$ *versus* $1/n_h^2$ has a y-intercept of R/4 and a slope of -R. Remember n_h must be a whole number.

In Part C, the wavelengths in the line spectrum of an unknown element are determined by comparing their position with the known wavelengths from the mercury line spectrum, Table 12.1. Then a match of the unknown's spectrum with the spectra for several known elements in Table 12.2 is made for its identification.

Procedure

A. Flame Tests

1. Dip the end of a platinum or nichrome wire into conc HCl (**Caution**: *Avoid skin contact. Flush affected area with large amounts of water*). Heat the wire in the flame's hottest region (Figure 1.2) until there is no visible color (Figure 12.3). Repeat this cleaning procedure as necessary.

2. In a watch glass, add a few crystals of $CaCl_2$ to 2 or 3 drops of conc HCl. Stir. Dip the clean wire into the solution and return it to the flame's hottest part. Note the color of the flame (view it through a spectroscope, if available). Correlate the color of the flame with the dominate wavelength of the sample's emission using the chart on the Data Sheet. Check the appropriate section.

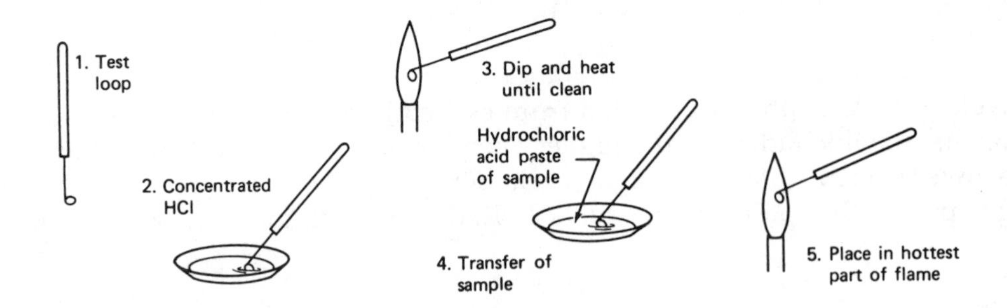

Figure 12.3
The Procedure for Performing a Flame Text

3. Repeat Steps 1 and 2 with $CuCl_2$, NaCl, $BaCl_2$, LiCl, $SrCl_2$, and KCl. For KCl view the flame through a cobalt glass plate.[2]

B. HYDROGEN SPECTRUM

1. In this experiment a photographic slide is used to view the hydrogen spectrum. Insert the H/Hg slide into the projector. Notice the three spectra on the slide. A continuous spectrum is at the bottom of the slide, next the H–spectrum, and at the top the Hg–spectrum. Align the visible lines of the Hg–spectrum with the wavelengths in Table 12.1, which are also marked on the tear–out wavelength scale.

Table 12.1
The Wavelengths of the Visible Lines in the Mercury Spectrum

Violet	404.7 nm
Violet	407.8 nm
Blue	435.8 nm
Green	546.1 nm
Yellow	577.0 nm
Yellow	579.1 nm

2. Mark the position of the hydrogen lines on the tear–out wavelength scale and record the wavelengths of the hydrogen spectrum. Calculate $\Delta E_{atom} = E_{photon}$, $1/\lambda$, n_h, $1/n_h^2$. Plot $1/\lambda$ *vs* $1/n_h^2$ to determine the Rydberg constant. Have your instructor approve your graph.

C. UNKNOWN SPECTRUM

Using an unknown/Hg slide (see instructor), identify the wavelengths for the lines in the spectrum of the unknown. Using Table 12.2, identify the element having the unknown spectrum.

---

[2]A cobalt glass plate filters all wavelengths other than those emitted by excited K^+ ions.

Table 12.2
Wavelengths and Relative Intensities of the Emission Spectra for Several Elements

Element	Wavelength (nm)	Relative Intensity
Argon	451.1	100
	560.7	35
	591.2	50
	603.2	70
	604.3	35
	641.6	70
	667.8	100
	675.2	150
	696.5	10000
	703.0	150
	706.7	10000
	706.9	100
Barium	435.0	80
	553.5	1000
	580.0	100
	582.6	150
	601.9	100
	606.3	200
	611.1	300
	648.3	150
	649.9	300
	652.7	150
	659.5	3000
	665.4	150
Cadmium	467.8	200
	478.0	300
	508.6	1000
	610.0	300
	643.8	2000
Cesium	455.5	1000
	459.3	460
	546.6	60
	566.4	210
	584.5	300
	601.0	640
	621.3	1000
	635.5	320
	658.7	490
	672.3	3300

Element	Wavelength (nm)	Relative Intensity
Helium	388.9	500
	396.5	20
	402.6	50
	412.1	12
	438.8	10
	447.1	200
	468.6	30
	471.3	30
	492.2	20
	501.5	100
	587.5	500
	587.6	100
	667.8	100
Neon	585.2	500
	587.2	100
	588.2	100
	594.5	100
	596.5	100
	597.4	100
	597.6	120
	603.0	100
	607.4	100
	614.3	100
	616.4	120
	618.2	250
	621.7	150
	626.6	150
	633.4	100
	638.3	120
	640.2	200
	650.7	150
	660.0	150
Potassium	404.4	18
	404.7	17
	536.0	14
	578.2	16
	580.1	17
	580.2	15
	583.2	17
	691.1	19

Element	Wavelength (nm)	Relative Intensity
Rubidium	420.2	1000
	421.6	500
	536.3	40
	543.2	75
	572.4	60
	607.1	75
	620.6	75
	630.0	120
Sodium	466.5	120
	466.9	200
	497.9	200
	498.3	400
	568.2	280
	568.8	560
	589.0	80000
	589.6	40000
	616.1	240
Thallium	377.6	12000
	436.0	2
	535.0	18000
	655.0	16
	671.4	6
Zinc	468.0	300
	472.2	400
	481.1	400
	507.0	15
	518.2	200
	577.7	10
	623.8	8
	636.2	1000
	647.9	10
	692.8	15

Tearout Wavelength Scale to be Used for Calibration of the Unknown Emission Spectra

Hg Reference

375 400 425 450 475 500 525 550 575 600 625 650 675

Wavelength (nanometers)

SPECTROSCOPY–LAB PREVIEW

Date________Name____________________________________Lab Sec._____Desk No._____

1. Distinguish between an absorption and an emission spectrum.

2. The most intense line in the indium emission–line spectrum occurs at 451.1 nm.

   a. Calculate the energy of these photons.

   b. What color do these lines exhibit?

3. a. Using the equation, $\frac{1}{\lambda} = R\left[\frac{1}{n_l^2} - \frac{1}{n_h^2}\right]$, calculate the wavelength for an electron transition from the $n_h = 4$ to the $n_l = 1$ energy level in the hydrogen atom. Assume $R = 1.1 \times 10^{-2}$ nm.

   b. Where (visible, ultraviolet, or infrared) does this line appear in the electromagnetic spectrum?

   c. Calculate the energy of the photon resulting from this electron transition.

4. A large number of hydrogen atoms have electrons excited to the $n_h = 4$ state. How many possible spectral lines can appear in the emission spectrum as a result of the electron reaching the ground state ($n_l = 1$)? Remember a spectral line appears when an electron de–excites from a higher to a lower energy state. Diagram all possible pathways for de–excitation from $n_h = 4$ to $n_l = 1$.

n = 4________________________

n = 3________________________

n = 2________________________

n = 1________________________

5. What dominant color appears for the mercury street bulb? the sodium street lamp? an incandescent bulb?

SPECTROSCOPY–DATA SHEET

Date_________Name_______________________________________Lab Sec. _____Desk No._____

A. FLAME TESTS

λ(nm)	400	450	500	550	600	650	700	750
$CaCl_2$								
$CuCl_2$								
NaCl								
$BaCl_2$								
LiCl								
$SrCl_2$								
KCl								

ultraviolet violet blue green yellow orange red infrared

B. HYDROGEN SPECTRUM

λ(nm)	color	$\Delta E_{atom} = E_{photon}$	1/λ	n_h (calc)	n_h(integer)	$1/n_h^2$
____	____	____________	____	____	____	____
____	____	____________	____	____	____	____
____	____	____________	____	____	____	____
____	____	____________	____	____	____	____

Show sample calculation of E_{photon} and n_h.

2. Plot 1/ λ (ordinate) *vs* $1/n_h^2$ (abscissa) on linear graph paper.

3. Instructor's approval of graph. __________________

4. From the plot determine

   a. the value of R, the Rydberg constant ____________________

   b. the value of $R/2^2$ ____________________

5. a. The accepted value of the Rydberg constant is (see instructor) ________________

   b. The percent error in R is ____________%

C. UNKNOWN SPECTRUM ____________

Lines in the spectrum (nm) ___________, ___________, ___________, ___________,

___________, ___________, ___________, ___________.

Unknown element: __________.

Questions

1. Why does a mercury light appear blue, even though yellow and green lines appear in the spectrum?

2. Explain why the color for ions in the flame tests differ.

3. Will the use of a spectroscope change the results of this experiment? Explain.

EXPERIMENT 13
CHEMICAL PERIODICITY

Objectives

- To observe the physical appearance of thirteen common elements
- To observe the chemical reactivity of elements in the third period and in Groups IA, IVA, and VIIA
- To predict the physical and chemical properties of other elements

Principles

Many chemical and physical properties for an element can be predicted from its location in the periodic chart. For example, elements at the left of the chart are shiny solids, conduct electricity, and are malleable; elements at the right of the chart are amorphous, nonconductors of electricity, and may even be gases; the metals at the left form salts with the nonmetals at the right. Elements of a **group** (vertical column) have similar chemical and physical properties with each successive element showing a predictable degree of reactivity. Successive elements in a **period** (horizontal row) show more dramatic differences in properties, progressing from metallic properties at the left of the period to nonmetallic properties at the right.

In this experiment, we will look at the similarities and differences in the chemical and physical properties of the elements of Period 3 and Groups IA (the alkali metals), IVA, and VIIA (the halogens). These elements and some of their parameters are listed in Table 13.1.

Definitions for the parameters in Table 13.1 are

- **Atomic radius:** the radius of an atom, expressed in angstroms, Å, where $1Å = 1 \times 10^{-10}m$
- **Ionization Energy:** energy required to remove one mole of electrons from a mole of the gaseous element, expressed in kJ/mol
- **Electronegativity:** an atom's relative attraction for the electrons used for bonding to another atom, expressed on a scale relative to fluorine being assigned a number of 4.1
- **Density:** the ratio of the mass of substance to its volume, expressed as g/cm^3 for solids and liquids and g/L at STP[1] for gases
- **Melting point:** the temperature at which the solid phase becomes a liquid
- **Boiling point:** the temperature at which the liquid phase becomes a gas[2]

---

[1]Standard temperature and pressure are 0°C (273K) and 1 atm (101.325 kPa).

[2]More specific definitions of melting and boiling points will be given later in the course.

Table 13.1
Measured Parameters of Some Elements

Periodic Properties	IA	IIA	IIIA	IVA	VA	VIA	VIIA
	Li			**C**			**F**
Atomic Radius (Å)	1.52			0.77			0.64
Ionization Energy (kJ/mol)	520			1086			1680
Electronegativity	1.0			2.5			4.1
Density (g/cm^3)	0.53			2.25[1]			1.696g/L
Melting Point (°C)	79			3570			-219.6
Boiling Point (°C)	1317			---			-188.1
	Na	**Mg**	**Al**	**Si**	**P**	**S**	**Cl**
Atomic Radius (Å)	1.86	1.60	1.43	1.17	1.10	1.04	0.99
Ionization Energy (kJ/mol)	496	737	577	786	1012	1000	1255
Electronegativity	1.0	1.3	1.5	1.8	2.1	2.4	2.9
Density (g/cm^3)	0.97	1.74	2.70	2.33	1.82	2.07	3.214g/L
Melting Point (°C)	97.8	651	660.2	1410	44.1[2]	114.5	-101.0
Boiling Point (°C)	892	1107	2467	2355	280	445	-34.6
	K						**Br**
Atomic Radius (Å)	2.27						1.14
Ionization Energy (kJ/mol)	419						1138
Electronegativity	2.8						0.9
Density (g/cm^3)	3.12						1.53
Melting Point (°C)	63.6						38.8
Boiling Point (°C)	774						701
	Rb			**Sn**			**I**
Atomic Radius (Å)	2.48			1.30			1.33
Ionization Energy (kJ/mol)	405			707			1004
Electronegativity	0.9			1.7			2.2
Density (g/cm^3)	1.53			7.30[3]			4.93
Melting Point (°C)	38.8			232			113.5
Boiling Point (°C)	701			2270			184.4
	Cs			**Pb**			
Atomic Radius (Å)	2.65			1.54			
Ionization Energy (kJ/mol)	376			715			
Electronegativity	0.9			1.6			
Density (g/cm^3)	1.87			11.35			
Melting Point (°C)	28.7			328			
Boiling Point (°C)	685			1750			

[1]Graphite [2]White phosphorus [3]White tin

Techniques

- Techniques 6a, c, page 7 — Heating Liquids
- Technique 14, page 18 — Testing with Litmus
- Technique 15, page 18 — Graphing Techniques

Procedure

A. Physical Properties

1. Samples of Na, Mg, Al, Si, P, and S are on the reagent table. Note that Na metal is stored under a nonaqueous liquid because of its reactivity with water and O_2 in air (Part B.1). Your lab instructor will cut a piece of Na–notice its luster and other metallic characteristics. Use steel wool to polish the pieces of Mg and Al for better viewing. Two forms of P exist: white P is so reactive with O_2 that it ignites in air; therefore, white P is stored under water. When stored under water for a long time, white P slowly changes to the more stable red P, which does not ignite in air. Record your observations on the Data Sheet.

2. Preparation of Cl_2. Place $^1/_2$–mL (10 drops) of 5% NaOCl (household laundry bleach) in a 75mm test tube and add 5 drops of toluene (**Caution**: *do not inhale*). Which layer is toluene and what is its color? Add 5 drops of 6M HCl (**Caution**: 6M HCl *is corrosive; wash it immediately from skin and clothes*). Agitate the solution by holding the upper part of the test tube with your thumb and index finger and tapping the lower part with your "pinky" finger (Figure 13.1). Observe the color of the toluene layer. What is the color of Cl_2?

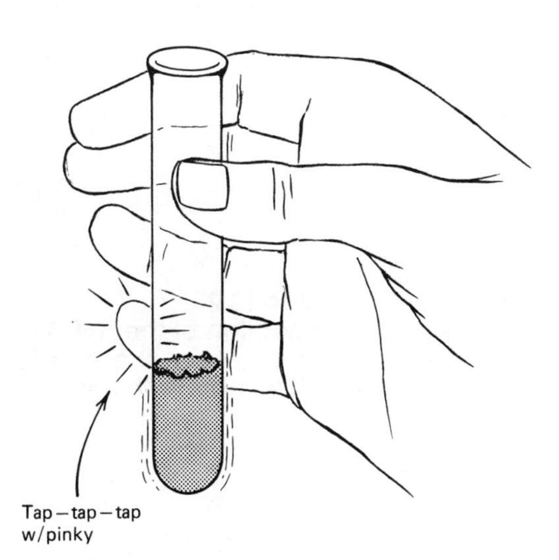

Figure 13.1
Agitating a Solution in a Test Tube with the "Pinky" Finger

3. Preparation of Br_2 and I_2. Mix equal portions of KBr and MnO_2; transfer a portion of the mixture to a dry , clean 75mm test tube until a depth of 3mm (≅1/8–inch) is reached.[3] Add 3 drops of conc H_2SO_4 (**Caution**: *don't let it touch your skin*). Gently warm the mixture over a low flame to initiate the reaction. What evidence of a reaction has occurred? Allow the test tube to cool, add 10 drops of water, and 5 drops of toluene, and agitate. What is the color of Br_2?

   Repeat the procedure in Part A.3b, substituting KI for the KBr. What is the color of I_2?

---

[3]Share the unused portion of the mixture with your neighbor–don't waste it!

4. Plot on graph paper the atomic radius (ordinate) *vs* atomic number (abscissa) and on the *same* graph, the ionization energy *vs* atomic number for the elements of Period 3. Label each axis and title the graph. Have your instructor approve your graph.

B. Chemical Properties

1. ***Demonstration Only.*** Na and K/H_2O. (**Caution**: *Never allow Na or K to touch the skin; each causes a severe skin burn*) Wrap a *pea–size* (no larger!) piece of freshly cut Na metal in aluminum foil. Fill a 200mm test tube with water and invert it into an 800mL beaker $^3/_4$–filled with water. Punch 5 pin–sized holes in the aluminum foil, grasp it with crucible tongs, and place it beneath the mouth of the water–filled test tube (Figure 13.2). After gas evolution has ceased, remove the test tube from the beaker, keeping it inverted. Place the mouth of the test tube over a Bunsen flame. What is the evolved gas? Don't be so alarmed as to drop the test tube–it may cost you 70¢ to replace it! Perform the litmus test on the water in the beaker. Is the water now acidic or basic? Based upon the properties of the gas and the acidity (or basicity) of the solution in the beaker, write a balanced equation for the reaction of Na with H_2O. Repeat the procedure with K metal.

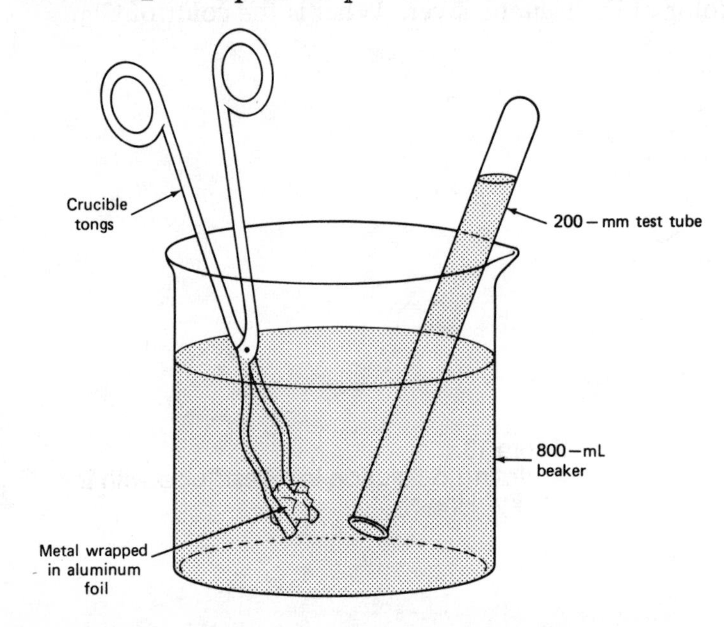

Figure 13.2
Collection of the $H_2(g)$ Evolved from the Reaction of Na and K with H_2O

2. ***Demonstration Only.*** Na/H_2O and CH_3OH (methanol). Set a 150mL beaker containing 20mL of methanol, CH_3OH, and a 150mL beaker containing 20mL of water behind a safety shield (Yes, *behind a safety shield*–if you don't have one, proceed to Part B.3). Cut two "BB"–size (no larger!) pieces of Na metal and, with tongs or tweezers, place one into each beaker. *Immediately* cover each beaker with a watch glass. Describe the reaction. How are water and CH_3OH similar?

3. Mg and Al/H_2O and HCl

a. Place 5mL of distilled (or de–ionized) water into separate 150mm test tubes and heat to boiling. Half–fill a 250mL beaker with water and heat to 95°C. Polish, with steel wool or sand paper, 2cm strips of Mg and Al to remove their oxide coatings.

b. Quickly add each metal to a separate test tube of hot water. Place the test tubes in the 250mm beaker, maintained at 85–95°C, for 10 minutes. What is observed?

c. Remove the test tubes containing any of the remaining Mg and Al samples. Perform a litmus test on the water in each test tube. Add 2mL of 6M HCl (**Caution**: *Avoid skin contact*). Record your observations. What is the evolved gas?

d. How does the reactivity of Na, Mg, and Al change in proceeding across Period 3?

4. C, Si, Sn, and Pb/ HCl

a. Place a small piece of carbon (a piece of "lead" from a pencil is satisfactory), silicon, mossy tin, and lead into four separate 150mm test tubes. Add 5mL of 6M HCl (**Caution**: *avoid skin contact*) to each test tube. Agitate each mixture.

b. Warm the mixtures. After 5 minutes, record any observations that characterize evidence of a chemical reaction. How does the reactivity of the elements in Group IVA change with increasing atomic number?

5. Reactivity of Cl_2, Br_2, and I_2

a. Set up six numbered 75mm test tubes. Place 1mL of the solution in each. Place a polished Mg strip in the first set of test tubes and a polished Al strip in the second set.

	Set #1			Set #2		
Test tube number	1	2	3	4	5	6
Solution	Cl_2/H_2O	Br_2/H_2O	I_2/H_2O	Cl_2/H_2O	Br_2/H_2O	I_2/H_2O
Metal	Mg	Mg	Mg	Al	Al	Al

b. Allow the mixtures to set for several minutes until differentiating evidence of chemical reactions has occurred. Record your observations. Account for the differences in the extent of the reactions.

6. Cl_2, Br^-, and I^-

a. Refer to Part A.2 to again prepare Cl_2 using the 5% NaOCl and 6M HCl solutions with the added toluene. Add a "pinch"[4] of KBr to the Cl_2/toluene/H_2O mixture and agitate. Account for your observation.

b. Repeat the preparation of Cl_2. Add a "pinch" of KI to the Cl_2/toluene/H_2O mixture and agitate. What chemical reaction has occurred?

---

[4]A "pinch" should be no larger than a grain of rice.

7. Cl^-, Br_2, AND I^-

a. Dissolve a "pinch" of KCl in $^1/_2$-mL of water in a 75mm test tube; add 5 drops of toluene. Add 5 drops of 2% Br_2/ H_2O (**Caution:** Br_2 *is corrosive and causes severe skin burns*) to the test tube and agitate. What happens? Does the color of the Br_2 in the toluene layer disappear?

b. Repeat Part B.7a, substituting KI for the KCl. Is occurrence of a chemical reaction evident? Write a balanced equation to represent your observation.

Periodic Classification of the Elements

IA	IIA	IIIB	IVB	VB	VIB	VIIB	VIII			IB	IIB	IIIA	IVA	VA	VIA	VIIA	0
1 H 1.0079																1 H 1.0079	2 He 4.00260
3 Li 6.941	4 Be 9.01218											5 B 10.81	6 C 12.011	7 N 14.0067	8 O 15.9994	9 F 18.99840	10 Ne 20.179
11 Na 22.98977	12 Mg 24.305											13 Al 26.98154	14 Si 28.086	15 P 30.97376	16 S 32.06	17 Cl 35.453	18 Ar 39.948
19 K 39.098	20 Ca 40.08	21 Sc 44.9559	22 Ti 47.90	23 V 50.9414	24 Cr 51.996	25 Mn 54.9380	26 Fe 55.847	27 Co 58.9332	28 Ni 58.70	29 Cu 63.546	30 Zn 65.38	31 Ga 69.72	32 Ge 72.59	33 As 74.9216	34 Se 78.96	35 Br 79.904	36 Kr 83.80
37 Rb 85.4678	38 Sr 87.62	39 Y 88.9059	40 Zr 91.22	41 Nb 92.9064	42 Mo 95.94	43 Tc 98.9062	44 Ru 101.07	45 Rh 102.9055	46 Pd 106.4	47 Ag 107.868	48 Cd 112.40	49 In 114.82	50 Sn 118.69	51 Sb 121.75	52 Te 127.60	53 I 126.9045	54 Xe 131.30
55 Cs 132.9054	56 Ba 137.34	57 La* 138.9055	72 Hf 178.49	73 Ta 180.9479	74 W 183.85	75 Re 186.207	76 Os 190.2	77 Ir 192.22	78 Pt 195.09	79 Au 196.9665	80 Hg 200.59	81 Tl 204.37	82 Pb 207.2	83 Bi 208.9804	84 Po (210)	85 At (210)	86 Rn (222)
87 Fr (223)	88 Ra 226.0254	89 Ac** (227)	104 (260)	105 (260)													

***Lanthanum Series**

58 Ce 140.12	59 Pr 140.9077	60 Nd 144.24	61 Pm (147)	62 Sm 150.4	63 Eu 151.96	64 Gd 157.25	65 Tb 158.9254	66 Dy 162.50	67 Ho 164.9304	68 Er 167.26	69 Tm 168.9342	70 Yb 173.04	71 Lu 174.97

****Actinium Series**

90 Th 232.0381	91 Pa 231.0359	92 U 238.029	93 Np 237.0482	94 Pu (244)	95 Am (243)	96 Cm (247)	97 Bk (247)	98 Cf (251)	99 Es (254)	100 Fm (257)	101 Md (258)	102 No (255)	103 Lr (256)

CHEMICAL PERIODICITY-LAB PREVIEW

Date________Name____Lab Sec. ____Desk No.____

1. a. What is a group of elements?

   b. What is a period of elements?

2. Refer to Table 13.1 to answer the following.

   a. Which element in Period 3 is most dense?________ least dense?________

   b. Which element in Group VIIA has the highest electronegativity?________ the lowest electronegativity?________

   c. Which element in Group IA has the highest melting point?________ the lowest boiling point?________

   d. In general, the densities of the elements in a group ________ as the atomic number increases.

   e. In general, the ionization energies of the elements in a period ________ as the atomic number increases.

   f. The electronegativity of the elements in a group ________ and in a period ________ as the atomic number increases.

3. The relative chemical reactivity for a number of elements is studied in today's experiment.

   a. List the elements.

   b. List the elements whose reactivities will be demonstrated by the instructor.

4. What commercially available compound is used to generate Cl_2 in the experiment?________

5. a. Describe the technique for testing a solution with litmus.

   b. Describe the *proper* technique for heating a solution in a test tube.

   c. What is the purpose of placing a glass stirring rod in a beaker that is being used to heat water?

CHEMICAL PERIODICITY-DATA SHEET

Date________Name____Lab Sec.____Desk No.____

A. PHYSICAL PROPERTIES

1.

Element	Symbol	Atomic Number	Atomic Mass	Physical State (g, l, s)	Color	Comments
sodium	____	_____	_____	_____	________	_______________
magnesium	____	_____	_____	_____	________	_______________
aluminum	____	_____	_____	_____	________	_______________
silicon	____	_____	_____	_____	________	_______________
phosphorus	____	_____	_____	_____	________	_______________
sulfur	____	_____	_____	_____	________	_______________

2. Preparation of Cl_2. Which is the toluene layer?________ ________What color is toluene?________________ What is the color of Cl_2?________

3. Preparation of Br_2 and I_2. Evidence for the preparation of Br_2.

Evidence for the preparation of I_2.

Color of Br_2 ________________ Color of I_2________________

4. Instructor's approval of graph.________________________

From the graph what generalized statement can you make about the relationship between atomic radii and ionization energies for a period of elements.

Is the same statement valid for a group of elements?________ If not, what statement holds true?

B. Chemical Properties

1. Na and K/H_2O

a. Gas evolved. Na: ______________; K: ______________

b. Litmus test, acidic or basic. Na: ______________; K: ______________

c. Write balanced equations for the reaction of Na and K with H_2O.

2. Na/H_2O and CH_3OH (Methanol)

a. What similarities exist in the reactions of Na with H_2O and CH_3OH?

b. Is Na more reactive in H_2O or CH_3OH?________

3. Mg and Al/H_2O and HCl

	Observation	Gas evolved	Litmus test
Mg/H_2O	------------------------	-----------	-----------
Al/H_2O	------------------------	-----------	-----------
Mg/HCl	------------------------	-----------	
Al/HCl	------------------------	-----------	

Compare the relative reactivity of Na, Mg, and Al with H_2O and with HCl.

4. C, Si, Sn, and Pb/ HCl

a. What general change in physical appearance occurs in this group of elements as the atomic number increases?

b. Which elements react with warm 6M HCl?______________

c. What is observed that indicates that a reaction has occurred?

C: ______________________; Si: ______________________

Sn: ______________________; Pb: ______________________

5. REACTIVITY OF Cl_2, Br_2, AND I_2

a. List in decreasing order the relative reactivity of Cl_2, Br_2, and I_2 with Mg.

________ > ________ > ________

b. List in decreasing order the relative reactivity of Cl_2, Br_2, and I_2 with Al.

________ > ________ > ________

c. Which metal, Mg or Al, is more reactive with the halogens?________

6. Cl_2, Br^-, and I^-

	Observation	Balanced equation for the reaction
$Cl_2 + Br^-$	________________________	________________________
$Cl_2 + I^-$	________________________	________________________

7. Cl^-, Br_2, and I^-

	Observation	Balanced equation for the reaction
$Cl^- + Br_2$	________________________	________________________
$Br_2 + I^-$	________________________	________________________

List the halogens in order of decreasing activity. ________ > ________ > ________
Explain your listing.

Questions

1. What tool did your lab instructor use to cut the Na metal?________________ What property does Na exhibit, one that we don't normally associate with a metal, that allowed the lab instructor to use that tool?

2. a. Metallic oxides, like Na_2O and BaO, dissolved in water turn red litmus blue. What chemical property does this indicate about metallic oxides?

   b. Predict the effect on litmus when nonmetallic oxides, like SO_2 and CO_2, are dissolved in water. Explain.

3. Cl_2 is used extensively as a bleaching agent and as a disinfectant. Without regard to any adverse effects, would Br_2 be a more or less effective as a bleaching agent and disinfectant? Explain.

4. Predict the reactivity of Cs in methanol relative to that of Na. Explain.

5. Predict the reactivity of Si in water relative to that of Na, Mg, and Al. Explain.

6. Predict the reactivity of Ge in warm 6M HCl relative to other Group IVA elements. Explain.

7. Predict for Ge its

    a. atomic radius________________

    b. ionization energy________________

    c. density________________

EXPERIMENT 14
SPECTROPHOTOMETRIC IRON ANALYSIS

Objectives

- To determine an unknown ferrous ion concentration in solution
- To develop the techniques for the use and operation of a spectrophotometer
- To determine the iron concentration in a vitamin tablet

Principles

The principle underlying a spectrophotometric method of analysis involves the interaction of electromagnetic (EM) radiation with matter. While the more common regions of the EM spectrum are the ultraviolet, visible, and the infrared, only visible light will be used in this experiment. Ths visible region has a wavelength range from about 400nm to 700nm; the 400nm limit approximates a violet color while the 700nm region has a red color.

Every chemical possesses a characteristic set of electronic, vibrational, and rotational energy states. Because they are characteristic, energy transitions between these states are often used to identify the presence and/or the concentration of a chemical in a mixture. This unique set of energy states for a chemical is therefore analogous to the unique set of fingerprints possessed by each person–each can be used for characteristic identification.

In the visible region of the EM spectrum, most absorptions result in the excitation of an electron from a lower to a higher state. When EM radiation (consisting of all energies) falls incident upon an atom or molecule, the only energy absorbed is that corresponding exactly to the difference between two electronic energy states. The atom or molecule that absorbs the radiation is in an **excited state**. The remainder of the EM radiation passes through the sample and an EM detector (either our own eye or an instrument) detects it. The absorbed energy, $\mathcal{E}$, is related to the wavelength, λ, of the EM radiation.

$$\mathcal{E} = h\frac{c}{\lambda}$$

h is Planck's constant and **c** is the speed of light.

When our eye is the detector, the color that we see is due to the EM radiation which the sample does not absorb. The appearance of the sample is that of the complementary color of the absorbed EM radiation. For example, if our sample solution absorbs energy in the yellow region of the visible spectrum, then the remaining wavelengths are transmitted and the sample appears violet. The greater the concentration of the yellow absorbing substance, the darker is the violet appearance. Table 14.1 list the wavelengths of the visible spectrum.

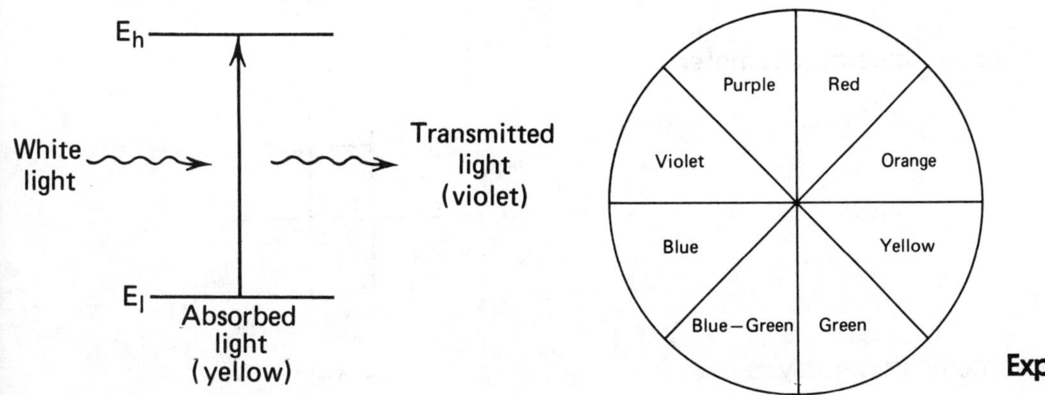

Table 14.1
Color and Wavelengths in the Visible Region of the Electromagnetic Spectrum

Color	Wavelength (nanometers*)	Color Transmitted
red	750-610	blue-green
orange	610-595	blue
yellow	595-580	violet
green	580-500	purple
blue	500-435	orange
violet	435-380	yellow

*1 nanometer = 1 x 10^{-9} meter

In this experiment a source of visible EM radiation is used to determine the concentration of ferrous, Fe^{2+}, ion in an aqueous solution. The wavelength at which a maximum absorption of EM radiation occurs is set on the spectrophotometer, an instrument that measures light intensities with a photosensitive cell (similar to our eye) at specific (but variable) wavelengths (Figure 14.1).

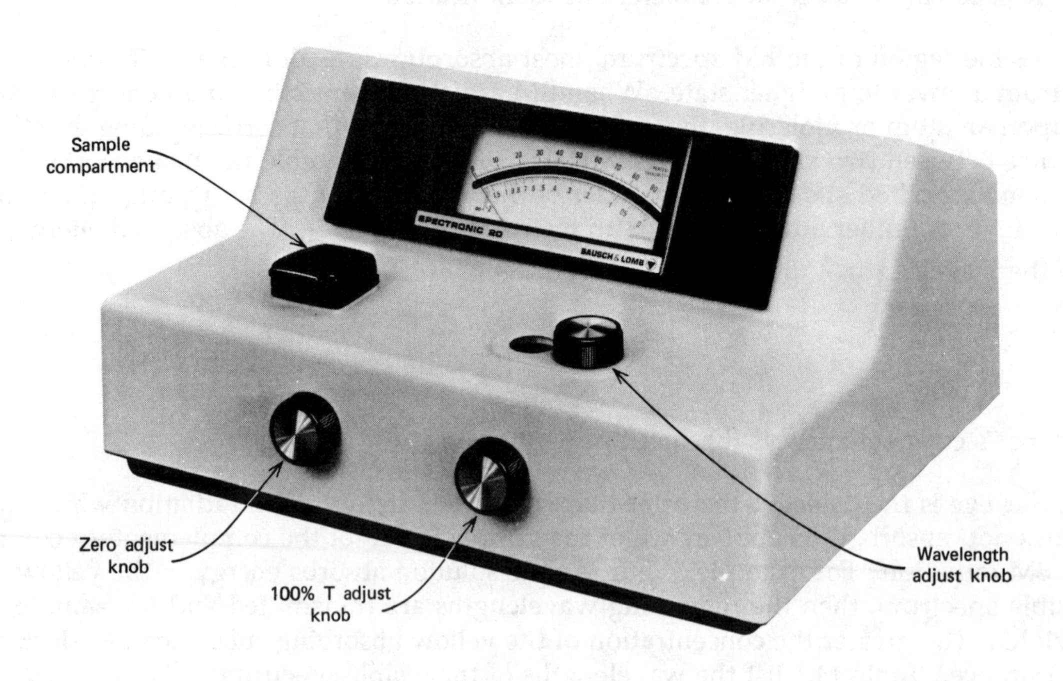

Figure 14.1
A Visible Spectophotometer

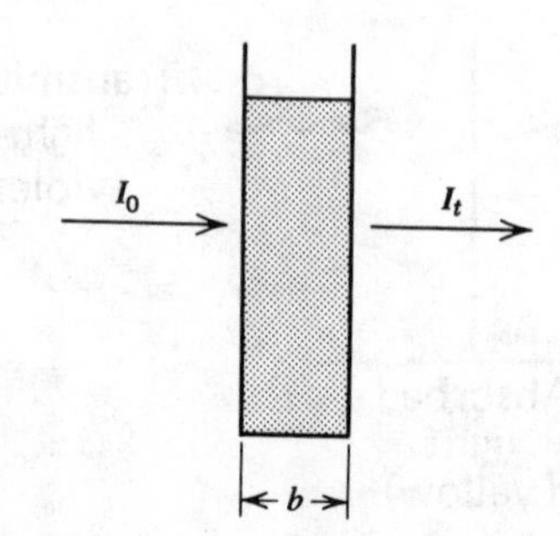

Several factors control the amount of EM radiation that the Fe^{2+} absorbs.

- the concentration of the Fe^{2+} in solution
- the thickness of the solution through which the EM radiation passes (this is determined by the size of the cuvet)
- the probability of the EM radiation being absorbed by the Fe^{2+} (this is called its molar absorptivity coefficient of extinction coefficient). This factor is constant for a given compound.

The ratio of the intensities of the transmitted EM radiation, I_t, to the incident EM radiation, I_o, is called the sample's **transmittance, T**, or expressed as a percent, %T

$$\%T = \frac{I_t}{I_o} \times 100$$

Frequently a chemist is more interested in measuring the amount of EM radiation that is absorbed by the Fe^{2+} rather than the amount that is transmitted. The Fe^{2+} concentration is proportional to the EM radiation absorbed (called the **absorbance, A**, of the solution); the Fe^{2+} concentration is also related to the incident and transmitted EM radiation and the absorbance of the solution by the equation

$$\log \frac{I_o}{I_t} = a \bullet b \bullet [Fe^{2+}] = A = -\log \frac{I_t}{I_o} = -\log \frac{\%T}{100}$$

a is called the extinction coefficient for Fe^{2+} and **b** is the thickness of the solution–both of which are constants at a given wavelength in a given cuvet. This equation is commonly referred to as Beer's Law, the important relationship being that $\mathbf{A \propto [Fe^{2+}]}$.

MEASURING THE Fe^{2+} CONCENTRATION

In this experiment the iron sample is dissolved in an aqueous solution and all of the iron is converted to Fe^{2+}. The Fe^{2+} forms a red–orange ion with 1,10–phenanthroline (also called ortho-phenanthroline and abbreviated o–phen), $[Fe(o\text{–}phen)_3]^{2+}$. The absorbance of this ion is determined at a wavelength where a maximum absorption of the EM radiation occurs, λ_{max}. The λ_{max} for $[Fe(o\text{–}phen)_3]^{2+}$ is determined in Part A from a plot of **A** *vs* λ for a standard $[Fe(o\text{–}phen)_3]^{2+}$ solution.

Ferric ion, Fe^{3+}, does not form the red–orange ion with o–phen; therefore, in the experiment all of the Fe^{3+} is the solution is reduced to Fe^{2+} with hydroxylamine hydrochloride, $NH_2OH \bullet HCl$.

$$2\,Fe^{3+}(aq) + 2NH_2OH \bullet HCl(aq) \rightarrow 2Fe^{2+}(aq) + N_2(g) + 2H_2O + 4H^+(aq) + 2Cl^-(aq)$$

The 1,10–phenanthroline is then added to the Fe^{2+} to form the red–orange ion.

$$Fe^{2+}(aq) + 3\text{ o–phen}(aq) \rightarrow [Fe(o\text{–}phen)_3]^{2+}(aq)$$

In addition, since the $[Fe(o\text{–}phen)_3]^{2+}$ ion is stable in the pH–range from 2 to 9, sodium acetate, $NaC_2H_3O_2$, is added, reacting with the HCl from the $NH_2OH \bullet HCl$ to from an $C_2H_3O_2^-$ / $HC_2H_3O_2$ combination that maintains the pH of the system between 4 and 6.

$$NH_2OH \bullet HCl(aq) + C_2H_3O_2^-(aq) \rightarrow HC_2H_3O_2(aq) + NH_2OH(aq) + Cl^-(aq)$$

Techniques

- Technique 2, page 2 — Transferring Liquid Reagents
- Technique 4b, page 5 — Separation of a Solid from a Liquid
- Technique 10, page 12 — Reading a Meniscus
- Technique 11, page 13 — Pipetting a Liquid or Solution
- Technique 13, page 16 — Using the Laboratory Balance
- Technique 15, page 18 — Graphing Techniques

In addition, you will develop the techniques for operating a spectrophotometer and handling cuvets. Also, you will gain additional experience in graphing data.

Procedure

You and a partner are using an expensive instrument in today's experiment. Follow your instructor's suggestions. You will need to obtain several pipets and 100mL volumetric flasks from the stockroom. Read ahead and be prepared.

A. Setting the λ_{max} on the Spectrophotometer

1. Obtain (and clean thoroughly with soap and water) a 10mL pipet, graduated at 1mL increments (a Mohr pipet), a 10mL graduated cylinder, and five 100mL volumetric flasks. Use the graduated 10mL pipet for the standard Fe^{3+} solution.

2. Pipet 5.00mL of the standard Fe^{3+} solution ($\cong$ 0.1mg Fe^{3+}/mL) into a clean 100mL volumetric flask[1]. Using the 10mL graduated cylinder, add 10mL of 10% $NaC_2H_3O_2$ and 10mL of 10% $NH_2OH \bullet HCl$. Agitate the solution and wait a few minutes to allow the complete reduction of Fe^{3+} to Fe^{2+}. Rinse the 10mL graduated cylinder after each use.

3. Use the 10mL graduated cylinder to add 10mL of 0.1% o-phen to the solution; dilute with distilled (or de–ionized) water to the "mark". Agitate the solution continuously for 10 minutes.

4. To prepare a blank solution[2], use your 10mL graduated cylinder to transfer 10mL of 10% $NaC_2H_3O_2$,10mL of 10% $NH_2OH \bullet HCl$, and 10mL of 0.1% o-phen to a 100mL flask. Dilute the mixture to 100mL with distilled (or de–ionized) water and agitate.

---

[1]It is suggested that a pair of students prepare this solution for the class. This avoids the waste of large quantities of the standard Fe^{3+} solution and a more efficient use of 100mL volumetric flasks.

[2]A **blank** solution is used to calibrate the spectrophotometer; the solution corrects for all substances that absorb EM radiation *except* the one of interest, in this case, $[Fe(o\text{–}phen)_3]^{2+}$.

5. Prepare two cuvets: rinse one cuvet twice with the blank solution and then fill; rinse a second cuvet (also twice) with the $[Fe(o\text{–}phen)_3]^{2+}$ solution (from Part A.3) and fill. Carefully dry the outside of each cuvet with a clean Kimwipe to remove fingerprints and water droplets. Thereafter, handle only the lip of the cuvets.[3]

6. Set the λ on the spectrophotometer at 400nm. Insert the blank solution into the sample holder and set the meter at 100%T with the 100%T knob. Remove the blank and (without any cuvet in the sample holder) set the meter to read 0%T with the zero adjust knob. Repeat until no further adjustments are necessary.

7. Place the cuvet containing the $[Fe(o\text{–}phen)_3]^{2+}$ solution in the sample holder, read the meter, and record the %T.

8. Repeat Steps 6 and 7 at 20nm intervals between 400nm and 600nm. Make additional %T measurements at 5nm intervals where the λ_{max} occurs.

9. Convert all of your %T readings to absorbance values. Plot the data as **A** (ordinate) *vs* λ (abscissa) on linear graph paper. Draw the best smooth curve through the data points. Have your instructor approve the graph.

B. Constructing the Calibration Curve

1. Set the spectrophotometer at the λ_{max} as determined from the graph in Part A.

2. Prepare the following solutions[4] in the same manner as described in Parts A.2 and A.3. Be sure to add the 0.1% o–phen after the Fe^{3+} has been reduced to Fe^{2+}. Use the 10mL pipet, calibrated at 1mL intervals, for transferring the standard Fe^{3+} solution.

test sol'n	0.1mg Fe^{3+}/mL sol'n (mL)	10% $NaC_2H_3O_2$ sol'n (mL)	10% $NH_2OH{\bullet}HCl$ sol'n (mL)	0.1% o-phen sol'n (mL)	volume (total mL)
1	8.00	10	10	10	100
2*	5.00	10	10	10	100
3	3.00	10	10	10	100
4	1.00	10	10	10	100
5*	0	10	10	10	100

*Prepared in Part A

3. Use the blank solution (from Part A.4) to check the calibration of the spectrophotometer.

4. Read and record the %T of the five solutions at the λ_{max}. Calculate the absorbance for each solution.

---

[3]Foreign material on the cuvet affects the intensity of the transmitted EM radiation.

[4]To increase the efficient use of 100mL volumetric flasks, the solutions may be used by other students in the laboratory.

5. Prepare a calibration curve by plotting on linear graph paper **A** (ordinate) *vs* Fe^{2+} concentration (mg/mL) (abscissa). Draw the best straight line through the five points on the graph. Have your instructor approve your graph.

C. Preparation of a Water Sample for Iron Analysis

1. Obtain at least 20mL of a water sample known to contain dissolved iron. It may be an unknown that has already been prepared for the experiment or it may be obtained from a drinking water supply, reservoir, or river. If there is any evidence of cloudiness, filter the sample.

2. Pipet 5.00mL of the water sample into a 100mL volumetric flask and dilute to the "mark". Pipet 5.00mL of this solution into a second 100mL volumetric flask, add the $NaC_2H_3O_2$, $NH_2OH{\bullet}HCl$, and o–phen solutions as in Parts A and B and dilute to the "mark". Proceed to Part E.

D. Preparation of a Vitamin Tablet for Iron Analysis

1. Crush a vitamin tablet with a mortar and pestle. Transfer the powder to a clean, dry, previously weighed (±0.01g) 100mL beaker. Reweigh the beaker and crushed tablet. Pipet 10mL of 6M HCl into the beaker and stir to dissolve (≅10 to 20 min).[5]

2. Quantitatively transfer the solution to a 100mL volumetric flask. Wash the solid remaining in the beaker with several portions of distilled (or de–ionized) water and add the washings to the flask. Dilute to the mark with distilled water and agitate for several minutes; if cloudiness persists, filter the solution.

3. Pipet 2mL of the (filtered) solution into a second 100mL volumetric flask, add the $NaC_2H_3O_2$, $NH_2OH{\bullet}HCl$, and o–phen solutions as in Parts A and B and dilute to the "mark". Proceed to Part E.

E. Determination of the Fe^{2+} Concentration in the Sample

1. Visually compare the color intensity of this solution with the standard solutions prepared in Part B; if the red–orange color is within the range of the standards, O.K.–if not, discard this solution and (quantitatively) adjust the second dilution so that the color intensity does lie within that of the standards. You will need to make some judgment on the amount to use.

2. Record the %T of the solution on the spectrophotometer and calculate its absorbance. It may be advisable at this point to check the calibration of the spectrophotometer with the blank solution; if it needs recalibration, the %T of the solution will need to be repeated.

3. Use the calibration curve prepared in Part B to determine the Fe^{2+} concentration (mg/mL) in the diluted sample. Calculate the quantity of iron in the original sample; be sure to account for the dilution of the original sample.

---

[5]Some of the tablet's binder may not dissolve; do not heat the solution in an attempt to dissolve the tablet, but merely continue with the experiment.

SPECTROPHOTOMETRIC IRON ANALYSIS–LAB PREVIEW

Date__________Name____Lab Sec. _____Desk No._____

1. List the purpose for each of these substances in the iron analysis.

   a. 1,10–phenanthroline

   b. $NaC_2H_3O_2$

   c. $NH_2OH{\bullet}HCl$ Write a balanced equation for its reaction with Fe^{3+}. The oxidation half–reaction of NH_2OH is

$$2NH_2OH \rightarrow N_2 + 2H_2O + 2H^+ + 2e^-$$

2. A concentration of 2ppm Fe^{2+} means 2g Fe^{2+} in 10^6g solution. Assuming the density of the solution is 1.0g/mL, express 2ppm Fe^{2+} in

   a. mg Fe^{2+}/mL solution

   b. mg Fe^{2+}/L solution

3. What is the color of a solution that absorbs 600nm EM radiation?

4. If $[Fe(o\text{–}phen)_3]^{2+}$ appears red–orange, what is the approximate wavelength for its maximum absorption?

5. A sample containing 18.0mg of iron dissolves in 100mL of solution. Five milliliters are then withdrawn and diluted to 250mL. What is the iron concentration in the diluted sample? Express your answer in mg Fe/mL and in ppm Fe.

6. A test solution was prepared by pipetting 2.0mL of an original sample and then diluting it to 250mL. The concentration of iron in the test solution was determined (from a calibration curve) to be 1.66 x 10^{-3}mg/mL or 1.66ppm. What was the iron concentration in the original sample. Express your answer in mg Fe/mL *and* in ppm Fe.

SPECTROPHOTOMETRIC IRON ANALYSIS-DATA SHEET

Date________Name____Lab Sec. _____Desk No._____

A. SETTING THE λ_{max} ON THE SPECTROPHOTOMETER

λ(nm)	%T	A	λ(nm)	%T	A	λ(nm)	%T	A
-----	-----	-----	-----	-----	-----	-----	-----	-----
-----	-----	-----	-----	-----	-----	-----	-----	-----
-----	-----	-----	-----	-----	-----	-----	-----	-----
-----	-----	-----	-----	-----	-----	-----	-----	-----
-----	-----	-----	-----	-----	-----	-----	-----	-----
-----	-----	-----	-----	-----	-----	-----	-----	-----

Plot the data, A (ordinate) *vs* λ(abscissa) on linear graph paper.

Instructor's Approval of graph.________________________________

B. CONSTRUCTING THE CALIBRATION CURVE

Test sol'n	Fe^{2+} Concentration (total volume = 100mL) mg Fe^{2+}/mL	ppm Fe^{2+}	%T	A
1	----------	----------	----------	----------
2	----------*	----------	----------	----------
3	----------	----------	----------	----------
4	----------	----------	----------	----------
5	----------	----------	----------	----------

* Sample calculation for Fe^{2+} (mg Fe^{2+}/mL) for Test Sol'n #2 (show work here).

Plot the data, A (ordinate) *vs* Fe^{2+} concentration (mg Fe^{2+}/mL) (abscissa) on linear graph paper.

C/E. Iron Concentration in Water Sample

	Trial 1	Trial 2
1. %T	----------	----------
2. Absorbance	----------	----------
3. Iron concentration in diluted sample from calibration curve (mg Fe/mL)	----------	----------
4. Iron concentration in original sample (mg Fe/mL)	----------	----------
(ppm Fe)	----------	----------

Show sample calculation.

D/E. Iron Concentration in Vitamin Tablet

	Trial 1	Trial 2
1. Mass of 100mL beaker (g)	----------	----------
2. Mass of 100mL beaker + crushed tablet	----------	----------
3. Mass of tablet (g)	----------	----------
4. %T	----------	----------
5. Absorbance	----------	----------
6. Iron concentration in diluted sample from calibration curve (mg Fe/mL)	----------	----------
7. Iron concentration in original sample (mg Fe/mL)	----------	----------
8. grams Fe/gram tablet	----------	----------
9. %Fe in tablet	----------	----------

Show sample calculation.

Questions

1. This experiment can measure the iron concentration in the 0.1ppm to 10ppm range. Explain how a solution having a higher iron concentration can be determined, and still remain within the limits of this sensitivity range.

2. How will an unfiltered sample affect the absorbance value for a solution?

3. If not enough time is allowed for the following reactions, how will it affect the reported iron concentration in the sample?

   a. the reduction of the Fe^{3+}.

   b. the formation of $[Fe(o\text{–}phen)_3]^{2+}$.

4. State the purpose for the blank solution in measuring the %T of the $[Fe(o\text{–}phen)_3]^{2+}$ solutions. Why is distilled water not a suitable blank?

5. a. If a 0.9cm cuvet is mistakenly substituted for a 0.8cm cuvet in one of the measurements, will the %T reading be higher or lower than it should be for that solution? Explain.

   b. Will the reported iron concentration be high or low for that solution?

EXPERIMENT 15
MOLECULAR GEOMETRY

Objectives

- To construct models for covalently–bonded molecules and polyatomic ions
- To apply theories of bonding to molecules and polyatomic ions
- To examine the various molecular shapes predicted by these theories

Principles

The structure of a molecule is basic in explaining its chemical and physical properties. For example, the facts that water is a liquid at room temperature, dissolves innumerable salts and sugars, is more dense than ice, boils at a relatively high temperature, and has a low vapor pressure can be explained through an understanding of its bonding and the bent arrangement of its atoms in the molecule.

Chemists have developed a number of bonding theories that account for physical and chemical properties. In 1916, G. N. Lewis postulated that *valence* electrons are most important in bonding atoms together. He proposed the **Octet Rule** in which atoms form bonds by losing, gaining, or sharing valence electrons to achieve an electron configuration that is the same as that of its nearest noble gas in the Periodic Table–with the exception of helium, this number of electrons is eight. The bond formed is ionic or covalent depending upon whether the electrons are transferred or shared (respectively). Lewis' Octet Rule is quite effective in explaining bonding, especially for bonds formed between the representative elements.

For water, the Lewis structure shows that when hydrogen and oxygen *share* valence electrons; each atom is isoelectronic[1] with a noble gas–hydrogen with helium and oxygen with neon.

O

H H

However, Lewis structures only account for the bonding of each atom and not the three–dimensional (3–D) structure of the molecule or polyatomic ion. Subsequent bonding theories also address the 3–D structure of a molecule; these theories, in addition, account for some of the chemical and physical properties.

VALENCE SHELL ELECTRON PAIR REPULSION (VSEPR) THEORY

VSEPR theory proposes that the geometry of a molecule or polyatomic ion is a result of a repulsive interaction of electron pairs in the valence shell of an atom; the most significant atom for determining its 3–D structure is the *central* atom. The orientation of the atoms is such that there is minimal interaction between the valence shell electron pairs; this, in turn, maximizes the distance between the electron pairs and also the atoms in the molecule.

---

[1]Isoelectronic means "the same electron configuration".

For methane, CH_4, the four valence electron pairs on the carbon atom (the central atom of the molecule) repel; this positions the electron pairs (and, for CH_4, the four hydrogen atoms) at the corners of a tetrahedron in a 3–D arrangement. This positioning can be generalized to include all molecular systems having four valence shell electron pairs. Since the oxygen atom in water also has four valence shell electron pairs, they, too, are arranged tetrahedrally–two electron pairs bond the hydrogen atoms to the oxygen and two electron pairs are nonbonding. This causes water to have a V–shaped arrangement of the H–O–H atoms in the molecule.

H
••
H : C : H
••
H

In summary, the arrangement of valence shell electron pairs, both bonding and nonbonding pairs, around the central atom gives rise to the corresponding 3–D structures of molecules and polyatomic ions. The 3–D structures and bond angles for molecules and polyatomic ions with a central atom having from two to six valence shell electron pairs are listed in Table 15.1, p. 166.

Valence Bond (VB) Theory

Covalent bonding between atoms according to VB theory is more than just a sharing of electrons; VB theory utilizes the valence shell atomic orbitals on each atom to form bonds. When two atomic orbitals on adjacent atoms overlap, a bond is formed–the greater the overlap, the stronger the bond. If the two overlapping orbitals point directly at one another, the bond is a **sigma– (σ-) bond**. In HF, when the 1s atomic orbital on hydrogen overlaps with a 2p atomic orbital of fluorine, a pair of electrons are shared between the two atoms to form a σ-bond.

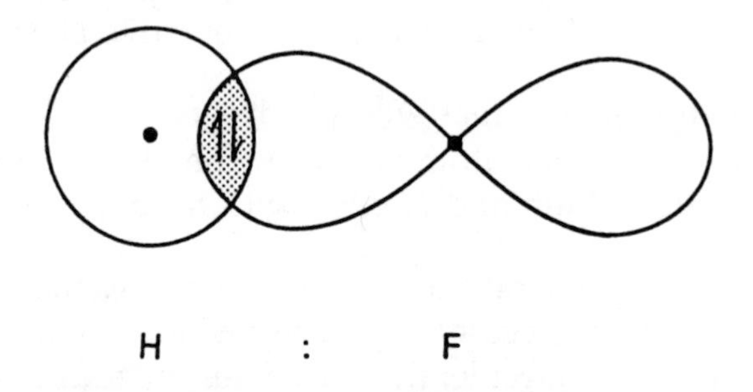

The orientation of the valence shell atomic orbitals involved in σ–bonding determines the geometry of the molecule or polyatomic ion. Oftentimes an atom reorients and reshapes its valence shell atomic orbitals to produce a molecular structure that is more stable; this "new" configuration of atomic orbitals is known as the **hybridization** of valence shell atomic orbitals. A close look at the bonding and structure of CH_4 can clarify our understanding of these concepts.

The valence shell electron configuration of the carbon atom is $2s^2 2p^2$–only two unpaired electrons appear to be available for bonding (Figure 15.1). However, for methane, four C–H bonds form: all the bonds have the same bond strength, all H–C–H bond angles are 109.5°, and its structure is tetrahedral. To account for these properties, VB theory states that the one 2s and the three 2p atomic orbitals (valence shell orbitals) mix and rearrange (*hybridize*) to form four equivalent "new" orbitals (Figure 15.2), each containing a single electron, which overlap with the 1s orbitals on the hydrogen atoms to form four C–H σ-bonds.

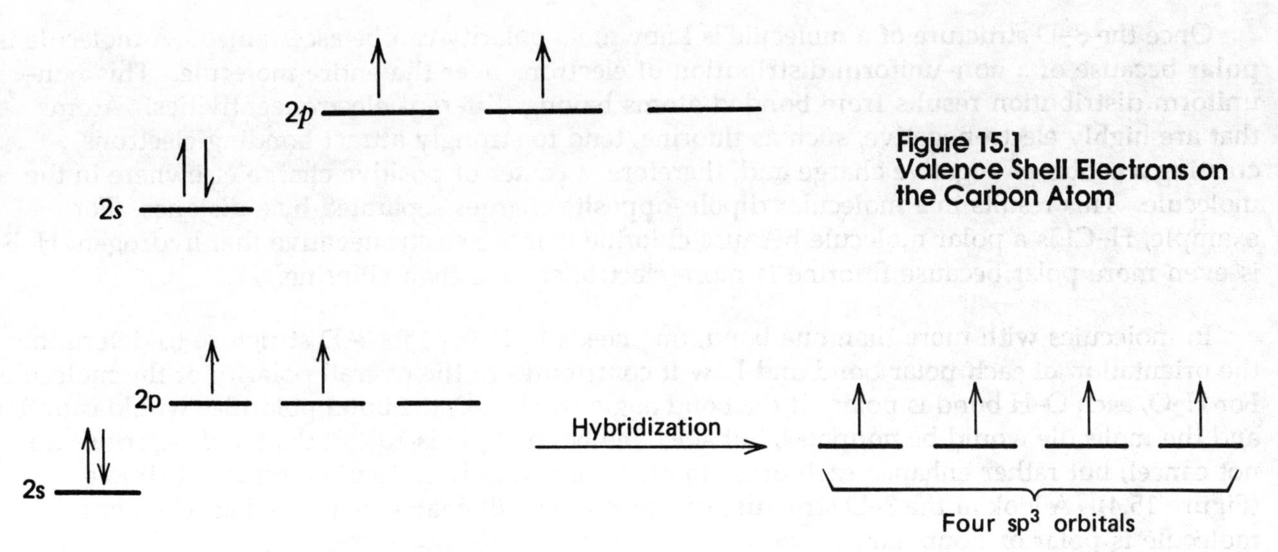

Figure 15.1 Valence Shell Electrons on the Carbon Atom

Figure 15.2 A Hybridization of the Valence Shell Orbitals on the Carbon Atom

This hybridization is called **sp^3** – one s– and three p– atomic orbitals form four sp^3 hybrid orbitals. The four sp^3 hybrid orbitals are directed toward the corners of a tetrahedron to minimize the electrostatic repulsion between hydrogen nuclei (Figure 15.3).

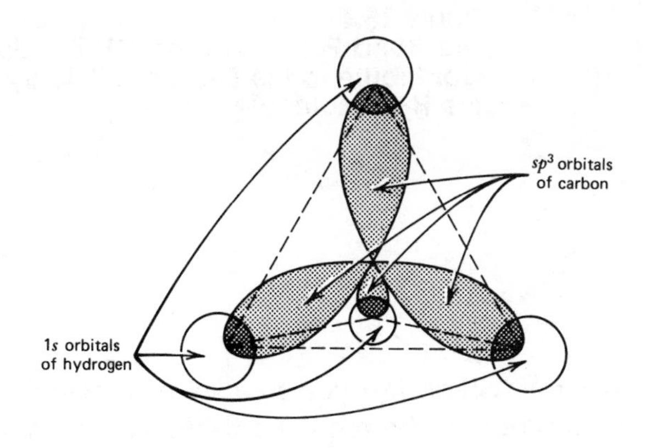

Figure 15.3
sp^3 Hybridized Valence Shell Orbitals on the Carbon in Methane

For some molecules or polyatomic ions a nonbonding pair of electrons in the valence shell of the central atom occupies a hybrid orbital. Because this hybrid orbital is the same as the hybrid orbitals forming σ-bonds, it contributes to the 3–D structure of the molecule. Therefore, the 3–D structure of a molecule is a result of the orientation of *all* hybrid orbitals–those forming σ-bonds and those containing nonbonding electron pairs.

A summary of the types of hybrid orbitals used by the valence shell of central atoms, the corresponding bond angles, and the geometries of the molecules is in Table 15.1, p. 166.

POLARITY AND DIPOLES

Once the 3–D structure of a molecule is known, its polarity can be ascertained. A molecule is **polar** because of a non–uniform distribution of electrons over the entire molecule. This non–uniform distribution results from bonded atoms having different electronegativities. Atoms that are highly electronegative, such as fluorine, tend to strongly attract bonding electrons, creating a center of negative charge and, therefore, a center of positive charge elsewhere in the molecule. This results in a molecular dipole–opposite charges separated by a distance. For example, H–Cl is a polar molecule because chlorine is more electronegative that hydrogen; H–F is even more polar because fluorine is more electronegative than chlorine.

In molecules with more than one bond, one needs to look at its 3–D structure to determine the orientation of each polar bond and how it contributes to the overall polarity of the molecule. For H_2O, each O–H bond is polar. If the bond angle were 180°, the bond polarities would cancel and the molecule would be nonpolar, but since the bond angle is 104.5°, the bond polarities do not cancel, but rather enhance each other to give the (overall) molecule a resultant dipole (Figure 15.4). A look at the 3–D structure of a molecule will enable us to predict whether a molecule is polar or nonpolar.

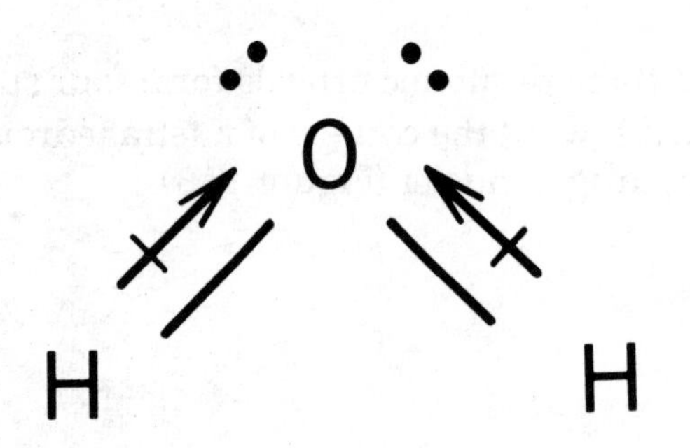

Figure 15.4
The Bond Polarity of the OH Bonds Contribute to the Resultant Polarity of the H_2O Molecule

Procedure

Check out a set of molecular models. Your instructor will assign you to a set of molecules or polyatomic ions. For each molecule or polyatomic ion, you will be required to do the following.

- Write its Lewis structure
- Determine the number of bonding electron pairs and the number of σ–bonds that exist
- Determine the number of nonbonding electron pairs on the central atom and the number of hybrid orbitals having nonbonding electron pairs
- Determine the type of hybrid orbitals used
- Predict the approximate bond angle
- Predict the 3–D structure
- Predict if the molecule is polar or nonpolar

MOLECULAR GEOMETRY-LAB PREVIEW

Date________Name____Lab Sec.____Desk No.____

1. a. A molecule that has two bonding and two nonbonding pairs of electrons has a

   ________________________ structure. Its approximate bond angle is ________°.

   b. A molecule that has four bonding and one nonbonding pairs of electrons has a

   ________________________ structure. Its approximate bond angle is ________°.

   c. A molecule that has three bonding and one nonbonding pairs of electrons has a

   ________________________ structure. Its approximate bond angle is ________°.

   d. A molecule that has six bonding and no nonbonding pairs of electrons has a

   ________________________ structure. Its approximate bond angle is ________°.

   e. A molecule that has two bonding and one nonbonding pairs of electrons has a

   ________________________ structure. Its approximate bond angle is ________°.

   f. A molecule that has two bonding and three nonbonding pairs of electrons has a

   ________________________ structure. Its approximate bond angle is ________°.

   g. A molecule that has four bonding and no nonbonding pairs of electrons has a

   ________________________ structure. Its approximate bond angle is ________°.

2. a. There are _____ equivalent atomic orbitals in an sp^3d hybridized system.

   b. In a _____ hybridized system there are two equivalent hybrid orbitals.

   c. There are _____ σ-bonds in an sp^2 hybridized molecule having three bonding orbitals.

   d. There are _____ σ-bonds in an sp^3d^2 hybridized molecule having six bonding orbitals.

   e. The number of hybrid orbitals on a sulfur atom that is sp^3d hybridized is _____. If four of its hybrid orbitals form σ-bonds, the geometry of the molecule is _____.

   f. The approximate bond angle for a molecule that is sp^2 hybridized is _____.°; for an sp^3 hybridized molecule, it is _____.°; for an sp^3d hybridized molecule, it is _____.°; for an sp hybridized molecule, it is _____.°; and for an sp^3d^2 hybridized molecule, it is _____°.

3. Sketch a representation, similar to Figure 15.2, for the hybridization of a $2s^2 2p^1$ electron configuration for carbon in CH_3^+. Remember the "+" indicates one less electron than the number of valence electrons in the atoms.

4. a. The number of σ-bonds in $BeCl_2$ is ________.

b. The number of bonding electron pairs in BrF_3 is ________.

c. The number of nonbonding electron pairs in BrF_3 is ________.

d. The number of bonding hybrid orbitals if CH_3^+ is ________.(See Question 3).

e. A molecule having bond angles of 120° and a triangular structure uses a ________set of hybrid orbitals in bonding.

f. The number of σ-bonds in SiF_4 is ________.

g. A molecule having bond angles of 90° and an octahedral structure uses a ________set of hybrid orbitals in bonding.

MOLECULAR GEOMETRY-DATA SHEET

Date_________ Name____ Lab Sec. _____ Desk No._____

Your instructor will assign molecules and/or ions from the following lists for you to characterize according to the format shown below. The central atom of the molecule/ion is in an italized bold face. The central atoms of the molecules/ions marked with an asterisk (*) do *not* obey the octet rule.

Column	Characteristics
A	Bonding electron pairs_____
B	σ–bonds_____
C	Nonbonding electron pairs_____
D	Hybrid orbitals having nonbonding electron pairs_____
E	Hybrid orbitals used_____
F	Approximate bond angle_____°
G	3-D structure_______________
H	Polar(P) or nonpolar(NP)__________

Set up a table on a separate sheet of paper, according to the following format and example.

Molecule or Ion	Lewis Structure	A	B	C	D	E	F	G	H
CH_4	H •• H : C : H •• H	4	4	0	none	sp^3	109.5°	tetra-hedral	NP

1. Complete the table (as outlined above) for the following molecules/ions.

a. CF_3Cl
b. NH_3
c. NH_4^+
d. H_2O
e. AsF_5*
f. AsF_3
g. H_3O^+
h. ClO_2^-
i. BF_4^-
j. XeF_4*
k. ICl_2^+
l. ICl_2^-*

2. Complete the table (as outlined above) for the following molecules/ions.

a. SF_2
b. SF_4*
c. SF_6*
d. SF_5^+*
e. BrF_2^+
f. BrF_2^-*
g. SbH_3
h. BrF_4^-*
i. BrF_4^+*
j. OF_2
k. $BeCl_2$
l. XeF_2*

3. Complete the table (as outlined above) for the following molecules/ions.

a. GaI_3*
b. CH_3^-
c. CH_3^+*
d. PF_3
e. PF_5*
f. PF_4^+
g. SbF_6^-*
h. SnF_4
i. SnF_2*
j. SiH_4
k. SnF_6^{2-}*
l. SO_4^{2-}

Table 15.1
The Likeness of VSEPR and Valence Bond Theories

Valance Shell Electron Pairs	Bonding Electron Pairs (VSEPR) or Bonding Orbitals (VB)	Nonbonding Electron Pairs (VSEPR) or Nonbonding Orbitals (VB)	Hybrid-ization	Bond Angle	Geometric Shapes	Examples
2	2	0	sp	180°	**linear**	$HgCl_2$, $BeCl_2$
3	3	0	sp^2	120°	**trigonal planar**	BF_3, $In(CH_3)_3$
	2	1	sp^2	<120°	V–shaped	$SnCl_2$, $PbBr_2$
4	4	0	sp^3	109.5°	**tetrahedral**	CH_4, $SnCl_4$
	3	1		<109.5°	trigonal pyramidal	NH_3, PCl_3, H_3O^+
	2	2		<109.5°	V–shaped	H_2O, OF_2, SCl_2
5	5	0	sp^3d	90°/ 120°	**trigonal bipyramidal**	PCl_5, $NbCl_5$
	4	1		>90°	irregular tetrahedral	SF_4, $TeCl_4$
	3	2		<90°	T–shaped	ClF_3
	2	3		180°	linear	ICl_2^-, XeF_2
6	6	0	sp^3d^2	90°	**octahedral**	SF_6
	5	1		>90°	square pyramidal	BrF_5
	4	2		90°	square planar	ICl_4^-, XeF_4

Questions

1. Distinguish, with a sketch, the shape of a p–orbital and an sp–orbital.

2. A π-bond (VB theory) or a double bond (VSEPR) does not affect the 3–D structure of a molecule.

   a. What mode of hybridization of valence shell atomic orbitals on the S and C atoms in SO_2 and CO_3^{2-} are predicted? Draw a Lewis structure for each.

   b. What 3-D structure do you predict for SO_2 and CO_3^{2-}?

Experiment 16
Formula Weight of a Volatile Compound

Objective

- To determine the formula weight and density of a volatile compound

Principles

The Dumas method (John Dumas, 1800-1884) for determining the formula weight of a volatile compound requires the use of the ideal gas law equation, **PV = nRT**.

In this experiment a compound is vaporized at a measured temperature, **T**, into the measured volume, **V**, of an Erlenmeyer flask. After the barometric pressure, **P**, is read from the laboratory barometer, the moles of gas, **n**, are calculated from the ideal gas law equation. When the pressure is recorded in atmospheres, the volume in liters, and temperature in kelvins, then the gas constant, **R**, equals 0.0821 L atm/mol K.

The mass of the vaporized compound, **m**, is measured from the difference between the empty flask and the vapor–filled flask. The formula weight, **FW**, is calculated from the equation,

$$\mathbf{FW} = \frac{\mathbf{m}}{\mathbf{n}}$$

The density of a gas is most often recorded when its temperature and pressure are at STP. A correction of the measured volume of the gas in the Erlenmeyer flask to STP, V_{corr}, must first be made before the density of the gas can be reported.

$$\textbf{density of gas} = \frac{\mathbf{m}}{\mathbf{V_{corr\ to\ STP}}}$$

Techniques

- Technique 1, page 1 — Inserting Glass Tubing through a Rubber Stopper
- Technique 6c, page 7 — Heating Liquids
- Technique 13, page 16 — Using the Laboratory Balance

Procedure

You are to complete two trials in today's experiment. Obtain no–more–than 15mL of an unknown liquid compound from your laboratory instructor and check out a 110°C thermometer from the stockroom. Record the number of your unknown.

1. Set up the apparatus shown in Figure 16.1. Fit a 125mL Erlenmeyer flask with a one–hole stopper. Insert a capillary tip piece of glass tubing, constructed in Experiment 1, into the rubber stopper. Clean the flask with soap and water, rinse thoroughly with distilled (or de-ionized) water, and *dry* either in a drying oven or by inverting the flask over a paper towel and allowing it to air–dry. Weigh (±0.001g) the *dry* flask and capillary–fitted stopper.

Figure 16.1
Flask and Capillary–fitted Stopper for Volatile Liquid

2. To prepare a hot water bath, fill a 600mL beaker half–full with tap water, add 2 or 3 drops of 6M HCl[1], support it on a wire gauze, and heat the water to boiling. While waiting for the water to boil, add approximately 6mL of your unknown liquid compound to the flask and insert the rubber stopper.

3. Remove the heat from the hot water bath. Insert and secure the Erlenmeyer flask, containing your unknown liquid compound, with a utility clamp in the hot water bath; be certain the flask does not touch the beaker wall (Figure 16.2). Adjust the water level high on the neck of the flask; you may need to add or remove water. Resume the heating of the hot water bath (**Caution**: *most unknowns are flammable; use a moderate flame for heating*).

4. As the unknown liquid compound is heated, some vapor escapes through the capillary tip. When vapors are no longer visible[2], continue heating for another 10 minutes. The flask should now be filled with the vapor of your unknown compound–no liquid should remain in the flask. Record the temperature (±0.1°C) of the boiling water.

---

[1] The HCl prevents the buildup of mineral deposits on the beaker as the water evaporates.

[2] To see vapors escaping through the capillary tip, look across the top of the capillary tip toward a lighted area. The vapor causes a diffraction of the lighted area.

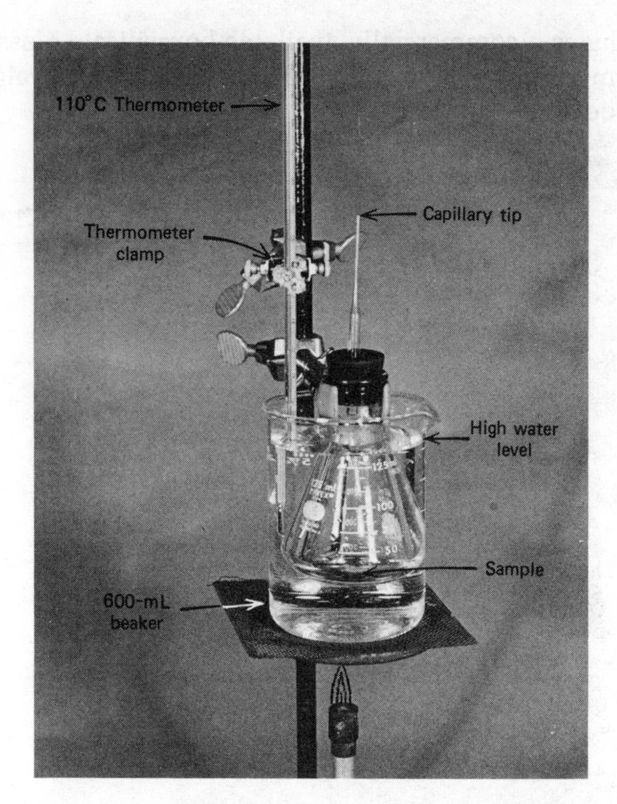

Figure 16.2
Setup for the Formula Weight Determination of a Volatile Liquid

5. Remove the flask; remove the heat from the hot water bath. Allow both to cool to room temperature. Sometimes the vapor remaining inside the flask condenses; that's O.K. Dry the outside of the flask and weigh it, its capillary–fitted stopper, and condensed vapor on the same balance that was used earlier.

6. To repeat the procedure, add 5mL of the unknown liquid compound to the Erlenmeyer flask and repeat Parts 3 through 5.

7. Discard the condensed vapor according to your instructor's instructions.

8. Fill the empty 125mL Erlenmeyer flask to the brim with tap water and insert the capillary–fitted stopper; be sure that water also fills the capillary tube. Remove the stopper. Measure the volume (±0.1mL) of the flask by transferring portions of the water to a 50mL graduated cylinder until it is all transferred. Add the volumes of water and record the total volume.

9. Read the laboratory barometer and record the barometric pressure.

This is a commercially available borosilicate glass bulb with a drawn tip that is used for determining the vapor density and the formula weight of volatile compounds by the Dumas method.

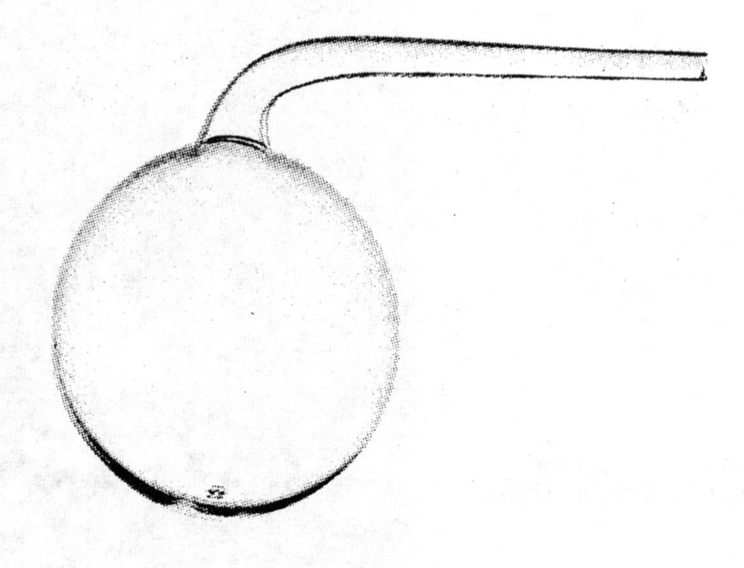

FORMULA WEIGHT OF A VOLATILE COMPOUND-LAB PREVIEW

Date__________Name____Lab Sec. ______Desk No._____

1. The vapor from an unknown compound occupies a 269mL Erlenmeyer flask at 98.7°C and 748 torr. The mass of the vapor is 0.791g.
   a. How many moles of vapor are present?

   b. What is the formula weight of the compound?

   c. What is the density of the vapor at standard temperature and pressure (STP)?

2. a. If the atmospheric pressure is mistakenly recorded at 760 torr in Question #1, what is the formula weight of the compound?

   b. What is the percent error caused by this assumed pressure?

3. Explain how the mass of the vaporized compound is measured in today's experiment?

4. a. How many trials are to be completed in this experiment? ________

   b. What volume of unknown compound should be obtained from your laboratory instructor?________

5. List the equipment that needs to be checked out from the stockroom.

6. a. When inserting a piece of glass tubing through a rubber stopper, what should be used as a lubricant?

   b. What is the purpose of placing a glass stirring rod into the beaker of a hot water bath?

   c. Specifically, where should the flame of a Bunsen burner be placed when heating water in a beaker?

FORMULA WEIGHT OF A VOLATILE COMPOUND-DATA SHEET

Date________Name____Lab Sec. ____Desk No.____

Unknown Number________________	Trial 1	Trial 2
1. Mass of dry flask + capillary-fitted stopper (g)	__________	__________
2. Mass of dry flask + capillary–fitted stopper, and condensed vapor (g)	__________	__________
3. Mass of vapor unknown, **m** (g)	__________	__________
4. Temperature of boiling water, **T** (°C)	__________	__________
5. Volume of 125mL flask ______+_______+_______ , **V** (mL)	__________	__________
6. Barometric pressure, **P** (atm)	__________	__________
7. Moles of vapor, **n** (mol)	__________	__________
8. Formula weight, **m/n** (g/mol)	__________*	__________
9. Average formula weight	__________	
10. Density of gas at STP (g/L)	__________	__________

*Show your calculations for Trial 1.

Questions

1. If the flask is not dried before weighing it after the liquid has been vaporized (in Part 5), will the formula weight of the unknown be too high or too low? Explain.

2. a. Suppose the barometric pressure during today's experiment is assumed to be 760 torr instead of the value you recorded. Would the formula weight of the unknown be reported too high or too low? Explain.

   b. What would be your percent error for the formula weight, if the assumption had been made?

3. If the volume of the vapor is assumed to be 125mL instead of the measured volume, what would be the percent error for the formula weight of the unknown liquid compound? Show your work.

4. If all of the unknown liquid compound does not vaporize into the 125mL Erlenmeyer flask in Part 4, will the reported formula weight be too high or too low? Explain.

Experiment 17
Acids, Bases, and Salts

Objective

- To become familiar with the chemical and physical properties of acids, bases, and salts.

Principles

Acids are are substances that, when dissolved in water, release hydrogen ion, H^+. For example HCl is a gas at room temperature and pressure, but when it is dissolved in water, it releases H^+; the H^+ then combines with the water to produce hydronium ion, H_3O^+.

$$HCl(g) + H_2O \rightarrow H_3O^+(aq) + Cl^-(aq)$$

Bases are substances that, when dissolved in water, produce hydroxide ion, OH^-. Barium hydroxide, $Ba(OH)_2$, is a solid at room temperature and pressure, but when added to water, it releases OH^-.

$$Ba(OH)_2(s) \xrightarrow{H_2O} Ba^{2+}(aq) + 2OH^-(aq)$$

When an acid is mixed with a base in an aqueous system, a reaction occurs producing water and a salt as products. This reaction is called a **neutralization reaction**. For example, when HCl and $Ba(OH)_2$ are mixed in solution, the products are water and the salt barium chloride, $BaCl_2$.

$$2HCl(g) + Ba(OH)_2(s) \xrightarrow{H_2O} 2H_2O + BaCl_2(aq)$$

Barium chloride is an ionic compound–this means that the compound consists of Ba^{2+} ions and Cl^- ions. In an aqueous solution an ionic compound tends to dissociate into its respective ions; barium chloride dissociates into Ba^{2+} and Cl^-. Therefore, a better representation of the reaction between HCl and $Ba(OH)_2$ is

$$2HCl(g) + Ba(OH)_2(s) \xrightarrow{H_2O} 2H_2O + Ba^{2+}(aq) + 2Cl^-(aq)$$

A **salt** therefore is any ionic compound that is a neutralization product of an acid–base reaction.

Substances that dissociate in water to produce ions are called **electrolytes**. Electrolytes conduct electrical current through aqueous solutions. Some compounds are called *strong* electrolytes–these substances when dissolved in water completely (100%) dissociate into ions. Hydrochloric acid, HCl, barium hydroxide, $Ba(OH)_2$, and barium chloride, $BaCl_2$, are strong electrolytes. Compounds that dissolve in water but do not (0%) dissociate are called **nonelectrolytes**. Sugar is a nonelectrolyte. The large number of compounds that only partially dissociate in an aqueous solution are called *weak* electrolytes. Acetic acid is a weak electrolyte.

Acidic solutions are produced from the action of water on (1) nonmetallic hydrides, such as HCl and HBr, (2) nonmetallic oxides, such as CO_2 and SO_3, or (3) compounds of hydrogen, oxygen, and one other element (usually a nonmetal), such as H_2SO_4 and HNO_3.

$$HCl(g) + H_2O \rightarrow H_3O^+(aq) + Cl^-(aq)$$
$$CO_2(g) + H_2O \rightarrow H_3O^+(aq) + HCO_3^-(aq)$$
$$H_2SO_4(l) + H_2O \rightarrow H_3O^+(aq) + HSO_4^-(aq)$$

Sulfuric acid, also called oil of vitriol, is perhaps the most versatile of all inorganic industrial chemicals. Presently, so many industries use it that sulfuric acid is often called the "Old Horse of Chemistry". There is hardly a chemical industry that does not use sulfuric acid in its manufacturing process. In 1986, when it ranked number one in chemical usage, nearly 74 billion pounds of H_2SO_4 were produced in the United States, nearly twice that of the second ranked chemical. Other acids that ranked among the "Top 50" chemicals in production were phosphoric acid, H_3PO_4, nitric acid, HNO_3, and hydrochloric acid, HCl (called muriatic acid). See Table 17.1.

In addition, many organic acids, such as acetic, adipic, and oleic acids, are useful and important to the chemical industry.

Basic solutions are produced from the action of water on (1) metallic hydroxides, such as $Ba(OH)_2$ and NaOH, (2) metallic oxides, such Na_2O, or (3) a select number of polyatomic anions, such as PO_4^{3-} and CO_3^{2-}.

$$Ba(OH)_2(s) \xrightarrow{H_2O} Ba^{2+}(aq) + 2OH^-(aq)$$
$$Na_2O(s) + H_2O \rightarrow 2Na^+(aq) + 2OH^-(aq)$$
$$PO_4^{3-}(aq) + H_2O \rightarrow HPO_4^{2-}(aq) + OH^-(aq)$$

The strong bases such as NaOH (called caustic soda or lye) and KOH (called caustic potash) are also known as **alkalis**.

A number of bases ranked in the "Top 50", most notably are ammonia, NH_3, sodium hydroxide, calcium oxide, CaO (called quicklime or, more simply, lime), and sodium carbonate, Na_2CO_3 (called soda ash). Again, refer to Table 17.1.

Table 17.1
Acids and Bases Ranked Among the "Top 50" in Production for 1986 in the United States from Chemical and Engineering News , April 13, 1987, p. 21.

Rank	Chemical	Formula	Billions of Pounds (1986)
1	sulfuric acid	H_2SO_4	73.64
5	lime	CaO	30.34
6	ammonia	NH_3	28.01
7	sodium hydroxide	NaOH	22.01
9	phosphoric acid	H_3PO_4	18.41
11	sodium carbonate	Na_2CO_3	17.20
13	nitric acid	HNO_3	13.12
23	hydrochloric acid	HCl	5.97
32	acetic acid	CH_3COOH	2.93
48	adipic acid	$C_4H_8(COOH)_2$	1.52

Techniques

- Technique 6a, page 7 — Heating Liquids
- Technique 9a,b, pages 10 & 11 — Handling Gases
- Technique 10, page 12 — Reading a Meniscus
- Technique 11, page 13 — Pipetting a Liquid or Solution
- Technique 13, page 16 — Using the Laboratory Balance
- Technique 14, page 18 — Testing with Litmus

Procedure

During this experiment, **STOP** at each numbered superscript (example, #1)and record your observation(s) or conclusion(s) on the Data Sheet.

Caution: *Dilute and concentrated (conc) acids and bases cause severe skin burns and irritation to mucous membranes. Be very careful in handling these chemicals. Clean up all spills immediately with excess water, followed by a covering of baking soda,* $NaHCO_3$. *Notify your instructor if a spill occurs. Read the "Laboratory Safety" section in Experiment 1.*

A. Conductivity

The apparatus shown in Figure 17.1 is used to determine the strength of various electrolytes. When the two electrodes are submerged in a solution that is a good conductor of current (a strong electrolyte), the circuit is completed and the bulb shines brightly; if the solution is a poor conductor the bulb burns dimly. Therefore the brightness of the bulb is a qualitative measure of the degree of dissociation of a substance into ions.

1. Connect the apparatus into an electrical outlet. Half–fill a 150mL beaker with distilled (or de–ionized) water and place in contact with the electrodes. Does the bulb glow?#1 Add about 0.5g of NaCl to the water and stir to dissolve. What happens to the conductivity of the solution?#2 Is NaCl a strong, weak, or nonelectrolyte?#3 Remove the NaCl solution and rinse the electrodes with distilled (or de–ionized) water from your wash bottle.

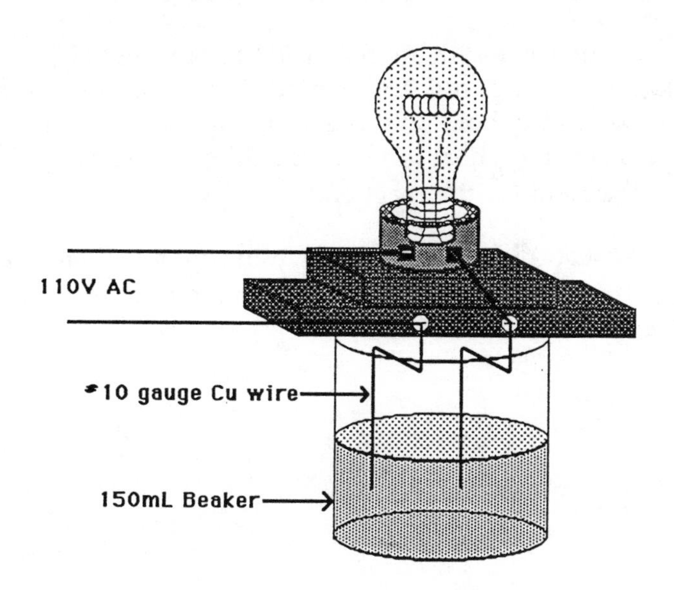

Figure 17.1
Apparatus for Testing the Conductivity of a Solution

2. Into a clean 150mL beaker place about 25mL of conc acetic acid (read the **Caution** for handling acids). Place the acid in contact with the electrodes. Classify its strength as an electrolyte.#4 While keeping the acetic acid in contact with the electrodes, add distilled (or de–ionized) from your wash bottle and swirl the solution. Observe. Continue to add water (up to 100mL), swirl, and observe the glow of the bulb. What happens to the conductivity of the acetic solution with dilution?#5 Rinse the electrodes with distilled (or de–ionized) water and discard the rinse.

3. Test the relative conductivity of the following solutions.* Be sure to rinse the electrodes after each test to avoid contamination of the solutions.

   0.1M $HC_2H_3O_2$, 0.1M HCl, 0.1M NH_3, 0.1M NaOH, 0.1M $NaNO_3$, 0.1M NH_4Cl, 0.1M $C_{12}H_{22}O_{11}$, and 0.1M $CO(NH_2)_2$. Record your observations on the Data Sheet. Determine if the solutions are strong, weak, or nonelectrolytes.#6

4. Place 25mL of 0.1M H_2SO_4 in a 150mL beaker; submerge the electrodes.#7 Swirl the solution and slowly add 0.2M $Ba(OH)_2$. What happens to the intensity of the glow? What is happening in the solution?#8 Continue to add 0.2M $Ba(OH)_2$ until the glow disappears. Now add additional $Ba(OH)_2$ solution. Explain your observations.#9

B. CONCENTRATED ACIDS

1. Several bottles of concentrated acids (purchased from the chemical supplies distributor) are on the reagent shelf. Select one of the acids and closely read the label. Answer the questions on the Data Sheet.

C. ACIDS. SULFURIC, NITRIC, HYDROCHLORIC, AND ACETIC ACIDS

1. Arrange four 75mm test tubes half–filled with water. Hold the first test tube in the palm of your hand and add 5 drops of conc HCl; does the test tube become warm?#10 Test the solution with litmus paper.#11 Repeat the tests with 5 drops of conc H_2SO_4, conc HNO_3, and conc $HC_2H_3O_2$ (**Caution**: *Do not allow the concentrated acids to touch the skin*).#12

2. Some acids can be prepared by the reaction of a salt with another acid. H_2SO_4 and H_3PO_4 are often used as the acids for the preparation. Place about 0.5g of NaCl in a 150mm test tube. Add 3 or 4 drops of conc H_2SO_4 to the solid salt; hold both red and blue moistened litmus papers at the mouth of the test tube. If necessary, heat to volatilize the acid. Explain the effect on litmus.#13 Write a balanced equation for the reaction.#14

3. Place a small piece of polished (with steel wool) Mg, Zn, Fe, and Cu in separate 75mm test tubes. Add about 10 drops of 6M HCl to each metal. Note any reaction in each test tube.#15 Repeat the test, substituting 6M HNO_3 for the 6M HCl.#16

---

* To avoid waste, the test solutions can be shared with other students, especially if care is taken to avoid contamination.

4. ***Instructor Demonstration.*** Test the dehydrating effects of conc H_2SO_4 by placing a few drops on a wood splint and, in an evaporating dish, on a small amount (≅1g) of sugar. Repeat the test with conc HCl and conc HNO_3. Record your observations for each acid on the Data Sheet.[#17-19]

5. Pipet 2mL of 0.1M NaOH into a 150mm test tube and add 1 drop of phenolphthalein indicator. Add drops of 0.5M HCl until a color change (from pink to colorless) occurs (shake after each drop). Record the number of drops.[#20]

   Repeat the determination, substituting 0.5M H_2SO_4 and 0.5M HNO_3 for the 0.5M HCl.[#21,22] What can you conclude about the available acidity of the three acids.[#23] Write balanced equations for the reactions.[#24]

D. Bases. Sodium Hydroxide, Calcium Oxide, and Sodium Carbonate and Ammonia

1. Place a small BB–size piece of NaOH in a 75mm test tube. Hold the test tube in the palm of your hand and add 1mL of water. Note the heat generated in the dissolving of the NaOH.[#25] Test the solution with litmus. Repeat the tests with oven cleaner or a solid drain cleaner, with a fresh sample of CaO, and with Na_2CO_3.[#26]

2. Set up the apparatus shown in Figure 17.2. Place 3g (±0.01g) of a 2:1 mixture (by volume) of NH_4Cl and $Ca(OH)_2$ into the 200mm test tube. Heat the mixture gently; begin at the mouth of the test tube and gradually extend the heat over the entire test tube. As soon as NH_3 gas evolves freely and the apparatus if free of air, collect *three* bottles of the NH_3 by air displacement.[§] Write balanced equations for the reaction of NH_4Cl with $Ca(OH)_2$ and for the reaction of HCl with NH_3.[#27] Note the color and odor of the NH_3.[#28]

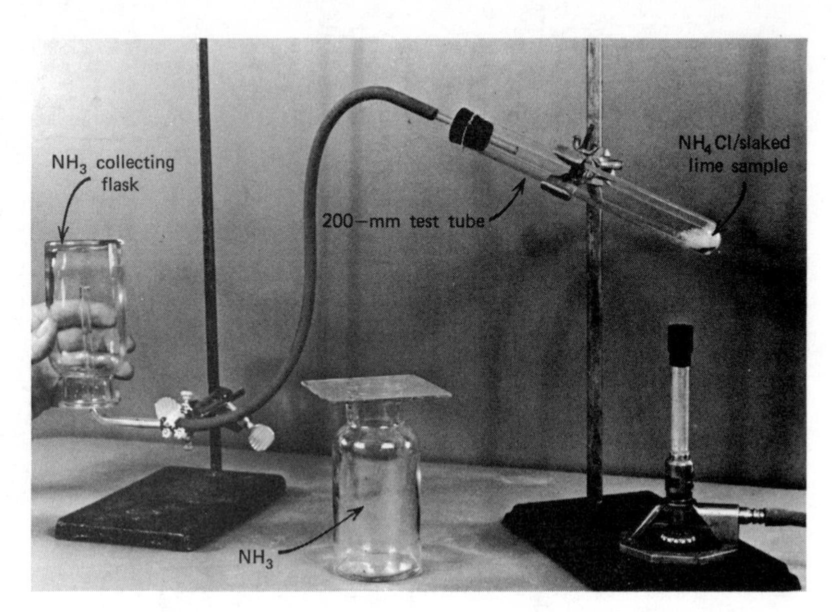

Figure 17.2
Collection of $NH_3(g)$ by Air Displacement

[§]The bottle is filled with NH_3 when a drop of conc HCl suspended at the tip of a glass rod fumes strongly when placed at the mouth of the bottle. Consult the instructor on this technique.

To test for the solubility of NH_3 in water, place about 300mL of water in an 800mL beaker. Add 1mL of phenolphthalein to the water. Bring the mouth of a bottle of NH_3 under the surface of the water (Figure 17.3). Allow it to remain there for 5 minutes; be sure that the mouth of the bottle stays submerged. What happens to the water level in the bottle?#29 What happens to the color of the solution? Why?#30

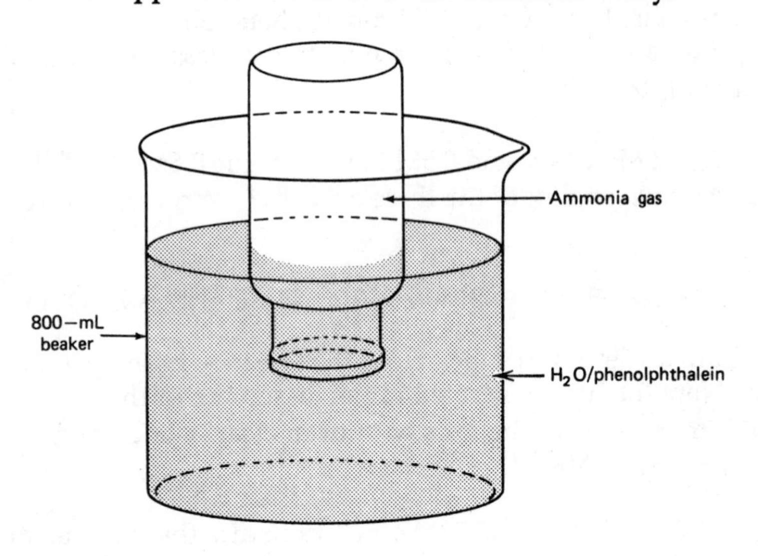

Figure 17.3
A Setup for Testing the Solubility of $NH_3(g)$ in Water

ACIDS, BASES, AND SALTS-LAB PREVIEW

Date__________Name____Lab Sec. _____Desk No._____

1. a. What is an acid?

   b. What is a base?

   c. What is a salt?

   d. What is an alkali?

2. Distinguish the difference between a strong electrolyte and a weak electrolyte.

3. a. What is a neutralization reaction?

   b. In what part of the experimental procedure is a neutralization reaction conducted?

4. a. What is the color of litmus paper in an acidic solution?_____________

   b. What is the color of litmus paper in a basic solution?_____________

   c. What is the color of phenolphthalein in an acidic solution?_____________

   d. What is the color of phenolphthalein in a basic solution?_____________

5. Briefly describe the technique for testing a solution with litmus paper.

6. How are gases that are less dense than air collected?

7. Describe the technique for heating a test tube with a direct flame.

8. Describe the technique for testing the odor of a chemical.

9. List the common (or commercial) names and the correct chemical names associated with these compounds.

	Common (commercial) Names	Chemical Names
a. Na_2CO_3	______________________	______________________
b. HCl	______________________	______________________
c. H_2SO_4	______________________	______________________
d. NaOH	______________________	______________________
e. CaO	______________________	______________________
f. $Ca(OH)_2$	______________________	______________________
g. $CaCO_3$	______________________	______________________
h. NH_3	______________________	______________________
i. $NaHCO_3$	______________________	______________________

10. How should acids spills be cleaned up in the laboratory?

ACIDS, BASES, AND SALTS-DATA SHEET

Date________Name____Lab Sec. _____Desk No._____

A. CONDUCTIVITY

	Compound	Observation	Electrolyte Strong	Weak	Non–
#1	H_2O	________________	_____	_____	_____
#2,3	NaCl	________________	_____	_____	_____
#4	conc $HC_2H_3O_2$	________________	_____	_____	_____
#5	dil $HC_2H_3O_2$	________________	_____	_____	_____
#6	$HC_2H_3O_2$	________________	_____	_____	_____
	HCl	________________	_____	_____	_____
	NH_3	________________	_____	_____	_____
	NaOH	________________	_____	_____	_____
	$NaNO_3$	________________	_____	_____	_____
	NH_4Cl	________________	_____	_____	_____
	$C_{12}H_{22}O_{11}$	________________	_____	_____	_____
	$CO(NH_2)_2$	________________	_____	_____	_____
#7	H_2SO_4	________________	_____	_____	_____
#8,9	$H_2SO_4 + Ba(OH)_2$	________________	_____	_____	_____

B. CONCENTRATED ACIDS

1. What is the percent (range) by weight of the acid in the bottle?______________

   What is its major impurity?________________________________

2. List two **Danger** warnings printed on the label.

   a. ____

   b. ____

3. List two **First Aid** remedies for the acid.

a. ______________________________

b. ______________________________

C. ACIDS

	Acid	Heat Change	Litmus Test
#10,11	HCl	____________	____________
#12	H_2SO_4	____________	____________
	HNO_3	____________	____________
	$HC_2H_3O_2$	____________	____________

#13 Litmus test on vapors______________________________

#14 Balanced equation: NaCl + $H_2SO_4 \rightarrow$ ______________________________

	Acid/Metal	Mg	Zn	Fe	Cu
#15	HCl	________	________	________	________
#16	HNO_3	________	________	________	________

	Observation	H_2SO_4	HCl	HNO_3
#17-19	wood splint	________	________	________
	sugar	________	________	________

	Acid	HCl	H_2SO_4	HNO_3
#20-22	Drops of NaOH	________	________	________

#23 Conclusion on available acidity

#24 Balanced Equations:

$NaOH + HCl \rightarrow$ ____

$NaOH + H_2SO_4 \rightarrow$ ______________________________________

$NaOH + HNO_3 \rightarrow$ ____

D. BASES

	Heat Change	Litmus Test
#25 NaOH	________________________	________________________
#26 Oven/Drain Cleaner	________________________	________________________
CaO	________________________	________________________
Na_2CO_3	________________________	________________________

#27 Balanced Equation:

$NH_4Cl + Ca(OH)_2 \rightarrow$ ____________________________

$HCl + NH_3 \rightarrow$ ____________________________

#28 Color of NH_3______________ ; Does NH_3 have an odor?______________

#29 Is NH_3 soluble in water?________________

#30 Why does the color of the solution change?

EXPERIMENT 18
HARD WATER ANALYSIS

Objectives

- To learn the cause and effects of hard water
- To determine the hardness of a water sample

Principles

Hard water contains the dissolved salts of calcium, magnesium, and iron. In low concentrations these ions are not considered harmful for domestic use. However at higher concentrations, these ions are considered undesirable for two reasons.

- Hardening ions form insoluble compounds with soaps. Soaps, sodium salts of fatty acids such as sodium stearate, $C_{17}H_{35}COO^- Na^+$, are very effective cleansing agents so long as they remain soluble; the presence of these hardening ions however causes the formation of a grey, insoluble soap scum, $(C_{17}H_{35}COO)_2Ca$.

$$2C_{17}H_{35}COO^-, Na^+(aq) + Ca^{2+}(aq) \rightarrow (C_{17}H_{35}COO)_2Ca\,(s) + 2Na^+(aq)$$

This grey precipitate appears as a "bathtub ring" and it also clings to clothes, causing white clothes to appear grey.

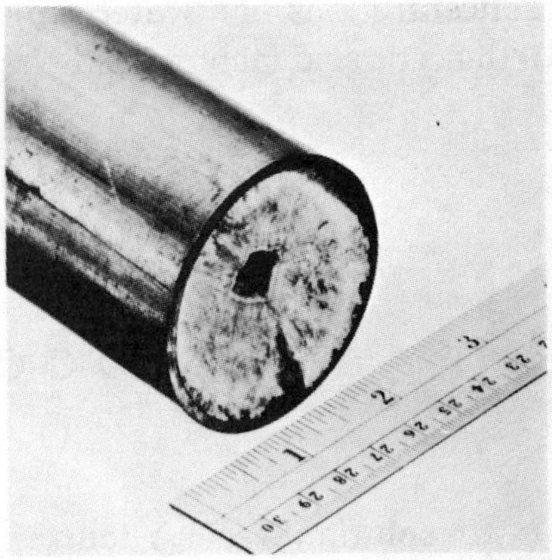

- Hard water is also responsible for the appearance and undesirable formation of "boiler scale" on tea kettles and pots used for heating water. The boiler scale is a poor conductor of heat and thus reduces the efficiency of transferring heat. Boiler scale can also build up on the inside of household hot water pipes and cause a reduction of water flow; in extreme cases, this buildup causes the pipe to break. Boiler scale consists primarily of the carbonate salts of the hardening ions and is formed according to

$$Ca^{2+}(aq) + 2HCO_3^-(aq) \xrightarrow{\Delta} CaCO_3(s) + CO_2(g) + H_2O$$

Ground water becomes hard water as a result of it flowing through underground limestone ($CaCO_3$) deposits; generally, the deeper water wells have a higher hardness than the shallow wells because of a longer time of contact with the $CaCO_3$. Surface water similarly accumulates hardening ions as a result of it flowing over limestone rock. In either case the CO_2 dissolved in rainwater[1] solubilizes limestone deposits.

---

[1] CO_2 dissolved in rainwater makes rainwater slightly acidic.

$$CO_2(g) + 2H_2O \rightarrow H_3O^+(aq) + HCO_3^-(aq)$$

$$CO_2(aq) + H_2O + CaCO_3(s) \rightarrow Ca^{2+}(aq) + 2HCO_3^-(aq)$$

Notice that this reaction is just the reverse of the reaction for the formation of boiler scale.

Because of the relative large natural abundance of limestone deposits and other calcium minerals, such as gypsum, $CaSO_4 \bullet 2H_2O$, it is not surprising that Ca^{2+} ion, in conjunction with Mg^{2+}, is a major component of the dissolved solids in water. A general classification of hard waters is listed in Table 18.1.

Table 18.1
Hardness Classification of Water

Hardness (ppm $CaCO_3$)	Classification
< 15ppm	very soft water
15ppm - 50 ppm	soft water
50ppm - 100ppm	medium hard water
100ppm - 200ppm	hard water
>200ppm	very hard water

The concentration of the hardening ions in a water sample is commonly expressed as though all of the hardness is due exclusively to $CaCO_3$. The units for hardness is mg $CaCO_3$/L, which is also ppm[2] $CaCO_3$.

In this experiment a titration technique is used to measure the combined Ca^{2+} and Mg^{2+} concentrations in a water sample. The titrant is the disodium salt of ethylenediamine-tetraacetic acid (abbreviated Na_2H_2Y)[3].

```
          O                                  O
          ||                                 ||
Na+O-C-CH2                            CH2-C-OH
           \                          /
            N-CH2-CH2-N
           /                          \
Na+O-C-CH2                            CH2-C-OH
          ||                                 ||
          O                                  O
```

In solution Na_2H_2Y ionizes to form $2Na^+$ and H_2Y^{2-}. H_2Y^{2-} reacts with the hardening ions, Ca^{2+} and Mg^{2+}, to form very stable complex ions, especially in a solution buffered at a pH of about 10. An ammonia–ammonium ion buffer is often used for this pH adjustment in the analysis.

---

[2] ppm means "parts per million"–1mg of $CaCO_3$ in 1 000 000mg solution is 1ppm $CaCO_3$. Assuming the density of the solution is 1g/mL, then 1 000 000mg solution = 1L solution.

[3] Ethylenediaminetetraacetic acid is often simply referred to as EDTA.

A special indicator is used to detect the endpoint in the titration. Called Eriochrome Black T (EBT), it also forms complex ions with the Ca^{2+} and Mg^{2+} ions, but more strongly with Mg^{2+}. Because only a small amount of EBT is added, only a small quantity of Mg^{2+}, and no Ca^{2+}, is complexed to EBT–most of the hardening ions are still "free" in solution. The EBT is sky–blue in solution and the $[Mg–EBT]^{2+}$ complex is wine–red.

$$\underset{\text{sky–blue}}{Mg^{2+} + EBT} \rightarrow \underset{\text{wine–red}}{[Mg–EBT]^{2+}}$$

Therefore even before any H_2Y^{2-} titrant is added in the titration, the solution is wine–red. As H_2Y^{2-} titrant is added, it complexes with the "free" Ca^{2+} and Mg^{2+}.

$$Ca^{2+}(aq) + H_2Y^{2-}(aq) \rightarrow CaY^{2-}(aq) + 2H^{+}(aq)$$
$$Mg^{2+}(aq) + H_2Y^{2-}(aq) \rightarrow MgY^{2-}(aq) + 2H^{+}(aq)$$

Once the H_2Y^{2-} complexes all of the "free" Ca^{2+} and Mg^{2+}, it then removes the Mg^{2+} from the $[Mg–EBT]^{2+}$; the solution turns from wine–red to sky–blue and the endpoint is reached.

$$\underset{\text{wine–red}}{[Mg^{2+}–EBT]^{2+}} + H_2Y^{2-}(aq) \rightarrow \underset{\text{sky–blue}}{MgY^{2-}(aq) + 2H^{+}(aq) + EBT}$$

For the endpoint to appear, Mg^{2+} must be present; therefore a small amount of MgY^{2-} is oftentimes added to the buffer solution. The added Mg^{2+} does not affect the amount of Na_2H_2Y because an equimolar amount of Na_2H_2Y is also added to the buffer.

From the balanced equations, it is apparent that once the molarity of the Na_2H_2Y solution is known, the moles of hardening ions in a water sample can be calculated.

$$\text{volume } H_2Y^{2-} \times \text{molarity } H_2Y^{2-} = \text{moles } H_2Y^{2-} = \text{moles hardening ions}$$

The hardening ions, for reporting purposes, are assumed to be exclusively Ca^{2+} from the dissolving of $CaCO_3$. Therefore the equivalent hardness expressed as mg $CaCO_3$ per liter of water sample is

$$\text{moles hardening ions} = \text{moles } Ca^{2+} = \text{moles } CaCO_3$$

$$\text{ppm } CaCO_3 \left(\frac{\text{mg } CaCO_3}{\text{L sample}}\right) = \frac{\text{mol } CaCO_3}{\text{L sample}} \times \frac{100.1\text{g } CaCO_3}{\text{mol}} \times \frac{\text{mg}}{10^{-3}\text{g}}$$

Techniques

- Technique 4b, page 5 — Separation of a Solid from a Liquid
- Technique 10, page 12 — Reading a Meniscus
- Technique 11, page 13 — Pipetting a Liquid or Solution
- Technique 12, page 14 — Titrating a Solution
- Technique 13, page 16 — Using the Laboratory Balance

Procedure

A. STANDARDIZATION OF A 0.01M DISODIUM ETHYLENEDIAMINETETRAACETATE, NA_2H_2Y, SOLUTION

Three trials should be completed for the standardization of the Na_2H_2Y solution. To save time, prepare the three Erlenmeyer flasks in Part A.3 at the same time.

1. Weigh (±0.01g) on weighing paper 1g of Na_2H_2Y; transfer it to a 250mL volumetric flask containing about 100mL of distilled (or de-ionized) water and stir to dissolve. Dilute to the "mark" with distilled (or de-ionized) water.

2. Prepare a buret for titration. Rinse the buret with the Na_2H_2Y solution and then fill. Record the volume (±0.01mL) of the solution.

3. Obtain about 80mL of the standard Ca^{2+} solution from the reagent shelf. Record the molarity of the solution. Pipet 25.0mL into a 125mL Erlenmeyer flask, add 1mL of the buffer (pH = 10) solution, and 2 drops of EBT indicator.

4. Titrate the standard Ca^{2+} solution with the Na_2H_2Y solution; stir continuously. Near the endpoint, slow the rate of addition to drops; the last few drops should be added at 3–5s intervals. The solution changes from wine–red to purple to blue–no tinge of the red–wine color should remain; the solution is *blue* at the endpoint.

5. Repeat the titrations on the other two samples and then calculate the molarity of the Na_2H_2Y solution.

B. ANALYSIS OF WATER SAMPLE

Complete three trials in your analysis. The first trial is your first indication of the hardness of your water sample–the hardness values may vary tremendously, depending upon your water sample. You may want to adjust the volume of water in the analysis for the second and third trials.

1. a. Obtain about 100mL of a water sample from your instructor. You may want to bring your own water sample or use the tap water in the laboratory.
   b. If the water sample is from a lake, stream, or ocean, you will need to gravity filter the sample before the analysis.
   c. If your sample is acidic, add drops of 1M NH_3 until it is basic to litmus.

2. Pipet 25.0mL of your (filtered, if necessary) water sample[4] into a 125mL Erlenmeyer flask, add 1mL of the buffer (pH = 10) solution, and 2 drops of EBT indicator.

3. Repeat Parts A.4 and A.5 to determine the hardness of your water sample.

---

[4]If your water is known to have a high hardness, decrease the volume of the water proportionally until it takes about 15mL of Na_2H_2Y titrant for your second and third trials. Similarly if your water sample is known to have a low hardness, increase the volume of the water proportionally.

HARD WATER ANALYSIS-LAB PREVIEW

Date_________Name____Lab Sec. _____Desk No._____

1. What ions are responsible for water hardness?

2. Which one of the hardening ions binds more tightly to (forms a stronger complex ion with) the anion of the disodium salt of ethylenediaminetetraacetate, H_2Y^{2-}?

3. a. How do hardening ions cause soap to be less effective?

   b. What is and what causes "boiler scale?"

4. a. While gravity filtering a solution, the funnel should be no more than _____ full at any time.

   b. Two folds are made on a piece of filter paper that is to be fit into a funnel. Describe the two folds.

5. The analysis procedure in this experiment requires that the titration be conducted at a pH = 10. How is the solution adjusted to this pH?

6. A 50.0mL water sample required 5.14mL of 0.01M Na_2H_2Y to reach the endpoint for the titration procedure described in this experiment.

   a. What is the name of the indicator used in the titration?

   b. What is the color change at the endpoint?

   c. Calculate the moles of hardening ions in the water sample.

   d. Assuming the hardness is due to $CaCO_3$, express the hardness concentration in mg $CaCO_3$/L sample.

   e. What is this hardness concentration expressed in ppm $CaCO_3$?

   f. Classify the hardness of this water according to Table 18.1.

7. The hardness of a water sample is known to be 500ppm $CaCO_3$. In analysis of a 50mL water sample, how many milliliters of 0.01M Na_2H_2Y are needed to reach the endpoint?

HARD WATER ANALYSIS-DATA SHEET

Date________Name______________________________________Lab Sec. _____Desk No.____

A. STANDARDIZATION OF A 0.01M DISODIUM ETHYLENEDIAMINETETRAACETATE, Na_2H_2Y, SOLUTION

	Trial 1	Trial 2	Trial 3
1. Mass of weighing paper + Na_2H_2Y (g)	__________	__________	__________
2. Mass of weighing paper (g)	__________	__________	__________
3. Mass of Na_2H_2Y (g)	__________	__________	__________
4. Volume of standard Ca^{2+} solution	_25.0mL_	_25.0mL_	_25.0mL_
5. Concentration of standard Ca^{2+} solution		__________	
6. Mol Ca^{2+} = mol Na_2H_2Y	__________	__________	__________
7. Buret reading, final (mL)	__________	__________	__________
8. Buret reading, initial (mL)	__________	__________	__________
9. Volume of Na_2H_2Y titrant (mL)	__________	__________	__________
10. Molarity of Na_2H_2Y solution (mol/L)	__________	__________	__________
11. Average molarity of Na_2H_2Y solution (mol/L)		__________	

B. ANALYSIS OF WATER SAMPLE

	Trial 1	Trial 2	Trial 3
1. Volume of water sample (mL)	__________	__________	__________
2. Buret reading, final (mL)	__________	__________	__________
3. Buret reading, initial (mL)	__________	__________	__________
4. Volume of Na_2H_2Y titrant (mL)	__________	__________	__________
5. Mol Na_2H_2Y = mol Ca^{2+} (mol)	__________	__________	__________
6. Mass of equivalent $CaCO_3$ (g)	__________	__________	__________
7. ppm $CaCO_3$ (mg $CaCO_3$/L sample)	__________	__________	__________
8. Average ppm $CaCO_3$		__________	

Questions

1. Because the chemist couldn't accurately detect the endpoint, the addition of Na_2H_2Y titrant was discontinued before the endpoint was reached. How will this affect the reported hardness of the water sample? Explain.

2. Explain what might happen in the analysis if the indicator had been omitted from the procedure.

3. Washing soda, $Na_2CO_3 \cdot 10H_2O$, is often used to "soften" hard water, i.e., to remove hardening ions. Assuming hardness is due to Ca^{2+}, the CO_3^{2-} ion precipitates the Ca^{2+}.

$$Ca^{2+}(aq) + CO_3^{2-}(aq) \rightarrow CaCO_3(s)$$

How many grams and pounds of washing soda are needed to remove the hardness from 500 gallons of water having a hardness of 150ppm $CaCO_3$?

EXPERIMENT 19
REDOX; ACTIVITY SERIES

Objectives

- To learn the rules for determining the oxidation numbers of the elements in compounds
- To develop a general understanding of redox reactions
- To determine the relative chemical reactivity of several metals

Principles

All elements have a relative chemical activity; most do not exist in their "elemental" form in nature, but rather they exist in chemical combination with other elements in their environment forming any number of compounds. When an element forms a compound, it no longer has the same physical or chemical properties; its properties have changed because it has undergone a fundamental change in its atomic structure–it has either lost, gained, or now shares valence electrons.

When an element combines to produce a compound, we account for what happens to the valence electrons by assuming that it has acquired an "apparent" charge. This apparent charge, the charge the atom would have if all the electrons in each bond were assigned to the more electronegative element, may be either positive (+) or negative (-). This apparent charge is called the **oxidation number** (or **oxidation state**) of the element in the compound. When the element appears to have *lost* valence electrons in the bond formation, the oxidation number of the element has increased and we say that it has been *oxidized*; but when the element appears to have *gained* valence electrons in the bond formation, the oxidation number of the element has decreased...it has been *reduced*.

Oxidation numbers are very handy bookkeeping devices for keeping track of electrons when elements combine to form compounds. If we can remember only a few generalizations concerning oxidation numbers, we can not only write the correct chemical formulas for a large number of compounds, but also analyze a chemical system to determine what is oxidized and what is reduced. These generalizations are summarized.

1. Any element in the "free" state (not combined with any other element) has an oxidation number of zero, regardless of the complexity of the molecule in which it occurs. Each atom in Ne, O_2, P_4, and S_8 has an oxidation number of zero.

2. The oxidation number of any ion (monoatomic or polyatomic) equals the charge on the ion. The Ca^{2+}, NH_4^+, S^{2-}, and PO_4^{3-} ions have oxidation numbers of 2^+, 1^+, 2^-, and 3^- respectively.

3. Oxygen is assigned an oxidation number of 2^- (except 1^- in peroxides, e.g., H_2O_2, and 2^+ in OF_2). The oxidation number of oxygen is 2^- in $KMnO_4$, Fe_2O_3, CaO, and N_2O.

4. Hydrogen has an oxidation number of 1^+ (except 1^- in metal hydrides, e.g., NaH or CaH_2). Its oxidation number is 1^+ in HCl, $HaHCO_3$, and NH_3.

5. Some elements exhibit only one oxidation number in certain compounds.

   a. Group IA elements (Li, Na, K, Rb, and Cs) always have a 1^+ oxidation number in compounds.

   b. Group IIA elements (Be, Mg, Ca, Sr, and Ba) always have a 2^+ oxidation number in compounds.

   c. Boron and aluminum always possess a 3^+ oxidation number in compounds.

   d. Group VIA elements (O, S, Se, and Te) exhibit a 2^- oxidation number in all *binary* compounds with *metals.*

   e. Group VIIA elements (F, Cl, Br, and I) exhibit a 1^- oxidation number in all *binary* compounds with *metals.*

6. The oxidation numbers of other elements can vary, depending upon with what elements it has combined in forming the compound. Generally the "other" elements in the compound have known oxidation numbers; since the compound must be neutral, its oxidation number can be calculated.

   Example 1. What is the oxidation number of Fe in Fe_2O_3?
   Since each oxygen is known to have a 2^- oxidation number (for a total charge of 6^-) and the charge of Fe_2O_3 is zero, the combined oxidation numbers of the two iron atoms must be 6^+, or *each* Fe atom must be 3^+.

   Example 2. What is the oxidation number of Fe in FeO?
   Since each oxygen is known to have a 2^- oxidation number and the charge of FeO is zero, the oxidation number of the iron atom must be 2^+.

7. The oxidation number of polyatomic ions is the charge of the ion. But what is the oxidation numbers of the constituent atoms in the ion?

   Example 3. What is the oxidation number of Cl in ClO_4^-, the perchlorate ion?
   Each oxygen has a 2^- oxidation number (for a total charge of 8^-) and the charge of ClO_4^- is 1^-. Therefore, the oxidation number of Cl must be 7^+. This is not a violation of the Rule 5e, because the Cl is *not* combined with a metal in a binary compound.

   Example 4. What is the oxidation number of Cl in ClO_2^-, the chlorite ion?
   Each oxygen has a 2^- oxidation number (for a total charge of 4^-) and the charge of ClO_2^- is 1^-. Therefore, the oxidation number of Cl must be 3^+.

   Example 5. What is the oxidation number of Cr in $K_2Cr_2O_7$?
   Each oxygen has a 2^- oxidation number (for a total charge of 14^-), each potassium has a 1^+ oxidation number (for a total charge of 2^+), and the charge of $K_2Cr_2O_7$ is zero. Since $2^+ + 2Cr + 14^- = 0$, the combined oxidation number of the two chromium atoms must be 12^+, or *each* Cr atom is 6^+.

   Additional practice is available on the Lab Preview.

Metallic and nonmetallic elements vary in their chemical activity. This experiment seeks to determine the relative chemical activity of several metals by measuring their tendencies to form a positive oxidation number by losing electrons (or being oxidized) and form cations. The metals, listed in order of increasing activity, comprise an abbreviated **activity series** in this experiment. Several tests are used to determine this relative activity.

Reaction of Metal with Water–Rate of Hydrogen Gas Evolution

Very reactive metals oxidize in water causing the reduction of water and the evolution of H_2 gas and heat.

$$\text{oxidation: } M \rightarrow M^{n+}(aq) + ne^-$$
$$\text{reduction: } 2H_2O + 2e^- \rightarrow 2OH^-(aq) + H_2(g)$$

The evolution rate of H_2 gas is not always an absolute criterion of chemical reactivity because the reaction involves several variables, e.g., state of a subdivision of metal. It is, however, a qualitative method for observing reactivity.

Reaction of Metal with Hot Water–Rate of Disappearance of Metal

Placing metals in hot water causes reactions similar to those occurring in cold water. In hot water the less reactive metals produce H_2 gas at such a slow rate that its evolution is not as detectable as the rate of disappearance of the metal. Although the particle size of the metal does influence the reaction rate, its rate of disappearance is a more certain qualitative method for comparison.

Reaction of Metal with a Nonoxidizing Acid–Rate of Hydrogen Gas Evolution

By comparison, the less reactive metals evolve H_2 gas only after the addition of a nonoxidizing acid, such as $HCl(aq)$ or $H_2SO_4(aq)$. The metal is oxidized and the H^+ is reduced.

$$\text{oxidation: } M \rightarrow M^{n+}(aq) + ne^-$$
$$\text{reduction: } 2H^+(aq) + 2e^- \rightarrow H_2(g)$$

Reaction of Metal with an Oxidizing Acid–Rate of Gas Evolution

Some metals can only be oxidized in the presence of oxidizing acids in forming their corresponding cations–these metals normally exist in the metallic state and are not easily oxidized. Common oxidizing acids include nitric acid, HNO_3, and perchloric acid, $HClO_4$.

$$\text{oxidation: } M \rightarrow M^{n+}(aq) + ne^-$$
$$\text{reduction: } 2H^+(aq) + NO_3^-(aq) + e^- \rightarrow NO_2(g) + H_2O \text{ for "conc" } HNO_3$$

Reaction of Metal with a More Reactive Metal Cation–A Displacement Reaction

The relative reactivity of two metals is established by the outcome of a reaction between a metal and the cation of the other. Where metal M is more reactive (a greater tendency to be oxidized) than metal R, M displaces R^{n+} from the aqueous solution. M goes into solution as M^{n+} and R^{n+} forms the metal, R.

$$M + R^{n+} \rightarrow M^{n+} + R$$

For example, when metallic iron is placed in a lead(II) solution, Fe metal is oxidized to Fe^{2+}; the Pb^{2+} is reduced to Pb metal.

$$\textbf{oxidation:} \quad Fe \rightarrow Fe^{2+} + 2e^-$$
$$\textbf{reduction:} \quad Pb^{2+} + 2e^- \rightarrow Pb$$

On the other hand, if Pb metal is placed in to an Fe^{2+} solution, no reaction occurs because of the two elements, Pb metal is less reactive and prefers the reduced state–Fe prefers the oxidized state, Fe^{2+}.

$$Pb + Fe^{2+} \rightarrow \text{no reaction}$$

These results show that Fe more readily oxidizes than does Pb. Iron, therefore, is said to have a greater activity that lead.

Techniques

- Technique 6a, d, page 7 — Heating liquids

Procedure

This experiment determines the relative activity of these metals: Ca, Na, K, Pb, Cu, Zn, Fe, Al, Mg, and Ni

You and your partner will need a number of solids and solutions for this experiment. Have them readily available so the testing can proceed quickly.

A. Reaction of Metal with Water–Rate of Hydrogen Gas Evolution

Demonstration Only. Ca, Na, and K. (**Caution:** *Never touch* Na, K, or Ca; *each causes a severe skin burn.*)[1] Wrap a pea-size (no larger!) piece of freshly cut metal in aluminum foil. Fill a 200mm Pyrex test tube with water and invert it in an 800mL beaker 3/4-filled with water. Set the beaker and test tube behind a safety shield. Punch 5 pin-size holes in the aluminum foil. With a pair of tongs or tweezers, place the wrapped metal in the mouth of the test tube, keeping it under water (Figure 19.1). Repeat the test for each metal and compare the rates of $H_2(g)$ evolution. Record these observations.

---

[1]It is strongly recommended that this be completed as a laboratory demonstration by the laboratory instructor.

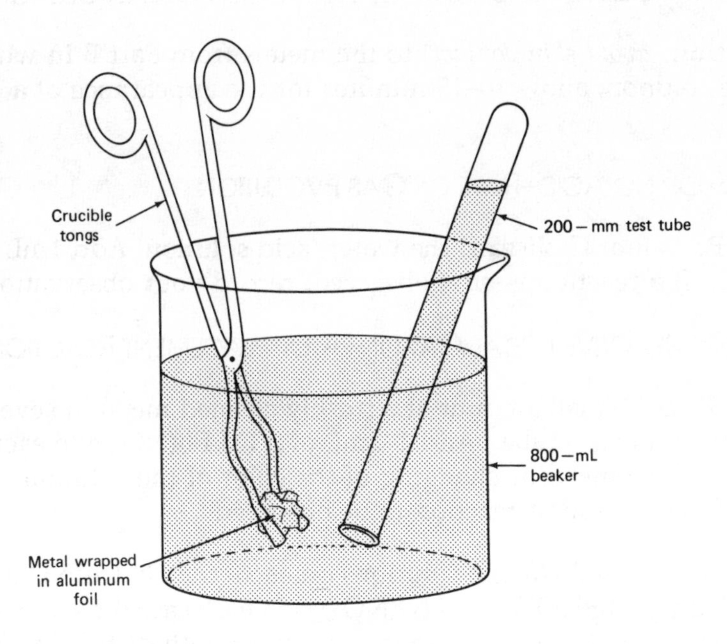

Figure 19.1
Collection of $H_2(g)$ from the Reaction of an Active Metal with Water

B. Reaction of Metal with Hot Water–Rate of Disappearance of Metal

Pb, Cu, Zn, Fe, Al, Mg, and Ni. Thoroughly clean each metal with steel wool to remove any oxide coating. This is especially critical for Al and Mg since they form tough, protective oxide coatings. After cleaning, place a small amount of each metal into 75mm test tubes (Figure 19.2). Quickly half-fill the test tubes with previously boiled, distilled (or de-ionized) hot water. Heat the test tubes in a hot water bath (85–90°C). Some reactions are not immediately apparent; maintain this temperature for at least 15 minutes. Look for $H_2(g)$ evolution, discoloration of the metal surface, or a disappearance of the metal. Record your observations.

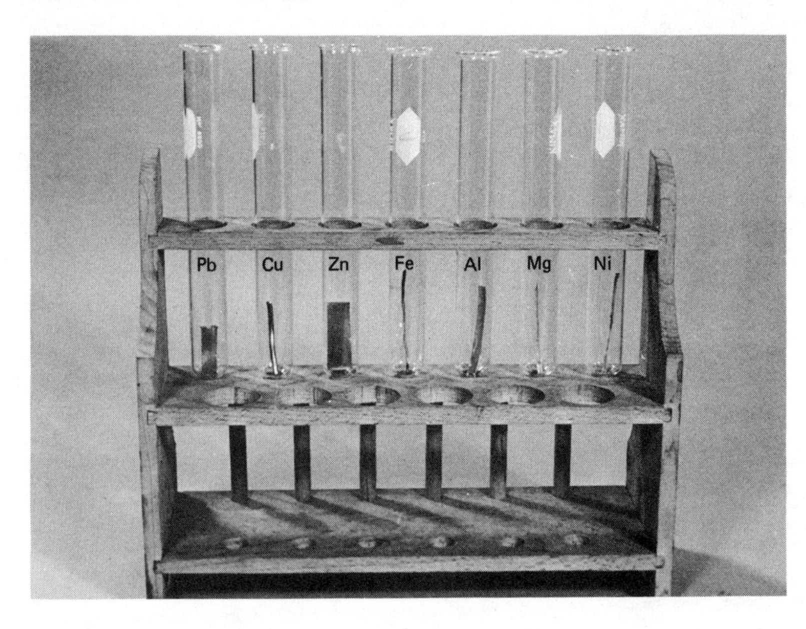

Figure 19.2
Arrangement of Metals for Testing Reactivity

C. Reaction of Metal with a Nonoxidizing Acid–Rate of Hydrogen Gas Evolution

Add 5 drops of conc HCl (**Caution**: *avoid skin contact*) to the metals from Part B in which no reaction was observed. Agitate the solution; allow 10–15 minutes for the appearance of a reaction. Record.

D. Reaction of Metal with an Oxidizing Acid–Rate of Gas Evolution

If a metal shows no reaction in Parts B or C, discard the water/acid solution. Add 1mL of 3M HNO_3 (**Caution**: *avoid skin contact*). If a reaction is now observed, record your observation.

E. Reaction of Metal with a More Reactive Metal Cation–A Displacement Reaction

1. Pb, Cu, Zn, Fe, Al, Mg, and Ni. Place a small amount of a freshly cleaned metal in seven labeled 75mm test tubes—one metal to each tube. Add 1–2mL of 0.1M $Pb(NO_3)_2$ to each test tube. Any tarnishing or dulling of the metal or changing of the color of the solution indicates a reaction. Allow 5–10 minutes for a reaction to be observed.

2. Repeat the procedure, using the following 0.1M test solutions on each metal: $Cu(NO_3)_2$, $Zn(NO_3)_2$, $Fe(NH_4)_2(SO_4)_2$, $Al(NO_3)_3$, $Mg(NO_3)_2$, and $Ni(NO_3)_2$. In each case the same metal sample may be reused if it is unreacted after the previous test, rinsed with distilled water, and cleaned with steel wool. Record.

F. Establishment of Activity Series

Using the observations from Parts A through E, list the 10 metals in order of increasing activity.

REDOX; ACTIVITY SERIES-LAB PREVIEW

Date________ Name______________________________________ Lab Sec. ____ Desk No.____

1. Indicate the oxidation number of the bold–faced element.

a. CO________ d. N_2O________ g. PCl_3________ j. PF_5________ m. SO_3^{2-}________

b. CO_2________ e. OF_2________ h. Cu_2SO_4________ k. HPO_3^-________ n. K_2MnO_4________

c. IF_7________ f. NI_3________ i. IO_3^-________ l. $AuCl_3$________ o. CrO_3________

2. From the following displacement reactions, arrange, at right, these hypothetical metals in order of increasing activity.

$A + B^+ \rightarrow A^+ + B$	________	least reactive
$C + B^+ \rightarrow$ no reaction	________	
$D + C^+ \rightarrow D^+ + C$	________	
$F + G^+ \rightarrow$ no reaction	________	
$A + F^+ \rightarrow$ no reaction	________	
$F + D^+ \rightarrow F^+ + D$	________	most reactive
$D + B^+ \rightarrow$ no reaction		
$G + A^+ \rightarrow G^+ + A$		

3. a. List four methods by which a metal's chemical activity can be determined.

__________________________ __________________________

__________________________ __________________________

b. Which method is used to determine the most reactive metals?

____

c. Which method is used to determine the least reactive metals?

____

4. Cesium is a very reactive metal, reacting with cold water to produce $H_2(g)$.

a. Write a balanced equation for the reaction.

b. The oxidizing agent is ________; the substance oxidized is ________.

5. Considering the relative ionization energies of the Group IA elements, what trend in their reactivity with cold water would you predict?

6. What reaction would you expect to occur if
   a. Zn metal is placed in a solution containing Ag^+? (Consider the relative reactivity of Zn *versus* Ag.)

   b. Au metal is placed in a solution containing Zn^{2+}?

7. Distinguish in a definition an oxidizing acid from a nonoxidizing acid. Give an example of each.

8. Explain, in brief, the correct technique for heating a liquid in a test tube using a direct flame.

9. What is Technique 6d?

REDOX; ACTIVITY SERIES-DATA SHEET

Date________Name____________________________________Lab Sec. _____Desk No._____

A. Reaction of Metal with Water–Rate of Hydrogen Gas Evolution

Ca, Na, and K

1. From the rate of $H_2(g)$ evolution, arrange Ca, Na, and K in order of increasing activity.______________________________

2. Write a balanced equation for each reaction of the metal with water.

B. Reaction of Metal with Hot Water–Rate of Disappearance of Metal

Pb, Cu, Zn, Fe, Al, Mg, and Ni

1. Which metals show a definite reaction with hot water?

____

2. If possible, arrange the metals that do react in order of increasing activity.

____

3. Write a balanced equation for each reaction.

C. Reaction of Metal with a Nonoxidizing Acid–Rate of Hydrogen Gas Evolution

Pb, Cu, Zn, Fe, Al, Mg, and Ni

1. Which metals show a definite reaction with HCl?

____

2. If possible, arrange these metals that do react in order of increasing activity.

____

3. Write a balanced equation for each reaction.

D. REACTION OF METAL WITH AN OXIDIZING ACID–RATE OF GAS EVOLUTION

1. Which of the remaining metals react with only HNO_3?_______________

E. REACTION OF METAL WITH A MORE REACTIVE METAL CATION–A DISPLACEMENT REACTION

Pb, Cu, Zn, Fe, Al, Mg, and Ni

Complete the table with NR (no reaction) or R (reaction) where appropriate.

For all reactions observed, write a balanced net–ionic equation. Use additional paper if necessary.

Test Reagents	Pb	Cu	Zn	Fe	Al	Mg	Ni
HCl	----	----	----	----	----	----	----
$Pb(NO_3)_2$	NR	----	----	----	----	----	----
$Cu(NO_3)_2$	----	NR	----	----	----	----	----
$Zn(NO_3)_2$	----	----	NR	----	----	----	----
$Fe(NH_4)_2(SO_4)_2$	----	----	----	NR	----	----	----
$Al(NO_3)_3$	----	----	----	----	NR	----	----
$Mg(NO_3)_2$	----	----	----	----	----	NR	----
$Ni(NO_3)_2$	----	----	----	----	----	----	NR

Balanced Equations

F. ESTABLISHMENT OF ACTIVITY SERIES

List the 10 metals studied in this experiment in order of increasing activity.

____

Questions

1. Tons of aluminum and magnesium are used annually where light–weight construction is required (e.g., for airplanes) even though each reacts rapidly with oxygen. Explain.

2. Zinc is used as the protective covering of steel on "galvanized iron". Explain its function in terms of chemical activity. Is zinc or iron more reactive?

3. a. In the laboratory can hydrochloric acid be added to copper metal to generate $H_2(g)$? Explain.

   b. Which metals do generate $H_2(g)$?

4. To reduce the corrosion of underground storage tanks (made of iron, actually steel which is an iron/carbon alloy), a sacrificial metal, such as magnesium, is attached to it. As a result the sacrificial metal corrodes instead of the iron. Explain.

5. Circle the metals that react with

   a. Na^+: Pb, Cu, Mg
   b. Fe^{2+}: Na, Mg, Cu
   c. Cu^{2+}: Ca, Zn, Pb

6. Write a balanced equation for only one (if any) of the reactions that occur in each part of Question 5.

   a. Na^+ + ______ →

   b. Fe^{2+} + ______ →

   c. Cu^{2+} + ______ →

EXPERIMENT 20
VITAMIN C ANALYSIS

Objectives

- To prepare and standardize a sodium thiosulfate solution
- To determine the vitamin C concentration in a vitamin tablet, a fresh fruit, or a fresh vegetable sample

Principles

The human body does not synthesize vitamins; therefore the vitamins that we need for catalyzing specific biochemical reactions are gained only from the food that we eat. Vitamin C, also called **ascorbic acid**, is one of the more abundant and easily obtained vitamins in nature. It is a colorless, water–soluble acid that, in addition to its acidic properties, is a powerful biochemical reducing agent, meaning it readily undergoes oxidation.

We are generally aware that vitamin C can be readily obtained from citrus fruits, but we do not often recognize the fact that it can also be obtained from fresh vegetables. However, storage and processing causes vegetables to lose their vitamin C content. Cooking (boiling or steaming) leaches the water–soluble ascorbic acid from the vegetables; in addition, the high temperatures accelerates its air oxidation. Therefore to maintain the vitamin C content, only freshly–harvested fruits or vegetables should be consumed; their natural protective coverings (e.g., orange peel) should be removed just before its consumption. Table 20.1 lists the concentration ranges of ascorbic acid for various vegetables.

Table 20.1
Ascorbic Acid in Foods

<10mg/100g	beets, carrots, eggs, milk
10-25mg/100g	asparagus, cranberries, cucumbers, green peas, lettuce, pineapple
25-100mg/100g	Brussels sprouts, citrus fruits, tomatos, spinach
100-350mg/100g	chili peppers, sweet peppers, turnip greens

Even though ascorbic acid is an acid, its *reducing* properties are used to analyze its concentration in various samples. There are many other acids present in foods (e.g, citric acid) that would interfere with the analysis and not permit us to selectively determine the ascorbic acid content. The equation for the oxidation of ascorbic acid is

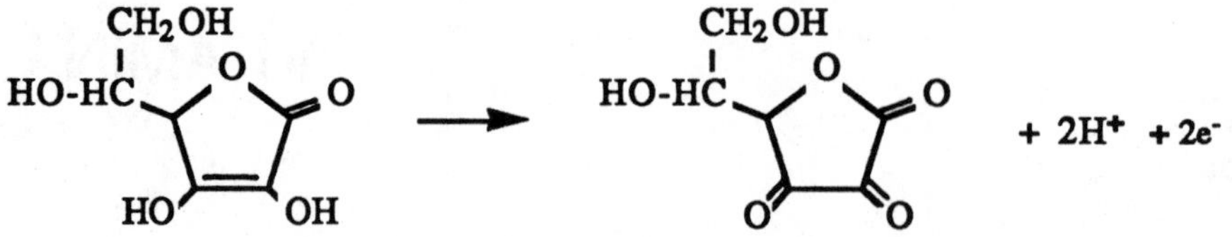

CH_2OH HO-HC O O HO OH → CH_2OH HO-HC O O O O $+ 2H^+ + 2e^-$

$$\text{or } C_6H_8O_6 \rightarrow C_6H_6O_6 + 2H^+ + 2e^-$$

Vitamin C Analysis

In analyzing for the ascorbic acid, the sample is dissolved in water and treated with a known excess of the oxidizing agent, IO_3^-; in an acidic solution containing an excess of I^-, IO_3^- converts to (red–brown) I_3^-, a milder oxidizing agent (Step 1).

$$\textbf{(Step 1)}\quad IO_3^-(aq) + 8I^-(aq) + 6H^+(aq) \rightarrow 3I_3^-(aq) + 3H_2O$$

$$3I_3^-(aq) + 3C_6H_8O_6(aq)\ \textbf{(Step 2)} \rightarrow 3C_6H_6O_6(aq) + 9I^-(aq) + 6H^+(aq)$$

$$3I_3^-(aq) + \textbf{(Step 3)}\ 6S_2O_3^{2-}(aq) \rightarrow 9I^-(aq) + 3S_4O_6^{2-}(aq) + 3H_2O$$

Some of the I_3^- then oxidizes the ascorbic acid (Step 2). The rest of the I_3^- (or the excess) is titrated with a standard thiosulfate, $S_2O_3^{2-}$, solution, producing the colorless I^- and $S_4O_6^{2-}$ ions (Step 3).

Therefore the difference between the I_3^- generated initially (from the IO_3^-) and that which is titrated in excess is a measure of the ascorbic acid content of the sample.

The stoichiometric point is detected using starch as an indicator. Just prior to the disappearance of the red–brown I_3^- in the titration, starch is added; this forms the deep–blue ion, $[I_3 \bullet \text{starch}]^-$.

$$I_3^-(aq) + \text{starch}(aq) \rightarrow [I_3 \bullet \text{starch}]^-(aq)$$

The addition of the $S_2O_3^{2-}$ titrant is continued until the $[I_3 \bullet \text{starch}]^-$ has been reduced to I^-; the solution appears colorless at the endpoint.

$$[I_3 \bullet \text{starch}]^-(aq) + 2S_2O_3^{2-}(aq) \rightarrow 3I^-(aq) + \text{starch} + S_4O_6^{2-}(aq)$$

STANDARDIZATION OF $Na_2S_2O_3$ SOLUTION

A standard $Na_2S_2O_3$ solution is prepared using solid KIO_3 as a primary standard. A quantitative amount of KIO_3 is dissolved in an acidic solution containing KI. The generated I_3^- (see equation above) is titrated with your prepared solution of $Na_2S_2O_3$ using starch as the indicator. Again, the endpoint will be a color change from deep–blue to colorless. The equations (Steps 1 and 3) and the analysis procedure are exactly the same as that described above for the vitamin C analysis, except that there is no vitamin C (Step 2) to react with any of the I_3^- in this standardization procedure.

Techniques

- Technique 4c, d, page 5 — Separation of a Solid from a Liquid
- Technique 10, page 12 — Reading a Meniscus
- Technique 11, page 13 — Pipetting a Liquid or Solution
- Technique 12, page 14 & 15 — Titrating a Solution
- Technique 13, page 16 — Using the Laboratory Balance

Procedure

In this experiment you are required to prepare and standardize a $Na_2S_2O_3$ solution using solid KIO_3 as a primary standard. Three trials are necessary. This solution is then used to analyze for ascorbic acid in a sample assigned by your instructor. Quantitative, reproducible data are objectives of this experiment; practice good laboratory techniques.

A. PREPARATION OF A (PRIMARY) STANDARD 0.01M KIO_3 SOLUTION

Weigh on weighing paper about 0.5g (±0.001) of KIO_3 (dried at 110°C), transfer the solid to a 250mL volumetric flask, dissolve and dilute to the mark. Calculate and record the molarity of the KIO_3 solution.

B. STANDARDIZATION OF A 0.1M $Na_2S_2O_3$ SOLUTION

1. This solution[1] should be prepared about one week in advance because of unavoidable decomposition that occurs. Dissolve about 12g (±0.01g) of $Na_2S_2O_3 \cdot 5H_2O$ with freshly boiled, distilled (or de-ionized) water and dilute to 500mL. Agitate until the salt dissolves.

2. Properly prepare a clean, 50mL buret for titration. Fill it with your $Na_2S_2O_3$ solution, drain the tip of air bubbles, and, after 30s, read and record the volume (±0.01mL).

3. Pipet 25mL of the standard KIO_3 solution into a 125mL Erlenmeyer flask and add about 2g (±0.01g) of solid KI. Add about 10mL of 0.5M H_2SO_4 and 0.5g of $NaHCO_3$.[2]

4. Immediately begin titrating with the $Na_2S_2O_3$ solution. When the red–brown solution (due to I_3^-) changes to a pale yellow color, add 2mL of starch solution. Stirring constantly, continue titrating slowly until the blue color disappears.

---

[1] Ask your instructor about this solution; it may have already been prepared.

[2] The $NaHCO_3$ reacts in the acidic solution to produce $CO_2(g)$, providing an inert atmosphere above the solution and minimizing the possibility of an air oxidation of the I^- ions (and in Part D, the ascorbic acid).

5. Repeat the procedure twice by rapidly adding the $Na_2S_2O_3$ titrant until 1mL before the stoichiometric point. Add the starch solution and continue titrating until the solution is colorless.

6. Refill the buret with the $Na_2S_2O_3$ solution in preparation for the sample analysis in Part D. Read and record the volume (±0.01mL) of $Na_2S_2O_3$ titrant in the buret.

C. Sample Preparation

1. Vitamin C tablet. Weigh (±0.001g) a portion of a vitamin tablet that is equivalent to about 100mg of ascorbic acid. Dissolve it in a 250mL Erlenmeyer flask with 40mL of 0.5M H_2SO_4[3] and then add about 0.5g $NaHCO_3$. Kool–Aid™, Tang™, or Gatorade™ may be substituted as dry samples, even though their ascorbic acid concentrations are much lower. Proceed immediately to Part D.

2. Fresh fruit sample. Filter 125mL to 130mL of freshly squeezed juice through several layers of cheesecloth (or vacuum–filter). Weigh (±0.01g) a clean, dry 250mL Erlenmeyer flask. Add about 100mL of filtered juice and reweigh. Add 40mL of 0.5M H_2SO_4 and 0.5g $NaHCO_3$. Concentrated fruit juices may also be used as samples. Proceed immediately to Part D.

3. Fresh vegetable sample. Weigh 100g (±0.01g) of a fresh vegetable. Transfer to a mortar[4] and grind. Add 5mL of 0.5M H_2SO_4 and continue to pulverize the sample. Add another 15mL of 0.5M H_2SO_4, stir, and filter through several layers of cheesecloth (or vacuum–filter). Add 20mL of 0.5M H_2SO_4 to the mortar, stir, and pour through the same filter. Repeat the washing of the mortar with 20mL of freshly boiled, distilled (or de-ionized) water. Combine all of the washings in a 250mL Erlenmeyer flask and add 0.5g $NaHCO_3$. Proceed immediately to Part D.

D. Vitamin C Analysis

1. Pipet 25.0mL of the standard KIO_3 solution (from Part A) into the sample solution and add 2g of KI. Add about 10mL of 0.5M H_2SO_4. Titrate the excess I_3^- in the sample with the standard $Na_2S_2O_3$ solution as described in Part B.4. Read and record the final buret reading (±0.01mL).

2. Repeat the analysis twice in order to complete the three trials.

---

[3]Remember that vitamin tablets contain binders and other material that may be insoluble in water–do not heat in an attempt to dissolve the tablet!

[4]A blender may be substituted for the mortar and pestle.

VITAMIN C ANALYSIS-LAB PREVIEW

Date__________Name____Lab Sec. ______Desk No.______

1. Explain why cooked fruits and vegetables have a lower vitamin C content than fresh fruits and vegetables.

2. A 25.0mL volume of 0.010M KIO_3, having an excess of KI, is added to a 0.246g sample of a Real Lemon™ solution containing vitamin C. The red–brown solution, caused by the presence of excess I_3^-, is titrated to a colorless starch endpoint with 10.7mL of 0.100M $Na_2S_2O_3$.

   a. How many moles of KIO_3 were added to the Real Lemon™ solution?

   b. Calculate the moles of I_3^- that are generated from this KIO_3?

   c. How many moles of I_3^- reacted with the 0.100M $Na_2S_2O_3$ in the titration?

   d. How many moles (of the total moles) of I_3^- had reacted with the vitamin C in the Real Lemon™ sample?

   e. Calculate the moles and grams of vitamin C, $C_6H_8O_6$, in the sample.

   f. Calculate the percent (by weight) of vitamin C in the Real Lemon™ sample.

3. Six ounces (1 fl. oz. = 29.57mL) of a well–known vegetable juice contains 35% of the recommended daily allowance of vitamin C (equal to 60mg). How many milliliters of the vegetable juice will provide 100% of the recommended daily allowance?

4. a. Which finger should be used to control the delivery of a solution from a pipet?

   b. How much time should elapse between the time that the stopcock on a buret is closed and the time that a volume reading should be made and recorded?

   c. What criterion is used to determine if a buret is clean?

5. a. What is the oxidizing agent in Part B of the procedure?

   b. What will be the color change of the starch indicator that signals the end of the titration in today's experiment?

   c. Vitamin C is an acid (ascorbic acid) and a reducing agent. Which property is utilized for its analysis in the experiment?

VITAMIN C ANALYSIS-DATA SHEET

Date________Name____Lab Sec. ____Desk No.____

A. PREPARATION OF A (PRIMARY) STANDARD 0.01M KIO_3 SOLUTION

1. Mass of KIO_3 + weighing paper (g) __________
2. Mass of weighing paper (g) __________
3. Mass of KIO_3 (g) __________
4. Moles of KIO_3 (mol) __________
5. Molarity of standard KIO_3 solution (mol/L) __________

B. STANDARDIZATION OF A 0.1M $NA_2S_2O_3$ SOLUTION

	Trial 1	Trial 2	Trial 3
1. Volume of KIO_3 solution (mL)	___25.0___	___25.0___	___25.0___
2. Moles of KIO_3 titrated (mol)	_________	_________	_________
3. Moles of I_3^- generated (mol)	_________	_________	_________
4. Buret reading, final (mL)	_________	_________	_________
5. Buret reading, initial (mL)	_________	_________	_________
6. Volume of $Na_2S_2O_3$ added (mL)	_________	_________	_________
7. Moles of $Na_2S_2O_3$ added (mol)	_________	_________	_________
8. Molarity of $Na_2S_2O_3$ solution (mol/L)	_________	_________	_________
9. Average molarity of $Na_2S_2O_3$ solution (mol/L)		__________	

C. SAMPLE PREPARATION

SAMPLE NAME: ____________________

1. Mass of sample (g)	_________	_________	_________

D. Vitamin C Analysis

	Trial 1	Trial 2	Trial 3
1. Volume of KIO_3 added (mL)	___25.0___	___25.0___	___25.0___
2. Moles of IO_3^- added (mol)	_________	_________	_________
3. Moles of I_3^- generated, total (mol)	_________	_________	_________
4. Buret reading, final (mL)	_________	_________	_________
5. Buret reading, initial (mL)	_________	_________	_________
6. Volume of $Na_2S_2O_3$ added (mL)	_________	_________	_________
7. Moles of $S_2O_3^{2-}$ added (mol)	_________	_________	_________
8. Moles of I_3^- titrated with $S_2O_3^{2-}$ (mol)	_________	_________	_________
9. Moles of I_3^- reduced by $C_6H_8O_6$ (mol)	_________	_________	_________
10. Moles of $C_6H_8O_6$ in sample (mol)	_________	_________	_________
11. Mass of $C_6H_8O_6$ in sample (g)	_________	_________	_________
12. Percent of $C_6H_8O_6$ in sample (%)	_________	_________	_________
13. Average percent of $C_6H_8O_6$ in sample (%)		_________	

Questions

1. How will the addition of 3g of KI to the sample in Part B.3, instead of the 2g, affect the molarity of the $Na_2S_2O_3$ solution? Explain.

2. If the blue color does not appear when the starch solution is added during the titration, should you continue titrating or discard the sample (or leave for a cup of coffee)? Explain (but with a serious response).

3. After adding the KIO_3 solution and KI to the sample (in Part D) and stirring, the sample solution remains colorless; what modification of the procedure can be made to correct for this unexpected observation in order to complete the analysis.

EXPERIMENT 21
BLEACH ANALYSIS

Objectives

- To prepare and standardize a sodium thiosulfate solution
- To determine the percent "available Cl_2" in a bleach

Principles

Most commercial bleaching agents contain the hypochlorite ion, ClO^-, as the "active ingredient" for making clothes whiter and/or removing stains. This ion is generally in the form of the sodium salt, NaClO, or the calcium salt, $Ca(ClO)_2$, in the bleach.

Normally when the strengths of various bleaches are compared, the standard oxidant for bleaching is assumed to be Cl_2. The oxidizing strength of the bleaching agent is rated according to an equivalent mass of Cl_2 per unit volume of the solution (or mass of the powder). This rating is called the "**available Cl_2**" in the bleach. Occasionally, its strength is expressed as percent chlorine, $\%Cl_2$, by weight.

Liquid laundry bleach, generally a 5.25% (by weight) NaOCl solution, is prepared by the electrolysis of a cold, stirred NaCl solution, producing Cl_2 at the anode and OH^- ion at the cathode. Cl_2 and OH^- react to form the ClO^- ion.

oxidation, anode:	$2Cl^-(aq)$	$\rightarrow Cl_2(g) + 2e^-$
reduction, cathode:	$2e^- + 2H_2O$	$\rightarrow 2OH^-(aq) + H_2(g)$
stirred solution:	$Cl_2(g) + 2OH^-(aq)$	$\rightarrow ClO^-(aq) + Cl^-(aq) + H_2O$
net overall reaction:	$Cl^-(aq) + H_2O$	$\rightarrow ClO^-(aq) + H_2(g)$

Bleaching powder–a mixture of $CaCl_2$, $Ca(ClO)Cl$, and $Ca(ClO)_2$–is prepared by the reaction of Cl_2 and slaked lime $Ca(OH)_2$.

In both types of bleaches, the ClO^- ion is the oxidizing agent, removing electrons from the substance that absorbs the visible light.

$$ClO^-(aq) + H_2O + 2e^- \rightarrow Cl^-(aq) + 2OH^-(aq)$$

Since visible light is no longer absorbed (the electrons are removed), the material appears whiter.

BLEACH ANALYSIS

In this experiment the oxidation-reduction analysis of a bleach involves the reaction of the ClO^- ion in the bleach with an excess of iodide ion.

$$ClO^-(aq) + 3I^-(aq) + H_2O \rightarrow I_3^-(aq) + Cl^-(aq) + 2OH^-(aq)$$

The solution is acidified and titrated with a standardized sodium thiosulfate $Na_2S_2O_3$, solution until the yellow color of I_3^- (or $I_2 \bullet I^-$)[1] nearly disappears.

$$I_3^-(aq) + 2S_2O_3^{2-}(aq) \rightarrow 3I^-(aq) + S_4O_6^{2-}(aq)$$

Just prior to the stoichiometric point, the addition of $Na_2S_2O_3$ is stopped and a starch solution is added. Starch forms a soluble, deep–blue $[I_3 \bullet starch]^-$ ion.

$$I_3^-(aq) + starch \rightarrow [I_3 \bullet starch]^-(aq) \quad (deep\text{–}blue)$$

The addition of $Na_2S_2O_3$ is resumed until all of the I_2 is reduced and the blue color disappears.

$$2e^- + [I_3 \bullet starch]^-(aq) \rightarrow 3I^-(aq) + starch \quad (colorless)$$

To summarize the analysis: a measured volume of $Na_2S_2O_3$ reacts with a generated quantity of I_3^-–this I_3^- is generated from the ClO^- in the bleach; therefore, since the moles of $Na_2S_2O_3$ that reacts is known, the moles of ClO^- in the bleach can be determined.

The "available Cl_2" is calculated as if free Cl_2 per unit volume (or mass, if the bleach is a powder) reacts with the I^- ion which is subsequently analyzed with $S_2O_3^{2-}$.

$$Cl_2 + 3I^-(aq) \rightarrow 2Cl^-(aq) + I_3^-(aq)$$
$$I_3^-(aq) + 2S_2O_3^{2-}(aq) \rightarrow 3I^-(aq) + S_4O_6^{2-}(aq)$$

$$Cl_2 + 2S_2O_3^{2-}(aq) \rightarrow 2Cl^-(aq) + S_4O_6^{2-}(aq)$$

This equation serves as the basis for determining the strength of bleaching agents in this experiment. Notice that both 1mol ClO^- and 1mol Cl_2 react with 2mol $S_2O_3^{2-}$.

STANDARDIZATION OF A $Na_2S_2O_3$ SOLUTION

A standard $Na_2S_2O_3$ solution is prepared using solid KIO_3 as a primary standard. A measured amount of KIO_3 is dissolved in an acidic solution containing an excess of KI

$$IO_3^-(aq) + 8I^-(aq) + 6H^+(aq) \rightarrow 3I_3^-(aq) + 3H_2O$$

The generated I_3^- is titrated with your prepared solution of $Na_2S_2O_3$ using starch as the indicator.

---

[1]The yellow color is due to the presence of I_2; the excess I^- combines with I_2 to form a soluble I_3^- ion. It is common to refer to $I_2(aq)$ and $I_3^-(aq)$ as the same substance.

$$3I_3^-(aq) + 3\text{ starch} \rightarrow 3[I_3\bullet\text{starch}]^-(aq) \quad \text{(deep–blue)}$$
$$3[I_3\bullet\text{starch}]^-(aq) + 6S_2O_3^{2-}(aq) \rightarrow 9I^-(aq) + 3S_4O_6^{2-}(aq) + 3H_2O + 3\text{ starch (colorless)}$$

Again, the endpoint is from a deep–blue color to colorless. The equations and the analysis procedure are exactly the same as that described above for the bleach analysis.

Techniques

- Technique 5, page 7 — Flushing a Precipitate from a Beaker
- Technique 10, page 12 — Reading a Meniscus
- Technique 11, page 13 — Pipetting a Solution
- Technique 12, page 14 & 15 — Titrating a Solution
- Technique 13, page 16 — Using the Laboratory Balance

Procedure

The strengths of two bleaching agents are determined in this experiment; two trials are required for each determination. Therefore, obtain about 12mL of each liquid bleach (or 5g of each powdered bleach) from your instructor.

A. Preparation of a (Primary) Standard 0.01M KIO_3 Solution

Weigh on weighing paper about 0.5g (±0.001) of KIO_3 (dried at 110°C), transfer the solid to a 250mL volumetric flask, dissolve and dilute to the mark. Calculate and record the molarity of the KIO_3 solution.

B. Standardization of a 0.1M $Na_2S_2O_3$ Solution

1. This solution[2] should be prepared about 1 week in advance because of unavoidable decomposition that occurs. Dissolve about 12g (±0.01g) of $Na_2S_2O_3\bullet5H_2O$ with freshly boiled, distilled (or de-ionized) water and dilute to 500mL. Agitate until the salt dissolves.

2. Properly prepare a clean, 50mL buret for titration. Fill it with your $Na_2S_2O_3$ solution, drain the tip of air bubbles, and, after 30s, read and record the volume (±0.01mL).

3. Pipet 25mL of the standard KIO_3 solution into a 125mL Erlenmeyer flask and add about 2g (±0.01g) of solid KI. Add about 10mL of 0.5M H_2SO_4 and 0.5g of $NaHCO_3$.[3] Immediately begin titrating with the $Na_2S_2O_3$ solution. When the red–brown solution (due to I_3^-) changes to a pale yellow color, add 2mL of starch solution. Stirring constantly, continue titrating slowly until the blue color disappears.

4. Repeat the procedure twice by *rapidly* adding the $Na_2S_2O_3$ titrant until 1mL before the stoichiometric point. Add the starch solution and continue titrating until the solution is colorless.

---

[2]Ask your instructor about this solution; it may have already been prepared.

[3]The $NaHCO_3$ reacts in the acidic solution to produce $CO_2(g)$, providing an inert atmosphere above the solution and minimizing the possibility of an air oxidation of the I^- ions.

5. Refill the buret with the $Na_2S_2O_3$ solution in preparation for the sample analysis in Part E. Read and record the volume (±0.01mL) of $Na_2S_2O_3$ titrant in the buret.

C. Preparation of Liquid Bleach

1. Pipet 10.0mL of bleach[4] into a 100mL volumetric flask and dilute to the "mark" with boiled, distilled (or de-ionized) water. Mix thoroughly. Pipet 25.0mL of this diluted bleach solution into a 250mL Erlenmeyer flask and add 20mL of distilled water, 2g of KI, and 10mL of 3M H_2SO_4. A yellow color indicates the presence of I_2 as I_3^-. Proceed to Part E.

D. Preparation of a Powdered Bleach

1. Place about 5g of powdered bleach into a mortar and grind. Weigh (±0.01g) on weighing paper the pulverized sample and transfer it to a 100mL volumetric flask fitted with a funnel. Dilute to 100mL with distilled (or de-ionized) water. Mix the solution thoroughly.[5] Pipet 25.0mL of this solution into a 250mL Erlenmeyer flask and add 20mL of distilled (or de-ionized) water, 2g of KI, and 15mL glacial acetic acid. A yellow color indicates the presence of I_2 as I_3^-. See footnote #5. Proceed to Part E.

E. Titration of Bleach

1. ***Immediately*** titrate the liberated I_2 with your standardized $Na_2S_2O_3$ solution until the yellow color is *almost* gone. Consult your instructor at this point if you are uncertain. Add 2mL of starch solution; this will form the deep–blue $[I_3 \bullet starch]^-$ ion.[6] While swirling the flask, continue titrating *slowly* until the deep–blue color disappears. Record the final volume of the titrant in the buret.

2. Repeat the experiment with the same bleaching agent. Analyze similar bleaching agents to determine the amount of "available Cl_2" and to determine the best buy.

---

[4]The density of liquid bleach is 1.084g/mL.

[5]All of the powder may not dissolve because of the insoluble abrasive in the bleach.

[6]If you are analyzing a powdered bleach, the 2mL of starch may be added after the glacial acetic acid solution in Part C.

BLEACH ANALYSIS-LAB PREVIEW

Date_________Name____Lab Sec. _____Desk No._____

1. a. Write the formula for the hypochlorite ion.

   b. In today's experiment, what substance is oxidized by the hypochlorite ion?

   c. Write the equation for the reaction in **b**.

2. Define "available Cl_2."

3. a. The number of moles of ClO^- reacting with 1mol of I^- is ______________.

   b. The number of moles of Cl_2 reacting with 1mol of I^- is ______________.

   c. The number of moles of ClO^- equivalent to 1mol of "available Cl_2" is

   \-\-\-\-\-\-\-\-\-\-\-\-\-\-\-

4. a. What is the color change at the endpoint in today's analysis?

   b. What is the cause of the color change?

5. Sodium thiosulfate is the titrant in the bleach analysis.

   a. Write the formula for sodium thiosulfate.______________

   b. Does sodium thiosulfate serve as an oxidizing agent or a reducing agent?

   .\-\-\-\-\-\-\-\-\-\-\-\-\-\-

   c. What does sodium thiosulfate oxidize (or reduce)?______________

   d. Write a balance equation for the reaction.

   e. Each mole of the sodium thiosulfate titrant added in the analysis is equivalent to ______ moles of ClO^- and ______ moles of Cl_2 in the bleach.

6. A 0.684g sample of a powdered bleach is analyzed according to the procedure in this experiment. A volume of 31.7mL of 0.100M $Na_2S_2O_3$ is required to reach the endpoint.
   a. How many moles of I_2 (or I_3^-) did the powdered bleach produce?

   b. Calculate the moles and grams of "available Cl_2" in the bleach.

   c. What is the percent (by weight) of available Cl_2 in the bleach?

7. A 10.0mL bleach sample is diluted to 100mL in a volumetric flask. A 25.0mL portion of this solution is analyzed according to the procedure in this experiment. If 11.3mL of 0.30M $Na_2S_2O_3$ is needed to reach the endpoint, what is the percent (by weight) of available Cl_2 in the *original* sample? Assume that the density of the bleach solution is 1.084g/mL.

8. Explain the technique for transferring an insoluble solid in one beaker into a second beaker or Erlenmeyer flask.

9. a. Where on a meniscus should the volume of solution be read in a buret?

   b. When reading a meniscus, how does the use of a black mark on a white card better define the volume of a solution?

10. When a solution is delivered from a pipet, some of the solution invariably remains in the tip. What should be done with that solution: should you force it out or leave it in the pipet? Explain.

BLEACH ANALYSIS-DATA SHEET

Date________ Name____ Lab Sec. _____ Desk No._____

A. PREPARATION OF A (PRIMARY) STANDARD 0.01M KIO_3 SOLUTION

1. Mass of KIO_3 + weighing paper (g) __________
2. Mass of weighing paper (g) __________
3. Mass of KIO_3 (g) __________
4. Moles of KIO_3 (mol) __________
5. Molarity of standard KIO_3 solution (mol/L) __________

B. STANDARDIZATION OF A 0.1M $NA_2S_2O_3$ SOLUTION

	Trial 1	Trial 2	Trial 3
1. Volume of KIO_3 solution (mL)	___25.0___	___25.0___	___25.0___
2. Moles of KIO_3 titrated (mol)	__________	__________	__________
3. Moles of I_3^- generated (mol)	__________	__________	__________
4. Buret reading, final (mL)	__________	__________	__________
5. Buret reading, initial (mL)	__________	__________	__________
6. Volume of $Na_2S_2O_3$ added (mL)	__________	__________	__________
7. Moles of $Na_2S_2O_3$ added (mol)	__________	__________	__________
8. Molarity of $Na_2S_2O_3$ solution (mol/L)	__________	__________	__________
9. Average molarity of $Na_2S_2O_3$ solution (mol/L)		__________	

C. PREPARATION OF LIQUID BLEACH

Sample and unit price (¢/g)	__________		__________	
	Trial 1	Trial 2	Trial 1	Trial 2
1. Volume of original liquid bleach titrated (mL)	__2.5mL__	__2.5mL__	__2.5mL__	__2.5mL__

D. PREPARATION OF A POWDERED BLEACH

Sample and unit price (¢/g)	________		________	
1. Mass of weighing paper and bleach (g)	--------	--------	--------	--------
2. Mass of weighing paper (g)	--------	--------	--------	--------
3. Mass of bleach (g)	--------	--------	--------	--------

E. TITRATION OF BLEACH

1. Buret reading, final (mL)	--------	--------	--------	--------
2. Buret reading, initial (mL)	--------	--------	--------	--------
3. Volume of $Na_2S_2O_3$ added (mL)	-----	--------	--------	--------
4. Moles of $Na_2S_2O_3$	--------	--------	--------	--------
5. Moles of "available Cl_2"	--------	--------	--------	--------
6. Mass of "available Cl_2"	--------	--------	--------	--------
7. Percent (by weight) "available Cl_2". (assume the density of liquid bleach to be 1.084g/mL)	--------	--------	--------	--------
8. Average % "available Cl_2	________		________	
9. Best buy (%Cl_2/unit price)	________			

Questions

1. The starch solution can be added at the same time as the KI solution in Part C. However, the $[I_3 \bullet starch]^-$ ion does not readily dissociate when the $Na_2S_2O_3$ titrant is added. What difficulties might you have encountered if the starch solution had been added in Part C?

2. If an air bubble initially trapped in the tip of the buret is released during the titration, will the reported percent "available Cl_2" be too high or too low? Explain.

EXPERIMENT 22
ALUM FROM SCRAP ALUMINUM

Objective

- To prepare an alum from scrap aluminum

Principles

Aluminum is the most abundant metal in the earth's crust (8.1% by weight), ranking just ahead of iron (5.0% by weight). Because of its low density, high tensile strength, and resistance to corrosion, it is widely used in the manufacture of airplanes and automobiles.

Its isolation from an ore is very energy intensive, requiring heat for dissolving the ore in aqueous NaOH and for drying the Al_2O_3 salt, and electricity for reducing molten Al^{3+} to Al metal. As a result, major aluminum companies are asking the public to save and return used aluminum cans, pans, and window screens. The "public's ore" is much cheaper to reprocess than nature's ore.

In this experiment, a scrap piece or aluminum is used to prepare an **alum**, a hydrated double salt with the formula, $KAl(SO_4)_2 \cdot 12H_2O$. This alum is used for water purification, for sewage treatment and in fire extinguishers.

Aluminum metal rapidly reacts in a hot aqueous KOH solution producing the soluble potassium aluminate, $KAl(OH)_4$, salt.

$$2Al\,(s) + 2KOH(aq) + 6H_2O \rightarrow 2KAl(OH)_4(aq) + 3H_2(g)$$

When this potassium aluminate solution is treated with sulfuric acid, insoluble $Al(OH)_3$ initially forms, but then it dissolves with heat and the addition of a slight of excess of H_2SO_4.

$$KAl(OH)_4(aq) + 2H_2SO_4(aq) \rightarrow Al(OH)_3(s) + 2K_2SO_4(aq) + 2H_2O$$
$$Al(OH)_3(s) + 2H_2SO_4(aq) \xrightarrow{\Delta} 2Al_2(SO_4)_3(aq) + 6H_2O$$

The alum, called potassium aluminum sulfate dodecahydrate, crystallizes when the nearly saturated solution is cooled.

$$K_2SO_4(aq) + Al_2(SO_4)_3(aq) + 12H_2O \rightarrow 2KAl(SO_4)_2 \cdot 12H_2O(s)$$

Techniques

- Technique 4b,c, page 5 — Separation of a Solid from a Liquid
- Technique 6b, page 7 — Heating Liquids
- Technique 13, page 16 — Using the Laboratory Balance

Procedure

For the synthesis of the alum you will need to obtain a 110°C thermometer and a vacuum–filter apparatus from the stockroom. Be sure to follow the laboratory safety suggestions in today's experiment.

A. Preparation of the Alum

1. Polish a small piece of scrap aluminum with steel wool. Cut about 1g (±0.01g) of the scrap aluminum into very small pieces. Place these into a 250mL Erlenmeyer flask.[1]

2. Add 50mL of 20% KOH to the Al pieces to dissolve the aluminum. (**Caution**: *Wear safety glasses. Do not splatter the solution.* KOH *is caustic–do not permit skin contact.*) Heating the flask gently with a small flame may be necessary. Since $H_2(g)$ is evolved, do this step in a well–ventilated area. Continue heating until all of the Al reacts.[2]

3. Heat the solution to reduce the liquid level to about 25mL. When evidence of reaction is no longer visible, gravity filter the warm solution into a 100mL beaker through very porous, low retention filter paper[3], through a thin layer of glass wool placed in a funnel, or by vacuum filtration.

---

[1]The smaller the Al pieces, the more rapid is the reaction. Aluminum foil may be used instead of scrap aluminum.

[2]Impurities, such as the plastic coating or paint covering, may remain as floating insoluble particles.

[3]Whatman No.4 or Fisher*brand* P8 are low retention filter paper brands.

4. Allow the clear filtrate (solution) to cool. While stirring, slowly add 30mL of 6M H_2SO_4 (**Caution**: *Avoid skin contact*). The solution should contain insoluble $Al(OH)_3$ after the H_2SO_4 addition.[4]

5. Gently reheat the mixture until the $Al(OH)_3$ dissolves. After the solution is clear, remove the heat, and filter, but *only* if any solids are present. Cool in an ice bath for about 20 minutes. Alum crystals should from in the beaker. For better results, allow the crystallization to continue overnight. If crystals do not form, reheat the mixture to reduce the volume of the solution to about 50mL or less and then cool again.

6. Vacuum–filter the crystals from the solution. Wash them on the filter paper with four 5mL portions of a 50/50 ethanol–water mixture.[5] Continue the vacuum suction until the crystals appear dry. Weigh (±0.01g) and show them to the laboratory instructor.

B. Melting Point of the Alum

1. Place finely ground alum to a depth of about 0.5cm in the bottom of a melting point capillary tube (Figure 22.1). Attach the capillary tube to the thermometer with a rubber band or tubing and position it in a hot–water bath (Figure 22.2).

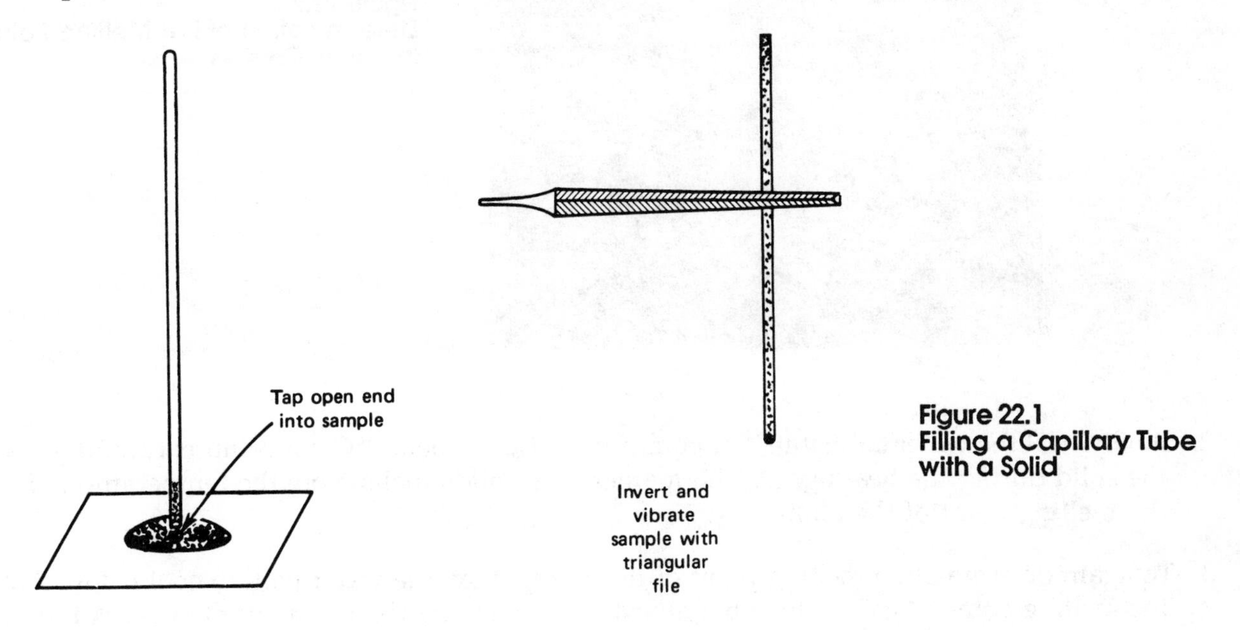
Tap open end into sample
Invert and vibrate sample with triangular file

Figure 22.1
Filling a Capillary Tube with a Solid

---

[4]You will need to use some judgment here; if a precipitate has not yet formed, add a small amount of $H_2SO_4(aq)$ until it does; if the precipitate has redissolved, add a small amount of KOH(aq) until the $Al(OH)_3$ reforms.

[5]Alum crystals have a very low solubility in a 50/50 ethanol–water mixture.

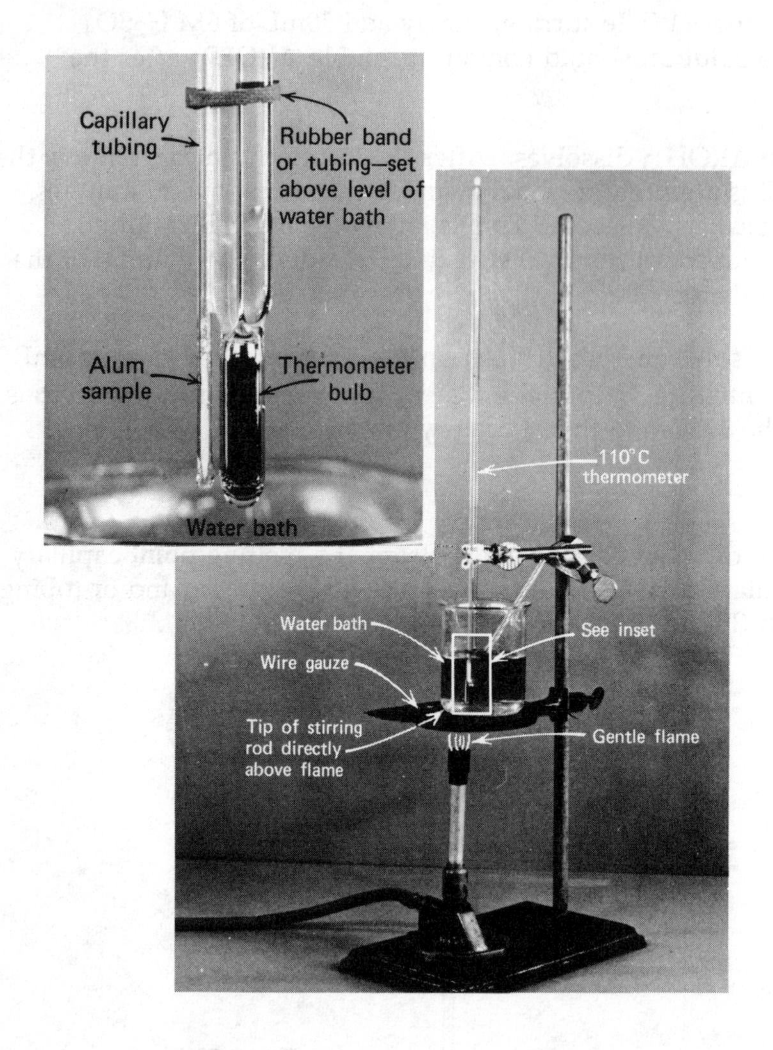

Figure 22.2
Determination of the Melting Point of the Alum Crystals

2. Slowly heat the water so that its temperature increases about 3°C per minute; carefully watch the solid during the heating. At the moment the solids melts, note the temperature. This is the melting point of the alum.

3. To again determine the melting point of the alum, allow the water bath to cool below that of the melting point. Again, reheat but at a slower rate (e.g., 1°C per minute) to check the temperature at which the alum changes from a solid to a liquid.

ALUM FROM SCRAP ALUMINUM–LAB PREVIEW

Date________Name____Lab Sec. _____Desk No._____

1. What does it mean when a salt is said to be a dodecahydrate?

2. An alum is a **double salt** with the general formula $M^{+}M^{3+}(SO_4)_2 \bullet 12H_2O$. M^{+} is commonly Na^{+}, K^{+}, Tl^{+}, NH_4^{+}, or Ag^{+}; M^{3+} is Al^{3+}, Fe^{3+}, Cr^{3+}, Ti^{3+}, or Co^{3+}.

   a. The alum commonly used for pickling cucumbers is ammonium aluminum sulfate dodecahydrate. Write its formula.

   b. Sodium aluminum sulfate dodecahydrate is the acid commonly used in baking powders. What is its formula?

   c. Potassium chromium(III) sulfate dodecahydrate is used in tanning leather and in waterproofing fabrics. Write its formula.

3. The Al^{3+} ion is amphoteric. What does this mean?

4. Explain the procedure for filling a capillary tube with a solid.

5. A mass of 13.02g $(NH_4)_2SO_4$ is dissolved in water. After the solution is heated, 27.22g of $Al_2(SO_4)_3 \bullet 18H_2O$ is added. Calculate the theoretical yield of the resulting alum (see Question 2 for the formula of the alum). Hint: this is a limiting reactant–type problem.

6. How many milliliters of H_2 gas at STP are evolved in the reaction of 1.02g Al with excess KOH?

7. a. While gravity filtering a solution, what is the maximum level to which a funnel should be filled with the solution?

   b. Explain how filter paper is sealed into a Buchner funnel when used for vacuum filtration.

   c. How should an Erlenmeyer flask containing a solution be held while heating it over an open flame?

ALUM FROM SCRAP ALUMINUM–DATA SHEET

Date________Name____Lab Sec. ____Desk No.____

A. PREPARATION OF THE ALUM

1. Mass of aluminum metal (g) _________________
2. Mass of alum, $KAl(SO_4)_2 \cdot 12H_2O$ (g) _________________
3. Theoretical yield of alum (g)* _________________
4. Percent yield (%) _________________

*Show calculation based upon pure aluminum metal.

B. MELTING POINT OF THE ALUM

1. Melting point (°C) _________________
2. Melting point, second determination (°C) _________________

Questions

1. Why is a 50/50 ethanol–water mixture used for washing the alum crystals after preparation rather than distilled (or de–ionized) water?

2. How can the reaction rate between Al and KOH be increased in Part A.1 of the procedure?

3. In the Lab Preview, it was noted that $Al(OH)_3$ is amphoteric. Where was this property observed in the experiment?

4. A greater yield with nearly perfect crystals is obtained when the alum solution is allowed to cool in a refrigerator overnight or for a few days. Explain.

5. How can you determine the mass of water hydrated to the alum crystals? See Experiment 8.

EXPERIMENT 23
MOLECULAR WEIGHT DETERMINATION

Objective

- To determine the molecular weight of a nonvolatile solute by observing the difference between the freezing points of a solvent and a solution

Principles

The addition of nonvolatile solute to a solvent produces several characteristic changes in the physical properties of a solvent. For example, salt is used to freeze ice cream because a salt–ice–water mixture exists at a lower temperature than an ice–water mixture; anti–freeze (ethylene glycol) is added to the cooling system of an automobile to prevent any freeze–up during the wintertime. Anti–freeze also reduces the probability of the coolant from boiling in the summertime; we might also predict (and we would predict correctly) that a salt–water mixture boils at a higher temperature than pure water.

From these observations we can conclude that a solute added to a solvent must therefore *reduce* the solvent's freezing point and *increase* the solvent's boiling point; for water as the solvent, a salt or ethylene glycol solution has a freezing point less than 0°C and a boiling point greater than 100°C.

The freezing point change, ΔT_f, and boiling point change, ΔT_b, are proportional to the molality, **m**, of the solute in solution. The proportionality is made an equality by inserting a constant; K_f and K_b are called the molal freezing and boiling constants for the solvent. These constants for several solvents are listed in Table 23.1

$$\Delta T_f = K_f m = K_f\left[\frac{\text{mol solute}}{\text{kg solvent}}\right] = K_f\left[\frac{(\text{g/MW})\ \text{solute}}{\text{kg solvent}}\right]$$

$$\Delta T_b = K_b m = K_b\left[\frac{\text{mol solute}}{\text{kg solvent}}\right] = K_b\left[\frac{(\text{g/MW})\ \text{solute}}{\text{kg solvent}}\right]$$

Table 23.1
Molal Freezing Point and Boiling Point Constants for Solvents

Substance	Freezing Point (°C)	K_f (°C•kg/mol)	Boiling Point (°C)	K_b (°C•kg/mol)
water	0.0	1.86	100.0	0.512
benzene	5.45	4.90	80.2	2.53
cyclohexane	.?*	20.0	80.7	2.79
naphthalene	80.2	6.9	..	..
camphor	178.4	37.7	..	..
acetic acid	16.6	3.90	118.3	3.07

.?* The freezing point of cyclohexane is measured in this experiment

The temperature changes, ΔT_f and ΔT_b, are dependent only on the concentration of the solute dissolved in the solvent, but independent of the solute itself. This means that a 1m aqueous sugar, $C_{12}H_{22}O_{11}$, solution has the same effect on ΔT_f and ΔT_b as a 1m aqueous ethylene glycol solution. It is important to note that two moles of solute, whether they be molecules or ions, have nearly twice the effect on ΔT_f and ΔT_f as does one mole. This means that a 2m ethylene glycol solution (2 moles of solute particles) and a 1m NaCl solution ($NaCl(aq) \rightarrow Na^+(aq) + Cl^-(aq)$–2 moles of solute particles) have a ΔT_f and ΔT_b approximately twice that of a 1m sugar solution (no dissociation into ions–1 mole of solute particles). Properties of solutions, such as a freezing point decrease and a boiling point increase, that are proportional to the number (and moles) of solute particles irregardless of their nature (ions or molecules) are called **colligative properties** of a solution.

In this experiment you will use the freezing point colligative property of cyclohexane to determine the molecular weight of a nonvolatile compound dissolved in cyclohexane. A mass of the unknown solute, added to a known mass of cyclohexane, causes a freezing point change, ΔT_f, which is measured. Since ΔT_f is proportional to the moles of solute added, the molecular weight of the unknown is calculated.

$$\mathbf{MW\ (g/mol) = \frac{mass\ of\ solute\ (g)}{moles\ of\ solute\ (mol)}}$$

Techniques

• Technique 1, page 1	Inserting a Thermometer Through a Rubber Stopper
• Technique 13, page 16	Using the Laboratory Balance
• Technique 15, page 18	Graphing Techniques

Procedure

You and a partner should complete at least 2 trials in today's experiment. A 110°C thermometer, a two–hole rubber stopper to fit your 200mm test tube, and a wire stirrer are needed from the stockroom. One of you should fill a 400mL beaker with ice.

A. Freezing Point of Cyclohexane

1. Place a clean, dry 200mm test tube in a 250mL beaker and weigh (±0.01g). Add approximately 15g (15–20mL) of cyclohexane to the test tube and reweigh, using the same balance (Figure 23.1).

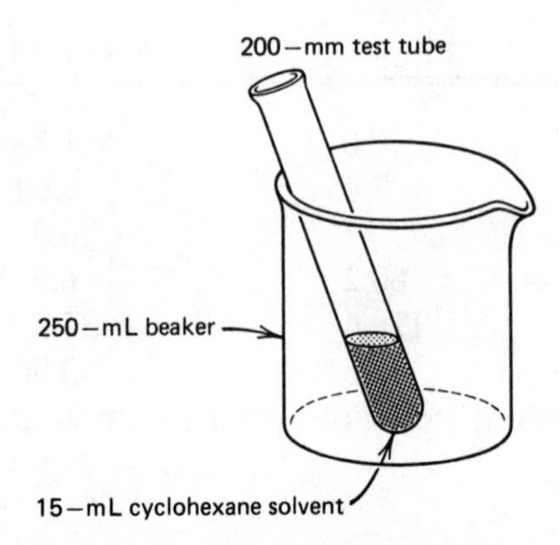

Figure 23.1
Weigh the Cyclohexane in a Beaker

2. Prepare about 300mL of an ice–water slurry in a 400mL beaker. Place the test tube containing the cyclohexane in the ice–water bath. Insert a thermometer, placed through one hole of a two–hole stopper, into the solvent to measure the temperature (Figure 23.2).

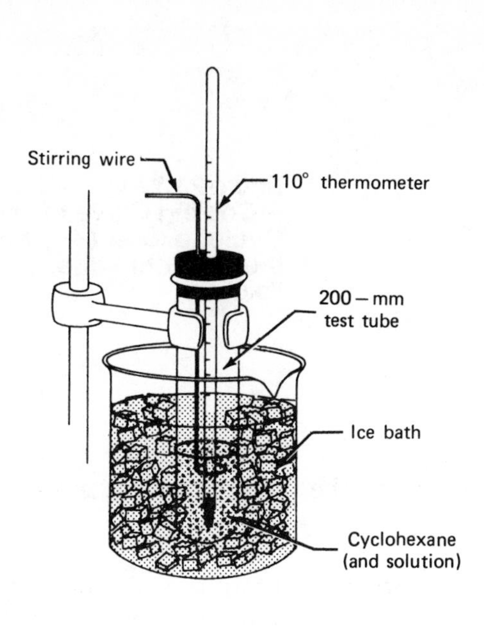

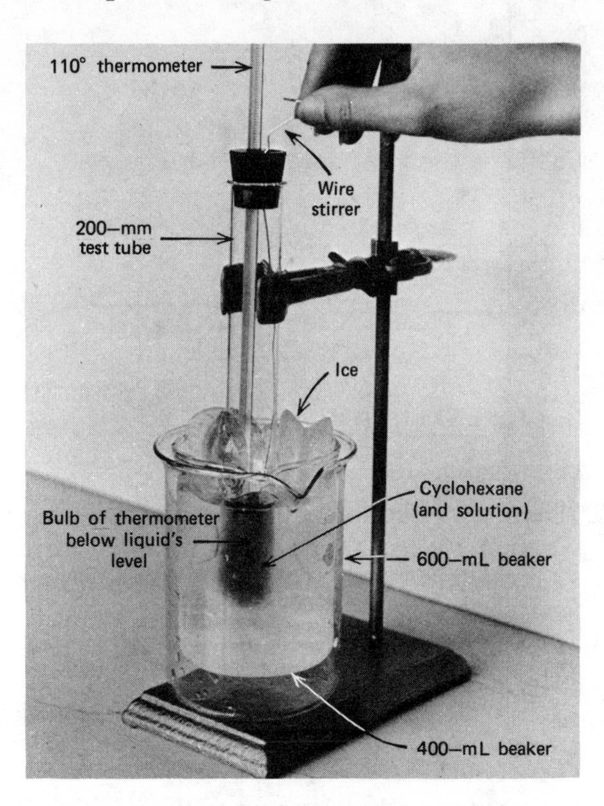

Figure 23.2 Apparatus for Determining the Freezing Point of Cyclohexane and a Cyclohexane Solution

3. While stirring with the wire stirrer, inserted through the other hole of the two–hole stopper, record on the Data Sheet the temperature (±0.1°C) at timed intervals (30s or 60s). The temperature remains virtually constant at the freezing point until the solidification is almost complete. Continue recording until the temperature begins to drop again.

4. On linear graph paper plot the temperature (°C, ordinate) *versus* time (seconds, abscissa) to obtain the "cooling curve" for cyclohexane. (See the solid line in Figure in 23.3.) Determine the freezing point of cyclohexane from the plot. Obtain your instructor's approval for your graph.

B. FREEZING POINT OF THE SOLUTION AND MOLECULAR WEIGHT OF AN UNKNOWN SOLUTE

1. Dry the outside of the test tube containing the cyclohexane and reweigh in the same 250mL beaker using the same balance as in Part A. Obtain a solid unknown from your instructor. Weigh approximately 0.2g to 0.5g of the unknown (ask your instructor for the approximate

mass to use).[1] Quantitatively transfer the unknown[2] to the cyclohexane in the 200mm test tube.

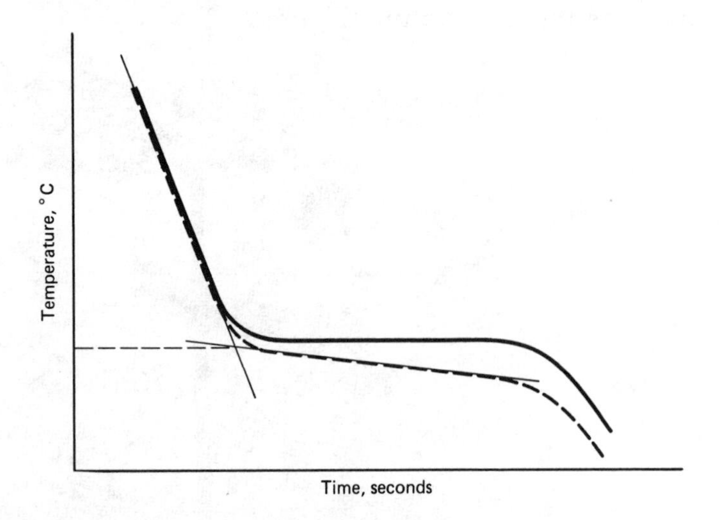

Figure 23.3
A Cooling Curve for the Cyclohexane (—) and for the Cyclohexane Solution (- - - -)

2. Determine the freezing point of the (now) solution in the same way as that for cyclohexane in Part A. When the solution nears its freezing point, record the temperature at more frequent intervals (15s to 20s).

3. Plot the temperature vs time on the *same* graph as in Part A to obtain the cooling curve for the solution. (See the broken line in Figure 23.3.) The curve will show a "break" at the temperature where freezing began; this is only an approximate freezing point and is not as well defined as that for pure cyclohexane. To determine the freezing point of the solution

   - draw a line tangent to the curve *prior* to the freezing point.
   - draw a line tangent to the curve *after* the freezing point is reached.
   - at the intersection point of the two drawn tangents, draw a line parallel to the abscissa (time axis) until it intersects the ordinate (temperature axis)–this temperature value is the freezing point of the solution.

   Again obtain your instructor's approval for your graph.

4. Warm the solution to about room temperature; repeat the temperature–time measurements and repeat the graphing until a reproducibility of the freezing point is achieved. After completing the collection of the experimental data, discard the solution into a jar marked "waste solution."

5. If time permits, repeat the procedure, starting with a new sample of cyclohexane (Part A) and a new mass of unknown (Part B).

---

[1] If the unknown is a **solid**, weigh on weighing paper. If the unknown is a **liquid**, weigh about 3mL of it in a 10mL graduated cylinder and pour this into the cyclohexane; reweigh the empty cylinder and calculate the mass difference for the mass of the unknown.

[2] In the transfer, be certain that *none* of the solute adheres to the wall of the test tube. If some does, roll the test tube so that the cyclohexane contacts and dissolves the solute.

MOLECULAR WEIGHT DETERMINATION-LAB PREVIEW

Date________Name____Lab Sec. _____Desk No._____

1. What is the freezing point of a 1m sugar (aqueous) solution?

2. Define a colligative property.

3. a. To insert a thermometer through a hole in a rubber stopper, what lubricant should be used and where should it be placed?

   b. At what (maximum) distance should your hands be placed from the rubber stopper when pushing the thermometer through the hole?

4. a. A 0.224g sample of a nonvolatile solute is dissolved in 15.0g of benzene. The solution freezes at 4.85°C. What is the molecular weight of the solute?

b. What would be the boiling point of this solution?

c. What would be the freezing point change, ΔT_f, of a solution if the same mass of solute had been added to 15.0g of naphthalene?

5. a. The molecular weight of p–nitrotoluene is 137g/mol. How many grams must be dissolved in 15.0g of benzene to lower its freezing point by 2.0°C?

b. How many grams of p-nitrotoluene must be dissolved in 15.0g of cyclohexane to lower its freezing point by 2.0°C?

6. a. In plotting data, which axis is the abscissa?_______________

b. In a plot of pressure *vs* volume, which values are customarily plotted along the y–axis?

7. Any given solution does not have a constant freezing point. Explain.

MOLECULAR WEIGHT DETERMINATION-DATA SHEET

Date________Name____Lab Sec. _____Desk No.____

A. FREEZING POINT OF CYCLOHEXANE

1. Mass of beaker, test tube, + cyclohexane (g) __________
2. Mass of beaker, test tube (g) __________
3. Mass of cyclohexane (g) __________
4. Freezing point, from cooling curve (°C) __________
5. Instructor's approval of graph ____________________

B. FREEZING POINT OF THE SOLUTION AND MOLECULAR WEIGHT OF AN UNKNOWN SOLUTE

	Trial 1	Trial 2
1. Mass of beaker, test tube + cyclohexane (g)	__________	__________
2. Mass of cyclohexane in solution (g)	__________	__________
3. Mass of unknown solute in solution (g)	__________	__________
4. Freezing point of solution, from cooling curve (°C)	__________	__________
5. Instructor's approval of graphs	__________	__________
6. K_f for cyclohexane	20.0°C kg/mol	
7. Freezing point change, ΔT_f (°C)	__________	__________
8. Molecular weight of unknown solute (g/mol)	__________	__________

9. Average molecular weight (g/mol) ________________

10. Percent (elemental) composition of unknown (Obtain this from your instructor) ________________

11. Empirical formula of unknown ________________

12. Molecular formula of unknown ________________

13. Actual molecular weight of unknown ________________

14. Percent deviation, $\frac{\text{actual MW-expt'l MW}}{\text{actual MW}} \times 100 =$ ________________

Questions

1. a. If the freezing point of the solution is recorded 0.2°C lower than its actual freezing point, will the molecular weight of the unknown be too high or too low? Explain.

   b. If the freezing point of cyclohexane is recorded 0.2°C lower that its actual freezing point, will the molecular weight of the unknown be too high or too low? Explain.

2. How will the change in the freezing point, ΔT_f, of cyclohexane be affected by
   a. the presence of a nonvolatile solute that dissociates? Explain.

   b. the presence of two solutes that react according to the equation, A + B→ C? Explain.

3. If a thermometer is miscalibrated to read 0.5°C higher than the actual temperature over its entire scale, how will it affect the *reported* molecular weight of the solute? Explain.

4. Suppose the "pure" cyclohexane in today's experiment was initially contaminated with a nonvolatile solute.
   a. How would the ΔT_f have been affected? Explain.

   b. Would the molecular weight of the unknown have been reported as being too high, too low, or unchanged? Explain.

5. Cyclohexane has a relatively high vapor pressure. If some of the cyclohexane evaporates during the experiment, how will this affect
   a. the ΔT_f in the experiment?

   b. the reported molecular weight of the unknown solute? Explain.

Experiment 24
LeChatelier's Principle

Objectives

- To study the effects of concentration and temperature changes on the equilibrium position in various chemical systems
- To study the pH effect of strong acid and strong base addition on buffered and unbuffered systems

Principles

Most chemical reactions do not produce a 100% yield of products according to the stoichiometry of the system, not because of experimental technique or design, but because of the reaction's chemical characteristics. While the products are produced initially at a relatively rapid rate, that production rate slows with time until an apparent cessation of the reaction occurs.

This apparent cessation of the reaction before a 100% yield is obtained implies that the chemical system has reached a state of **dynamic equilibrium**, a condition that is fulfilled for all *reversible* reactions–the rate at which the reactants combine to form the products equals the rate at which the products combine to re–form the reactants. For the reaction,

$$2\,SO_2(g) + O_2(g) \rightleftharpoons 2\,SO_3(g) + 197kJ$$

chemical equilibrium is reached when the reaction rate between SO_2 and O_2 to form SO_3 is the same as that between SO_3 molecules re–forming SO_2 and O_2.

If the concentration of one of the species in the *equilibrium* system changes or the temperature changes, then the equilibrium position of the system tends to shift compensating for the change. For example, assume that the above system has reached a state of dynamic equilibrium: if more SO_2 is added, it will (stoichiometrically) react with the O_2, increasing the moles of SO_3 until a *new* dynamic equilibrium is established. The new equilibrium will have more moles of SO_2 (because it was added) and SO_3 (because it formed from the reaction), but less moles of O_2 (because it reacted with the SO_2). The addition of the SO_2 resulted in a *shift* in the position of the original equilibrium to the *right*, a shift to compensate for the added SO_2.

A general statement governing all chemical systems at a state of dynamic equilibrium is ***"if an external stress (change in concentration, temperature, ...) is applied to a chemical system at equilibrium, the equilibrium shifts in the direction which minimizes the effect of that stress"***– this is **LeChatelier's Principle**, proposed by Henri Louis LeChatelier in 1888.

Often the concentrations of the species in a state of dynamic equilibrium can be quantitatively determined. From this, an equilibrium constant is calculated–its magnitude indicates the position of the equilibrium. Equilibrium constants are determined in Experiments 25 and 26.

In this experiment, we will observe the effects of creating a stress on a number of reactions that exist at equilibrium. These stresses will involve concentration and temperature changes.

A. Changes in Concentration

Chromate-dichromate equilibrium. The position of some systems at equilibrium can be easily shifted by adjusting the solution's acidity (or basicity). One such system is the chromate, CrO_4^{2-}/dichromate, $Cr_2O_7^{2-}$, system.

$$\underset{}{2\,H^+(aq)} + \underset{\text{(yellow)}}{CrO_4^{2-}(aq)} \rightleftharpoons \underset{\text{(orange)}}{Cr_2O_7^{2-}(aq)} + H_2O$$

A strongly acidic solution favors the formation of the orange $Cr_2O_7^{2-}$ ion, while a basic solution favors the formation of the yellow CrO_4^{2-} ion.

Many salts are only slightly soluble in water. The Ag^+ ion forms slightly soluble salts with many anions. This experiment studies several equilibria involving the relative solubility of the CO_3^{2-}, Cl^-, I^-, and S^{2-} silver salts; an explanation of these equilibria follows.

Silver carbonate equilibrium. Ag_2CO_3 precipitates in the presence of excess Ag^+ and CO_3^{2-} ions in aqueous solution to establish an equilibrium.

$$Ag_2CO_3(s) \rightleftharpoons 2\,Ag^+(aq) + CO_3^{2-}(aq)$$

Addition of HNO_3 causes Ag_2CO_3 to dissolve. The H^+ ions from the HNO_3 react with the free CO_3^{2-} ions, removing the CO_3^{2-} from the equilibrium. Because the system compensates for the removal of the CO_3^{2-} ions, the equilibrium shifts right. Ag_2CO_3 dissolves to form H_2CO_3 which, because of its instability at room temperature and pressure, decomposes to CO_2 and H_2O.

$$Ag_2CO_3(s) \rightleftharpoons 2\,Ag^+(aq) + CO_3^{2-}(aq)$$
$$\searrow 2H^+(aq) \text{ from } HNO_3$$
$$\rightarrow H_2CO_3(aq) \rightarrow H_2O + CO_2(aq)$$

The Ag^+ and NO_3^- remain in solution.

Silver chloride equilibrium. The addition of Cl^- ions to the Ag^+ ions remaining in solution causes insoluble, white AgCl to form.

$$Ag^+(aq) + Cl^-(aq) \rightleftharpoons AgCl(s)$$

Addition of NH_3 to the AgCl equilibrium "ties up" (i.e., forms a complex with) Ag^+ ion, forming soluble diamminesilver ion, $Ag(NH_3)_2^+$.

$$Ag^+(aq) + Cl^-(aq) \rightleftharpoons AgCl(s)$$
$$\searrow 2\,NH_3(aq)$$
$$\rightarrow Ag(NH_3)_2^+(aq) + Cl^-(aq)$$

The Ag^+ ion is removed from the AgCl equilibrium, causing a shift to the *left* and dissolving the AgCl.

Adding acid to the system causes a reaction with the NH_3 [$NH_3(aq) + H^+(aq) \rightarrow NH_4^+(aq)$]. This destroys the $Ag(NH_3)_2^+$ complex and releases the Ag^+, which then recombines with the Cl^- to reform AgCl(s).

$$Ag^+(aq) + Cl^-(aq) \rightleftharpoons AgCl(s)$$
$$\uparrow\ 2NH_3(aq) + 2H^+ \rightarrow NH_4^+$$
$$Ag(NH_3)_2^+(aq)$$

Silver iodide equilibrium. I^- ion from KI added to the $Ag(NH_3)_2^+$ equilibrium precipitates yellow AgI.

$$Ag(NH_3)_2^+(aq) \rightleftharpoons Ag^+(aq) + 2\,NH_3(aq)$$
$$\searrow I^-(aq) \rightarrow AgI(aq)$$

The I^- ion removes the Ag^+ ion causing a shift of the above $Ag(NH_3)_2^+$ equilibrium to the right.

Silver sulfide equilibrium. Silver sulfide, Ag_2S, is less soluble than AgI. Addition of S^{2-} ion (from Na_2S) precipitates Ag^+ from the AgI equilibrium.

$$2AgI(s) \rightleftharpoons 2Ag^+(aq) + 2I^-(aq)$$
$$\searrow S^{2-}(aq) \rightarrow Ag_2S(aq)$$

Common ion effect. The effect of adding an ion or ions common to those in an equilibrium system is called a **common ion effect**. This effect is observed for the following equilibria.

$$NH_4Cl(s) \rightleftharpoons NH_4^+(aq) + Cl^-(aq)$$
$$4Cl^-(aq) + Co(H_2O)_6^{2+}(aq) \rightleftharpoons CoCl_4^{2-}(aq) + 6H_2O$$
$$4Br^-(aq) + Cu(H_2O)_4^{2+}(aq) \rightleftharpoons CuBr_4^{2-}(aq) + 4H_2O$$
$$HC_2H_3O_2(aq) + H_2O \rightleftharpoons H_3O^+(aq) + C_2H_3O_2^-(aq)$$
$$Mg^{2+}(aq) + 2NH_3(aq) \rightleftharpoons 2Mg(OH)_2(s) + 2NH_4^+(aq)$$

B. Changes in Temperature

The reaction of SO_2 with O_2 producing SO_3 is exothermic by 197kJ.

$$2SO_2(g) + O_2(g) \rightleftharpoons 2SO_3(g) + 197kJ$$

To favor the formation of SO_3, the reaction vessel is kept cool. Removal of heat shifts the equilibrium to the *right* favoring the formation of SO_3. Added heat results in a shift in the direction that absorbs heat–in this case, to the *left*.

This experiment examines the effect of temperature on the equilibrium systems **$4Cl^-(aq) + Co(H_2O)_6^{2+}(aq) \rightleftharpoons CoCl_4^{2-}(aq) + 6H_2O$** and **$4Br^-(aq) + Cu(H_2O)_4^{2+}(aq) \rightleftharpoons CuBr_4^{2-}(aq) + 4H_2O$**. These systems involve an equilibrium between the "coordination spheres" about the cobalt(II) and copper(II) ions which are concentration *and* temperature dependent. The tetrachlorocobaltate(II) ion, $CoCl_4^{2-}$, and the tetraaquacopper(II) ion, $Cu(H_2O)_4^{2+}$, are more stable at the higher temperatures.

Techniques

- Technique 5d, page 7 — Separation of a Solid from a Liquid
- Technique 6a,d, page 7 — Heating Liquids

Procedure

At each numbered superscript (example #1)in the procedure, **STOP** , and record your observations.

A. Saturated NH_4Cl Solution

Add an excess of NH_4Cl to a 75mm test tube containing 2mL of H_2O and shake. Hold the test tube in the palm of your hand. Is the dissolution an endothermic or exothermic process?#1 Transfer the solution (none of the solid!) to a second 75mm test tube. Add 2-3 drops conc HCl (**Caution**: *conc* HCl *is a severe skin irritant*) to the saturated solution. Note the results.#2 Write a chemical equation to explain this. Gradually heat the test tube. What happens to the appearance of the system?#3

B. CrO_4^{2-}•$Cr_2O_7^{2-}$ Equilibrium

Place 2-3mL of 0.1M K_2CrO_4 solution in a 150mm test tube. Add 2-3 drops of 6M HNO_3 and note the color change.#4 Express the equilibrium system with an equation. Add 5-7 drops of 6M NaOH. Explain.#5 Reacidify with 6M HNO_3.#6

C. Silver Ion Equilibria

1. Add 5mL of 0.1M Na_2CO_3 to 5mL of 0.1M $AgNO_3$ in a 150mm test tube.#7 Now add 5 drops of 6M HNO_3 to the precipitate. What do you see?#8

2. After the HNO_3 dissolves the precipitate, add 5mL of 0.1M HCl.#9 Allow the precipitate to settle and decant the supernatant until approximately 5mL of solution remain in the 150mm test tube. Add *drops* of conc NH_3 (**Caution**: *avoid inhalation and skin contact with conc* NH_3) until the precipitate dissolves.#10 Reacidify the solution with 6M HNO_3 and record.#11 What happens if an excess of conc NH_3 is added? Try it.#12

3. After "trying it", add 5mL of 0.1M KI.#13

4. To the solution containing the yellow AgI precipitate, add 5mL of 0.1 M Na_2S. Record and explain your observations on the Data Sheet.#14

D. $Cu(H_2O)_4^{2+}$•$CuBr_4^{2-}$ Equilibrium (Concentration/Temperature Effects)

1. Place 0.5g of $CuBr_2$ in a dry 200mm test tube. Note its color.#15 Add 0.5mL (10 drops) of water,#16 then an additional 2mL of water. Observe the changes after each addition. Continue adding water, noting the color changes, until the total volume is about 15mL.#17 Write the ionic equation for the equilibrium.

2. Place about 1g of KBr in a 75mm test tube and add 2mL of water. What is the color of the K^+ and Br^- ions?#18

3. Place 2mL of the $CuBr_2$ solution prepared in Part D.1 in another test tube. Add about 1g of solid KBr. What happens?#19 Explain the color changes with ionic equations.

4. Place this solution (from Part D.3) in a boiling water bath. Write an equation that explains your observation.#20

E. $Co(H_2O)_6^{2+} \bullet CoCl_4^{2-}$ EQUILIBRIUM (CONCENTRATION/TEMPERATURE EFFECTS)

1. Place about 5mL of 1M $CoCl_2$ in a 150mm test tube. Record the solution's color.#21 Slowly and carefully add conc HCl (**Caution:** *avoid inhalation or skin contact*) until a color change occurs (about 3mL).#22 Write an equation that explains your observations. Slowly add water to the system.#23

2. Again place about 5mL of 1M $CoCl_2$ in a 150mm test tube; place in a boiling hot water bath. Compare the color to the original 1 M $CoCl_2$ solution.#24 Write an equation that explains your observation.

F. $HC_2H_3O_2 \bullet C_2H_3O_2^-$ EQUILIBRIUM (A BUFFER SYSTEM)

1. Add a few drops of universal indicator to 5mL of 0.1M $HC_2H_3O_2$ in a 150mm test tube; note the color.#25 Now add 2g of solid $NH_4C_2H_3O_2$ and agitate to dissolve the salt. Compare the solution's color with the pH color chart for the universal indicator.#26 Divide the solution into two equal volumes in separate 150mm test tubes labeled "1" and "2".

2. Place 2.5mL of distilled water into 150mm test tubes labeled "3" and "4". Add several drops of universal indicator to test tubes 3 and 4.

3. To test tubes 1 and 3, add 2.0mL of 0.1M NaOH; compare the colors of both solutions.#27 To test tubes 2 and 4, add 2.0mL of 0.1M HCl; compare the colors of these two solutions.#28 Controlling the pH change in a chemical system when an acid or base is added is termed **buffer action.** Explain the effect that a buffered system has on pH when a strong or base is added to it.#29

G. $Mg^{2+}(aq) \bullet Mg(OH)_2$ EQUILIBRIUM

To 5mL of distilled water in a 150mm test tube, add 2mL of 1 M $MgCl_2$ and 1mL of 3M NH_3. Note what occurs.#30 Add 1g of solid NH_4Cl and agitate the mixture. What happens to the precipitate?#31

LECHATELIER'S PRINCIPLE-LAB PREVIEW

Date__________Name____Lab Sec. _____Desk No._____

1. **Indicate the direction in which the equilibrium shifts in each of the chemical systems when the following stress is placed on each.**

   a. Ag^+ is added to $Ag^+(aq) + Cl^-(aq) \rightleftharpoons AgCl(s)$ ----------

   b. H_3O^+ is added to $Ag_2CO_3(s) \rightleftharpoons 2Ag^+(aq) + CO_3^{2-}(aq)$ ----------

   c. H_3O^+ is added to $Ag^+(aq) + 2NH_3(aq) \rightleftharpoons Ag(NH_3)_2^+(aq)$ ----------

   d. Ag^+ is removed from $Ag^+(aq) + I^-(aq) \rightleftharpoons AgI(s)$ ----------

   e. heat is added to $2NO(g) + Cl_2(g) \rightleftharpoons 2NOCl(g) + 77.1kJ$ ----------

   f. heat is removed from $2SO_2(g) + O_2(g) \rightleftharpoons 2SO_3(g) + 197kJ$ ----------

   g. Cl_2 is added to $2NO(g) + Cl_2(g) \rightleftharpoons 2NOCl(g) + 77.1kJ$ ----------

   h. Br^- is added to $4Br^-(aq) + Cu(H_2O)_4^{2+}(aq) \rightleftharpoons CuBr_4^{2-}(aq) + 4H_2O$ ----------

   i. $NaHCO_3$ is added to $CO_3^{2-}(aq) + H_2O \rightleftharpoons HCO_3^-(aq) + OH^-(aq)$ ----------

2. **Complete the following statements with "increase", "decrease", or "no change".**

   a. When Ag^+ is added to $Ag^+(aq) + Cl^-(aq) \rightleftharpoons AgCl(s)$, the Cl^- concentration __________.

   b. When Cl_2 is added to $2NO(g) + Cl_2(g) \rightleftharpoons 2NOCl(g) + 77.1kJ$, the temperature __________ and the NOCl __________ .

   c. The reaction, $CH_4(g) + 2H_2S(g) \rightleftharpoons CS_2(g) + 4H_2(g)$, is endothermic. If $CH_4(g)$ is added to the system, the H_2S concentration __________, the CS_2 concentration __________, and temperature __________.

   d. When CO_3^{2-} is added to the $Mg^{2+}(aq) + CO_3^{2-}(aq) \rightleftharpoons MgCO_3(s)$, the Mg^{2+} concentration __________ and the amount of insoluble $MgCO_3$ __________.

   e. Adding NH_3 to $4NH_3(aq) + Cu(H_2O)_4^{2+}(aq) \rightleftharpoons Cu(NH_3)_4^{2+}(aq) + 4H_2O$ causes the $Cu(H_2O)_4^{2+}$ concentration to __________ and the $Cu(NH_3)_4^{2+}$ concentration to __________ .

f. Adding H_3O^+ to $HCN(aq) + H_2O \rightleftharpoons H_3O^+(aq) + CN^-(aq)$ causes the CN^- concentration to ________. Adding OH^- to the same equilibrium causes the CN^- concentration to ________ and the HCN concentration to ________.

LECHATELIER'S PRINCIPLE-DATA SHEET

Date________ Name____ Lab Sec. ____ Desk No.____

A. SATURATED NH_4Cl SOLUTION

#1 Is the dissolution exothermic or endothermic?____________________

Equation for equilibrium system____________________________________

#2 Effect of HCl addition. Explain.____________________________________

#3 Effect of added heat. Explain.____________________________________

B. $CrO_4^{2-} \bullet Cr_2O_7^{2-}$ EQUILIBRIUM

Equation for equilibrium system.____________________________________

#4 Effect of 6M HNO_3. Explain.____________________________________

#5 Effect of 6M NaOH. Explain.____________________________________

#6 Effect of additional 6M HNO_3. Explain.____________________________________

C. SILVER ION EQUILIBRIA

#7 Equation for Ag_2CO_3 equilibrium system____________________________________

#8 Effect of 6M HNO_3. Explain.____________________________________

Equation____________________________________

#9 Effect of 0.1M HCl. Explain.____________________________________

Equation____________________________________

#10 Effect of conc NH_3. Explain.____________________________________

Equation____________________________________

#11 Effect of reacidification with 6M HNO_3. Explain.____________________________________

Equation____________________________________

#12 Effect of additional conc NH_3. Explain.____________________________________

#13 Effect of 0.1M KI. Explain.____________________________________

Equation______________________________

#14 Effect of 0.1M Na_2S. Explain.______________________________

D. $Cu(H_2O)_4^{2+}$•$CuBr_4^{2-}$ Equilibrium (Concentration/Temperature Effects)

#15 Color of $CuBr_2(s)$______________________________

#16 Color of $CuBr_2$ in $^1/_2$mL solution______________________________

#17 Color of $CuBr_2$ in 15mL solution______________________________

Ionic equation for equilibrium established with $CuBr_2$ in 15mL solution

____

#18 Color of $K^+(aq)$______________; color of $Br^-(aq)$______________

#19 Color after KBr addition (due to $CuBr_4^{2-}$)______________________________

Effect of KBr on the equilibrium______________________________

#20 Effect of added heat. Explain______________________________

E. $Co(H_2O)_6^{2+}$•$CoCl_4^{2-}$ Equilibrium (Concentration/Temperature Effects)

#21 Color of $CoCl_2(aq)$______________________________

#22 Effect of conc HCl. Explain______________________________

Equation for equilibrium______________________________

#23 Effect of added H_2O. Explain______________________________

#24 Effect of temperature increase. Explain.______________________________

F. $HC_2H_3O_2$•$C_2H_3O_2^-$ Equilibrium (A Buffer System)

Equation for ionization of $HC_2H_3O_2$ in water______________________________

#25 Color of universal indicator in $HC_2H_3O_2$______________________; pH =________

#26 Color of universal indicator in $NH_4C_2H_3O_2$ addition______________; pH =________

Effect of $NH_4C_2H_3O_2$ on the equilibrium______________________________

Test tube number	$HC_2H_3O_2 \bullet C_2H_3O_2^-$ 1	2	H_2O 3	4
#27 Color effect from 0.1M NaOH		•••••••		•••••••
Approximate pH		•••••••		•••••••
Approximate change in pH		•••••••		•••••••
#28 Color effect from 0.1M HCl	•••••••		•••••••	
Approximate pH	•••••••		•••••••	
Approximate change in pH	•••••••		•••••••	

#29 How does the magnitude of the pH change when HCl (a strong acid) or NaOH (a strong base) is added to a solution of $HC_2H_3O_2$ and $NH_4C_2H_3O_2$ as compared to the magnitude of the pH change that occurs when HCl or NaOH is added to pure water?

G. $Mg^{2+}(aq) \bullet Mg(OH)_2$ Equilibrium

#30 Addition of $NH_3(aq)$ to the $MgCl_2$ solution____________________

Equation____________________

#31 Effect of NH_4Cl____________________

Questions

1. Identify the color of each:

a. CrO_4^{2-}____________________ f. $CoCl_4^{2-}$____________________

b. $Cr_2O_7^{2-}$____________________ g. $Co(H_2O)_6^{2+}$____________________

c. AgI____________________ h. I^-____________________

d. $Ag(NH_3)_2^+$____________________ i. S^{2-}____________________

e. $Cu(H_2O)_4^{2+}$____________________ j. $CuBr_4^{2-}$____________________

2. Identify the color of the

a. $Co(H_2O)_6^{2+} \bullet CoCl_4^{2-}$ equilibrium at 80°C____________________

b. $Cu(H_2O)_4^{2+} \bullet CuBr_4^{2-}$ equilibrium at 80°C.____________________

3. a. Explain why the pH change was small when HCl was added to the $HC_2H_3O_2 \bullet C_2H_3O_2^-$ system, but much larger when added to water.

b. Explain why the pH change was small when NaOH was added to the $HC_2H_3O_2 \bullet C_2H_3O_2^-$ system, but much larger when added to water.

EXPERIMENT 25
AN EQUILIBRIUM CONSTANT

Objectives

- To develop the techniques for the use and operation of a spectrophotometer
- To determine the equilibrium constant for a soluble ionic system

Principles

The use of a spectrophotometer for measuring the concentration of an ion in solution is discussed in Experiment 14. If you did not complete that experiment, read the "Principles" section.

The magnitude of an equilibrium constant, K_c, expresses the equilibrium position for a chemical system. For example, a small K_c indicates that the equilibrium is far to the left whereas a large K_c indicates the equilibrium to be far to the right. The value of K_c is constant for a chemical system at a constant temperature.

This experiment determines K_c for an equilibrium system in which all species are ionic and soluble.

$$Fe(H_2O)_6^{3+}(aq) + SCN^-(aq) \rightleftharpoons Fe(H_2O)_5NCS^{2+}(aq) + H_2O$$

Because the concentration of H_2O is essentially constant in dilute aqueous solutions, we can ignore the waters of hydration and represent the equilibrium with the equation

$$Fe^{3+}(aq) + SCN^-(aq) \rightleftharpoons FeNCS^{2+}(aq)$$

Its equilibrium (or mass action) expression is

$$K_c = \frac{[FeNCS^{2+}]}{[Fe^{3+}][SCN^-]}$$

This equilibrium is prepared by mixing known concentrations of Fe^{3+} and SCN^-. Since the $FeNCS^{2+}$ ion shows a deep, blood–red color with an absorption maximum at about 447nm, its concentration is determined spectrophotometrically. By knowing the *initial* molar concentrations of Fe^{3+} and SCN^- and by measuring the $FeNCS^{2+}$ *equilibrium* molar concentration spectrophotometrically, the equilibrium molar concentrations of Fe^{3+} and SCN^- are calculated.

$$[Fe^{3+}]_{\text{at equilibrium}} = [Fe^{3+}]_{\text{initial}} - [FeNCS^{2+}]_{\text{at equilibrium}}$$
$$[SCN^-]_{\text{at equilibrium}} = [SCN^-]_{\text{initial}} - [FeNCS^{2+}]_{\text{at equilibrium}}$$

Using these equilibrium concentrations, the K_c for the equilibrium is calculated.

In Part A of this procedure, you will prepare a set of standard solutions of the $FeNCS^{2+}$ ion. The absorbance, A, for each solution is plotted against the [$FeNCS^{2+}$] to establish a calibration curve. This graph is then used to determine the [$FeNCS^{2+}$] at equilibrium for the systems in Part B. In preparing the standards for Part A, the [Fe^{3+}] is adjusted to be in a *large excess* relative to the [SCN^-]; this drives the equilibrium ($Fe^{3+}(aq) + SCN^-(aq) \rightleftharpoons FeNCS^{2+}(aq)$) far to the *right*. From this, we assume that the [$FeNCS^{2+}$] approximates the original [SCN^-], i.e., all of the SCN^- is in the form of the $FeNCS^{2+}$ ion.

The calculations for K_c are involved, but the Lab Preview should clarify most of these steps. The Data Sheet is also outlined in such detail as to help with the calculations.

Techniques

- Technique 10, page 12 — Reading a Meniscus
- Technique 11, page 13 — Pipetting a Liquid or Solution
- Technique 15, page 18 — Graphing Techniques

In addition, the proper technique for the handling of cuvets is necessary.

Procedure

A large number of 100mL volumetric flasks is used in this experiment. Ask your instructor about working with a partner(s) for Parts A and/or B.

A. A Set of Standard $FeNCS^{2+}$ Solutions

1. Pipet 0, 2, 4, 6, and 8mL of 0.002M NaSCN[1] into separate, labeled 100mL volumetric flasks. Pipet 25.0mL of 0.2M $Fe(NO_3)_3$[2] into each flask and dilute to the "mark" with 0.1M HNO_3. See Table 25.1. These solutions are used to establish a calibration curve for determining the equilibrium [$FeNCS^{2+}$], spectrophotometrically. Additional test solutions from 1mL and 10mL of 0.002M NaSCN are suggested.

Table 25.1
A Set of Standard $FeNCS^{2+}$ Solutions

Solution	0.2M $Fe(NO_3)_3$ (in 0.1M HNO_3)	0.002M NaSCN (in 0.1M HNO_3)	0.1M HNO_3
1	25.0mL	0.0mL	75.0mL
2	25.0mL	2.0mL	73.0mL
3	25.0mL	4.0mL	71.0mL
4	25.0mL	6.0mL	69.0mL
5	25.0mL	8.0mL	67.0mL

2. Listen carefully to your instructor's advice on the use and operation of the spectrophotometer. Solution #1 is to be used as the "blank". Stir each solution thoroughly with a clean, *dry* stirring rod.

---

[1]Record the *exact* concentration of the NaSCN on the Data Sheet.

[2]Record the *exact* concentration of the $Fe(NO_3)_3$ on the Data Sheet.

3. Rinse the cuvet with several portions of test solution and then fill it approximately three-fourths full. Carefully dry the outside of each cuvet with a clean Kimwipe to remove water droplets and fingerprints. Handle only the lip of the cuvet henceforth. Any foreign material on the cuvet affects the intensity of the transmitted light.

4. Set the wavelength on the spectrophotometer at 447nm. Calibrate the spectrophotometer with Solution #1: place the cuvet containing Solution #1 in the cuvet holder and set the spectrophotometer to read 100% T. Remove the cuvet and set the spectrophotometer to read 0%T. Repeat until no further adjustments are necessary.

5. Place the cuvet containing Solution #2 and read its %T; repeat with Solutions #3, #4, and #5. Convert your %T readings to absorbance values, A,[3] and plot them *versus* [$FeNCS^{2+}$]. Draw the best straight line through the five (or seven) points *and* the origin. Ask the instructor to approve your graph.

B. Set of Equilibrium Solutions

1. Prepare the test solutions in Table 25.2 in clean, *dry* 200mm test tubes. Use pipets for the volumetric measurements. If 0.002M $Fe(NO_3)_3$ is not available, dilute 1.0mL (measure with a 1.0mL pipet) of the 0.2M $Fe(NO_3)_3$ used in Part A to 100mL with 0.1M HNO_3 in a volumetric flask.

2. Stir each solution, read, and record its %T as in Part A.5. Again use Solution #1 as the blank for calibrating the spectrophotometer. Be careful in handling the cuvets–don't drop them!

3. From the calibration curve, determine the equilibrium [$FeNCS^{2+}$] for each solution.

Table 25.2
A Set of Equilibrium Solutions

Solution	0.002M $Fe(NO_3)_3$ (in 0.1M HNO_3)	0.002M NaSCN (in 0.1M HNO_3)	0.1M HNO_3
6	5.00mL	1.00mL	4.00mL
7	5.00mL	2.00mL	3.00mL
8	5.00mL	3.00mL	2.00mL
9	5.00mL	4.00mL	1.00mL
10	5.00mL	5.00mL	••••

[3] $A = -\log(\%T/100)$; also, A α [$FeNCS^{2+}$] where "α" means "proportional to"

An Equilibrium Constant-Lab Preview

Date__________Name____Lab Sec. _____Desk No._____

1. a. What is white light?

   b. What is monochromatic light?

2. Explain how 0%T and 100%T are set on the spectrophotometer for its calibration.

3. A maximum of visible light is absorbed at 447nm for the $FeNCS^{2+}$ ion. What is the dominant color of light absorbed?___________ and the dominant color of light transmitted?___________

4. What effect do fingerprints on a cuvet containing the test sample have on the transmittance (%T) of visible light through the solution?

5. A 5.0mL volume of 0.0200M $Fe(NO_3)_3$ is mixed with 5.0mL of 0.00200M NaSCN; the blood–red $FeNCS^{2+}$ ion forms and the equilibrium is established.

$$Fe^{3+}(aq) + SCN^-(aq) \rightleftharpoons FeNCS^{2+}(aq)$$

The equilibrium [$FeNCS^{2+}$], measured spectrophotometrically, is 7.0×10^{-4} mol/L. To calculate the K_c for the equilibrium system, proceed through the following steps.

a. moles of Fe^{3+}, initial _______________

b. moles of SCN^-, initial _______________

c. moles of $FeNCS^{2+}$ at equilibrium _______________

d. moles of Fe^{3+} reacted _______________

e. moles of SCN^- reacted _______________

f. moles of Fe^{3+} remaining unreacted, at equilibrium (a-d) _______________

g. moles of SCN^- remaining unreacted, at equilibrium (b-c) _______________

h. [Fe^{3+}] at equilibrium _______________

i. [SCN^-] at equilibrium _______________

j. [$FeNCS^{2+}$] at equilibrium $\underline{7.0 \times 10^{-4}\ mol/L}$

k. K_c of $FeNCS^{2+}$ _______________

6. Using the value of K_c in Question 5, determine the equilibrium [SCN^-] after 90.0mL of 0.100M Fe^{3+} are added to 10.0mL of a SCN^- solution; the equilibrium [$FeNCS^{2+}$] is 1.0×10^{-6} mol/L.

An Equilibrium Constant-Data Sheet

Date__________ Name____ Lab Sec. ____ Desk No.____

A. A Set of Standard FeNCS^{2+} Solutions

1. *Exact* molar concentration of NaSCN __________
2. *Exact* molar concentration of $Fe(NO_3)_3$ __________

3. Standard solutions	1	2	3	4	5
a. Volume of NaSCN (mL)	______	______	______	______	______
b. Moles of SCN^- (mol)	______	______	______	______	______
c. [SCN^-] (100mL solution)	______	______	______	______	______
d. [$FeSCN^{2+}$]	______	______	______	______	______
e. %T	______	______	______	______	______
f. Absorbance, A	______	______	______	______	______

g. Calibration curve, (A *vs* [$FeSCN^{2+}$]) Instructor's Approval ____________

B. Set of Equilibrium Solutions

1. Molar concentration of $Fe(NO_3)_3$ __________
2. Molar concentration of NaSCN __________

3. Solutions	6	7	8	9	10
a. Volume of $Fe(NO_3)_3$ (mL)	______	______	______	______	______
b. Moles of Fe^{3+}, initial (mol)	______	______	______	______	______
c. Volume of NaSCN (mL)	______	______	______	______	______
d. Moles of SCN^-, initial (mol)	______	______	______	______	______
e. %T	______	______	______	______	______
f. Absorbance, A	______	______	______	______	______

C. Determination of K_C

Solutions	6	7	8	9	10
1. [$FeSCN^{2+}$], calibration curve	______	______	______	______	______
2. [Fe^{3+}], <u>at equilibrium</u>					
a. moles of $FeSCN^{2+}$ in solution at equilibrium	______	______	______	______	______
b. moles of Fe^{3+}, reacted	______	______	______	______	______
c. moles of Fe^{3+}, unreacted	______	______	______	______	______
d. [Fe^{3+}], equilibrium (unreacted)*	______	______	______	______	______
3. [SCN^-], <u>at equilibrium</u>					
a. moles of SCN^-, reacted	______	______	______	______	______
b. moles of SCN^-, unreacted	______	______	______	______	______
c. [SCN^-], equilibrium (unreacted)**	______	______	______	______	______

---

*Show calculation for [Fe^{3+}], equilibrium

**Show calculation for [SCN^-], equilibrium

4. $K_c = \frac{[FeSCN^{2+}]}{[Fe^{3+}][SCN^-]}$ ------ ------ ------ ------ ------

5. Average K_c ----------

6. Standard deviation, $\sigma = \sqrt{\frac{d_1^2 + d_2^2 + d_3^2 + ...+ d_n^2}{(n-1)}}$ = ----------

d = difference between a single K_c value and the average K_c
n = number of K_c values determined

Questions

1. What effect will a dirty cuvet (caused by fingerprints, water spots, or lint) have on the value of K_c in this experiment?

2. In our calculations the solution's thickness and the probability of light absorption by the $FeNCS^{2+}$ were not considered. Explain.

3. Considering Solution #9, suppose the 1.00mL of 0.1M HNO_3 is not added.
   a. How will this affect the %T of the solution?

   b. How will this affect the reported the $FeNCS^{2+}$ concentration for the solution?

   c. Will the reported K_c be greater or less than it should have been? Explain.

4. Why is 0.2M $Fe(NO_3)_3$ in 0.1M HNO_3 (Solution #1) used as a "blank" for calibrating the spectrophotometer instead of distilled water?

5. Why should the spectrophotometer be periodically recalibrated during the course of the experiment?

EXPERIMENT 26
SOLUBILITY PRODUCT CONSTANT AND COMMON ION EFFECT

Objectives

- To determine the molar solubility and K_{sp} of $Ca(OH)_2$
- To determine the molar solubility of $Ca(OH)_2$ in the presence of added Ca^{2+}

Principles

Many common salts which have a very limited solubility in water are called **slightly soluble salts.** A saturated solution of a slightly soluble salt is a dynamic equilibrium between the solid salt and its ions in solution. Limestone rock appears to be insoluble in its natural environment; however, with the passage of time, the limestone slowly dissolves due to its very slight solubility. Limestone's major component is calcium carbonate, $CaCO_3$. Its equilibrium with the Ca^{2+} and CO_3^{2-} favors the solid $CaCO_3$, or the equilibrium lies far to the *left*.

$$CaCO_3(s) \rightleftharpoons Ca^{2+}(aq) + CO_3^{2-}(aq)$$

The mass action expression for this equilibrium equals an equilibrium constant, called the **solubility product constant, K_{sp},** for the salt. Since the concentration of a solid also remains a constant, the mathematical expression is

$$K_{sp} = [Ca^{2+}][CO_3^{2-}]$$

This equation states that the product of the Ca^{2+} ion and the CO_3^{2-} ion molar concentrations is a constant; it does **not** state that the molar concentrations of the Ca^{2+} ion and the CO_3^{2-} ion must be equal in solution. If a solution has a high $[Ca^{2+}]$, then its $[CO_3^{2-}]$ must be necessarily low–their product is experimentally found to be a constant.

According to the CRC's *Handbook of Chemistry and Physics,* the K_{sp} of $CaCO_3$ is 4.95×10^{-9} at 25°C. In a saturated solution, when the $[Ca^{2+}] = [CO_3^{2-}]$, the molar concentration of each ion is

$$K_{sp} = [Ca^{2+}][CO_3^{2-}] = 4.95 \times 10^{-9}$$

$$[Ca^{2+}] = [CO_3^{2-}] = \sqrt{4.95 \times 10^{-9}} = 7.04 \times 10^{-5} \text{ mol/L}$$

The **molar solubility** of $CaCO_3$ is also 7.04×10^{-5} mol/L because for each mole of $CaCO_3$ that dissolves 1 mol of Ca^{2+} and 1 mol of CO_3^{2-} are in solution.

What happens to the solubility of a slightly soluble salt when an ion, common to the salt, is added to a saturated solution? According to LeChatelier's Principle, its equilibrium shifts to compensate for the added (common) ions and favor the formation of more solid salt, reducing the salt's molar solubility.

Example. Suppose 0.0010 mol CO_3^{2-} is added to a saturated $CaCO_3$ solution. What happens? The molar solubility decreases because the added CO_3^{2-} shifts the equilibrium, $CaCO_3(s) \rightleftharpoons Ca^{2+}(aq) + CO_3^{2-}(aq)$, to the *left*, meaning that less moles of $CaCO_3$ can now dissolve. The new molar solubility of $CaCO_3$ equals the $[Ca^{2+}]$ in solution or

$$\text{molar solubility of } CaCO_3 = [Ca^{2+}] = \frac{K_{sp}}{[CO_3^{2-}]} = \frac{4.95 \times 10^{-9}}{0.0010} = 4.95 \times 10^{-6} \text{ mol/L}$$

The added CO_3^{2-} decreases the molar solubility of $CaCO_3$ from 7.04×10^{-5} mol/L to 4.95×10^{-6} mol/L.

In this experiment you will determine the K_{sp} of $Ca(OH)_2$ and its molar solubility in a saturated $Ca(OH)_2$ solution and in the presence of added Ca^{2+}. A saturated $Ca(OH)_2$ solution is decanted from solid $Ca(OH)_2$; the OH^- is titrated with a standard HCl solution to determine its concentration. According to the equation

$$Ca(OH)_2(s) \rightleftharpoons Ca^{2+}(aq) + 2OH^-(aq),$$

for each mole of $Ca(OH)_2$ that dissolves, 1 mol Ca^{2+} and 2 mol OH^- are present in solution. Thus by determining the $[OH^-]$, the $[Ca^{2+}]$, K_{sp}, and the molar solubility of $Ca(OH)_2$ can be calculated.

$$K_{sp} = [Ca^{2+}][OH^-]^2 = \left(\frac{1}{2}[OH^-]\right)[OH^-]^2$$

$$\text{The molar solubility of } Ca(OH)_2 = [Ca^{2+}] = \frac{1}{2}[OH^-]$$

Likewise we use the same procedure to determine the molar solubility of $Ca(OH)_2$ with Ca^{2+} added; Ca^{2+} is an ion common to the slightly soluble salt, $Ca(OH)_2$, equilibrium.

Techniques

- Technique 4a, page 4 — Separation of a Solid from a Liquid
- Technique 10, page 12 — Reading a Meniscus
- Technique 11, page 13 — Pipetting a Liquid or Solution
- Technique 12, page 14 & 15 — Titrating a Solution

Procedure

Three trials for Parts A and B are to be completed. To hasten the analysis, clean and label three 125mL or 250mL Erlenmeyer flasks. In Part A.3, pipet 25mL of the saturated $Ca(OH)_2$ solution into each flask. Titrate the three samples consecutively.

A. K_{SP} AND MOLAR SOLUBILITY OF $CA(OH)_2$

1. Prepare a saturated $Ca(OH)_2$ solution one week[1] before the experiment by adding about 3g of solid $Ca(OH)_2$ to 100mL of boiled, distilled (or de–ionized) water in a 125mL Erlenmeyer flask. Stir the solution and stopper.

---

[1] This solution may have been prepared for you. Ask your instructor.

2. Prepare a 50mL buret for titration. Rinse the clean buret and tip with two 5mL portions of a standard 0.02M HCl solution. Fill, read (±0.01mL), and record the volume of 0.02M HCl in the buret.

3. Allow the undissolved $Ca(OH)_2$ to remain settled on the bottom of the flask. Without disturbing the insoluble $Ca(OH)_2$, *carefully* decant the saturated solution into a second 125mL flask. Rinse a 25mL pipet with 1mL or 2mL of the saturated $Ca(OH)_2$ solution and discard. Pipet 25mL of the decantate into a clean 125mL Erlenmeyer flask and add 2 drops of methyl orange indicator. Record the solution's temperature. Titrate with the standard HCl solution. Record the volume (±0.01mL) needed to just turn the yellow color red.[2] Repeat the titration with two new samples of $Ca(OH)_2$ solution.

B. SOLUBILITY OF $CA(OH)_2$ IN THE PRESENCE OF CA^{2+}

1. Mix 3g of solid $Ca(OH)_2$ and 1g of solid $CaCl_2 \cdot 2H_2O$ with 100mL of boiled, distilled (or de–ionized) water in a 125mL Erlenmeyer flask one week[3] before the experiment, stir, and stopper the flask.

2. To complete the analysis, titrate a 25mL sample, that has been decanted from the original solution, with the standard 0.02M HCl solution, as was described in Parts A.2 and A.3.

---

[2]The color change at the endpoint for methyl orange is subtle. Keep the titrated solutions from the successive trials to compare the colors at the endpoint.

[3]This solution may have also been prepared for you. Ask your instructor.

The formation of stalagmites and stalagtites results from the limited solubility of $CaCO_3$. As the water evaporates, the $CaCO_3$ precipitates...but ever so slowly!

Solubility Product Constant and Common Ion Effect–Lab Preview

Date________Name____Lab Sec. _____Desk No._____

1. How is the molar concentration of OH^- measured in a saturated $Ca(OH)_2$ solution for this experiment.

2. a. How many times should a buret and/or pipet be rinsed before it is used to deliver a solution for an analysis?

   b. What criterion is used to determine whether or not a buret, pipet, or any other form of glassware is clean?

   c. Explain how to add less–than–a–drop of titrant from a buret.

3. a. What indicator is used in today's titration?

   b. What color change of the indicator is observed at the endpoint in the titration?

   Does this color change occur as a result of the indicator changing from an acidic to basic form or basic to acidic form?

4. Write the K_{sp} expression for these slightly soluble salt equilibria:

   a. $CuS(s) \rightleftharpoons Cu^{2+}(aq) + S^{2-}(aq)$ $K_{sp} =$

   b. $BaSO_4(s) \rightleftharpoons Ba^{2+}(aq) + SO_4^{2-}(aq)$ $K_{sp} =$

   c. $Ca_3(PO_4)_2(s) \rightleftharpoons 3Ca^{2+}(aq) + 2PO_4^{3-}(aq)$ $K_{sp} =$

5. a. Calculate the molar solubility of AgI. $K_{sp} = 1.5 \times 10^{-16}$

b. Calculate the molar solubility of AgI in the presence of 0.020M KI.

6. The pH of a saturated $Ni(OH)_2$ solution at equilibrium is 8.92.

a. Calculate the $[OH^-]$.

b. Calculate the $[Ni^{2+}]$.

c. Calculate the K_{sp} of $Ni(OH)_2$.

SOLUBILITY PRODUCT CONSTANT AND COMMON ION EFFECT-DATA SHEET

Date__________ Name____ Lab Sec. _____ Desk No._____

A. K_{SP} AND MOLAR SOLUBILITY OF $CA(OH)_2$

	Trial 1	Trial 2	Trial 3
1. Concentration of standard HCl solution (mol/L)		__________	
2. Buret reading, initial (mL)	________	________	________
3. Buret reading, final (mL)	________	________	________
4. Volume of standard HCl used (mL)	________	________	________
5. Moles of HCl added (mol)	________	________	________
6. Moles of OH^- in sat'd $Ca(OH)_2$ solution (mol)	________	________	________
7. Volume of sat'd $Ca(OH)_2$ solution (mL)	__25.0__	__25.0__	__25.0__
8. Temperature of $Ca(OH)_2$ solution (°C)	________	________	________
9. $[OH^-]$ at equilibrium (mol/L)	________	________	________
10. $[Ca^{2+}]$ at equilibrium (mol/L)	________	________	________
11. K_{sp} of $Ca(OH)_2$	________	________	________
12. Average K_{sp} at ___°C		__________	
13. Molar solubility of $Ca(OH)_2$ at ___°C		__________	

B. Solubility of $Ca(OH)_2$ in the Presence of Ca^{2+}

	Trial 1	Trial 2	Trial 3
1. Concentration of standard HCl solution (mol/L)		__________	
2. Buret reading, initial (mL)	________	________	________
3. Buret reading, final (mL)	________	________	________
4. Volume of standard HCl used (mL)	________	________	________
5. Moles of HCl added (mol)	________	________	________
6. Moles of OH^- in $Ca(OH)_2/Ca^{2+}$ solution(mol)	_______	________	________
7. Volume of $Ca(OH)_2/Ca^{2+}$ solution titrated (mL)	__25.0__	_25.0__	__25.0__
8. Temperature of $Ca(OH)_2/Ca^{2+}$ solution (°C)	________	________	________
9. $[OH^-]$ at equilibrium (mol/L)	________	________	________
10. $[Ca^{2+}]$ at equilibrium (mol/L)	________	________	________

11. Molar solubility of $Ca(OH)_2$ in $Ca(OH)_2/Ca^{2+}$ solution at ___°C ______________

Questions

1. How does the addition of $CaCl_2$ affect the molar solubility of $Ca(OH)_2$?

2. a. In Part A how will the transfer of some solid $Ca(OH)_2$ into the Erlenmeyer titrating flask affect the amount of standard HCl used to reach an endpoint?

b. How will this affect the reported K_{sp} value of $Ca(OH)_2$? Explain.

c. As a result of this transfer error, will the molar solubility of $Ca(OH)_2$ be higher or lower than the accepted value for $Ca(OH)_2$? Explain.

3. If the endpoint of the titration is surpassed in Part A, will the K_{sp} value for $Ca(OH)_2$ be higher or lower than the accepted value? Explain.

4. How will the calculated K_{sp} value for $Ca(OH)_2$ be affected if the original $Ca(OH)_2$ solution is not saturated?

5. Why was the temperature recorded in this experiment?

6. Does adding boiled, distilled water to the Erlenmeyer titrating flask, in order to wash the sides of the flask and the buret tip, affect the K_{sp} of $Ca(OH)_2$? Explain.

7. How will tap water instead of boiled, distilled water affect the K_{sp} of $Ca(OH)_2$ in Part A.

EXPERIMENT 27
PH, HYDROLYSIS, AND BUFFERS

Objectives

- To measure the pH of some acids and bases
- To measure the degree of hydrolysis of various ions
- To observe the effectiveness of a buffer system

Principles

PH

The acidity or basicity of most aqueous solutions is frequently due to small concentrations of H^+ or OH^-. Small concentrations of H^+ are often, and more conveniently, expressed as **pH**, the negative logarithm of the molar concentration of H^+ in an aqueous solution.

$$pH = -\log[H^+]$$

Water dissociates *very* slightly producing equal concentrations of H^+ and OH^-.

$$H_2O \rightleftharpoons H^+(aq) + OH^-(aq)$$

At 25°C, the $[H^+] = [OH^-] = 1.0 \times 10^{-7}$ mol/L, producing a pH = 7.00. Addition of an acid to water increases the relative $[H^+]$ (and reduces the relative $[OH^-]$) causing a pH *less than* 7. A base, on the other hand, increases the $[OH^-]$ in solution (and decreases the $[H^+]$) producing a pH *greater than* 7.

In this experiment we will measure the pH of some acids and bases using several acid–base indicators. An **indicator** is a weak organic acid, HIn; the undissociated acid has a color different than its conjugate base, In^-.

$$HIn(aq) \rightleftharpoons H^+(aq) + In^-(aq)$$

Therefore when a solution is acidic, the indicator will have the color of the HIn form of the organic acid, but will have the color of the In^- in a basic solution. Since indicators are of varying strengths, the dominant form of the indicator (HIn or In^-) is pH dependent. The pH dependency on the relative concentrations of HIn and In^- along with the corresponding colors for HIn and In^- are listed in Table 27.1.

Table 27.1
Acid-Base Indicators: Colors and pH Change

Indicator	Acid (HIn) Color	pH Range	Base (In^-) Color
thymol blue	red	1.2–2.8	yellow
methyl orange	red	3.2–4.4	orange/yellow
methyl red	red	4.8–6.0	yellow
litmus	red	4.7–8.3	blue
bromocresol purple	yellow	5.2–6.8	purple
bromothymol blue	yellow	6.0–7.6	blue
m-nitrophenol	colorless	6.8–8.6	yellow
thymol blue	yellow	8.0–9.6	blue
phenolphthalein	colorless	8.2–10.0	pink
alizarin yellow R	yellow	10.1–12.0	red

HYDROLYSIS

All salts dissolve to some extent in an aqueous solution to form their respective cations and anions. These ions are **hydrated** (surrounded and attracted to the solvent H_2O molecules) to varying degrees. For example, the cations Ba^{2+} and Na^+ are weakly hydrated while the Fe^{3+} and Al^{3+} are strongly hydrated in solution.

A strongly hydrated cation attracts the negative end of the polar H_2O molecule; this weakens the O-H bond and liberates free H^+ into the solution. This results in an acidic solution with a pH less than 7. The hydrolysis of Fe^{3+} produces an acidic solution, H^+, and $FeOH^{2+}$.

$$Fe^{3+} \text{--} {}^{\delta-}O(H^{\delta+})_2 \rightarrow [Fe \text{--} OH_2]^{3+} \rightarrow FeOH^{2+} + H^+$$

A strongly hydrated anion, on the other hand, attracts the positive end of the polar H_2O molecule; this again weakens the O–H bond and frees OH^- into the solution. This produces a basic solution with a pH greater than 7. The hydrolysis of PO_4^{3-} produces a basic solution, OH^-, and HPO_4^{2-}.

$$PO_4^{3-} \text{----} {}^{\delta+}H\text{–}O\text{–}H^{\delta+} \rightarrow [O_3P\text{–}O\text{–}H \cdots OH]^{3-} \rightarrow [O_3P\text{–}OH]^{2-} + OH^-$$

The result of this splitting of water molecules to produce an acidic/basic solution is called **hydrolysis.**

Cations and anions having a very weak attraction for the polar water molecules do not affect the pH of the solution. These ions, called **spectator ions**, include the following.

cations: Group IA (Na^+, K^+, Rb^+, Cs^+), Group IIA (Mg^{2+}, Ca^{2+}, Sr^{2+}, Ba^{2+}) and all other metal cations in the 1^+ oxidation state.

anions: Cl^-, Br^-, I^-, NO_3^-, ClO_4^-, and ClO_3^-

This experiment measures the pH of a number of salts in solution to determine the extent of ion hydrolysis. Even the nose is used as an indicator for the ammonium salts. The ion of the salt most strongly hydrated is identified and an equation is written to account for the observation.

BUFFERS

In many areas of research, chemists, biologists, and environmentalists need an aqueous solution that resists a large pH change when H^+ (from a strong acid) or OH^- (from a strong base) is added. An aqueous solution that adjusts to the H^+ or OH^- addition without a large resultant change in pH is called a **buffer solution.** (Figure 27.1)

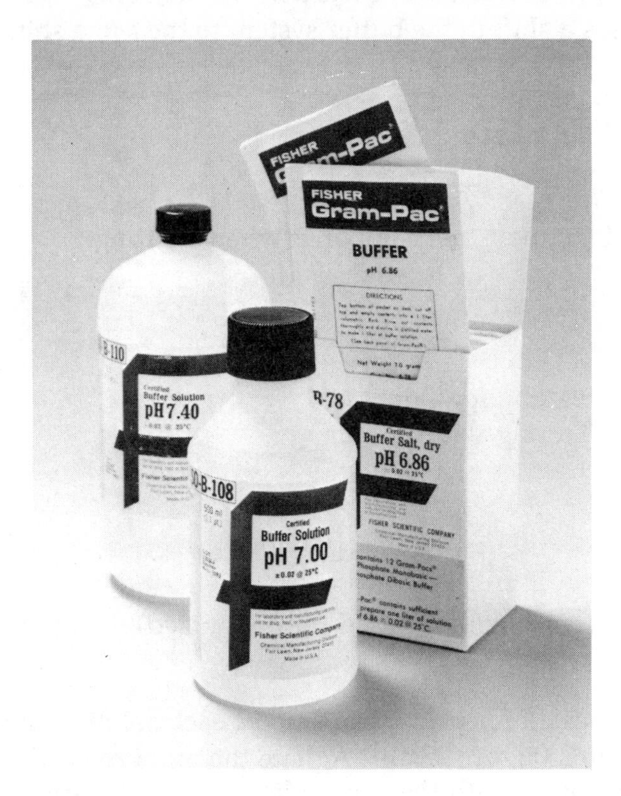

Figure 27.1
Commercial Buffer Solutions with a Desired pH

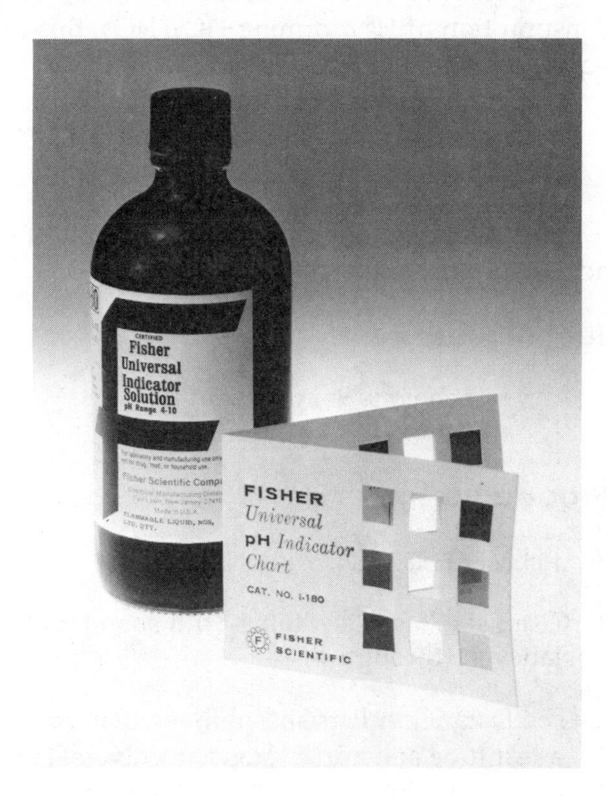

Figure 27.2
A Universal Indicator Solution and Color Chart for Determining pH

A buffer solution must have two components: one, is a substance that will react with H^+ (from a strong acid)–this must be a base, a proton acceptor; the other substance must be capable of reacting with OH^- (from a strong base)–this must be an acid, a proton donor. In either case, the effect of the addition is neutralized. The two components of a typical buffer system are a *weak* acid and it conjugate base (or a weak base along with its conjugate acid). A typical buffer system is the $HC_2H_3O_2/C_2H_3O_2^-$ system.

$$HC_2H_3O_2(aq) \rightleftharpoons H^+(aq) + C_2H_3O_2^-(aq) \quad K_a = 1.8 \times 10^{-5}$$

The addition of OH^- to the system results in its reaction with H^+, forming H_2O and more $C_2H_3O_2^-$ (a weak base).

$$HC_2H_3O_2(aq) \rightleftharpoons H^+(aq) + C_2H_3O_2^-(aq)$$
$$H^+(aq) + OH^-(aq) \rightarrow H_2O$$

The system shifts *right*, the extent of the shift equals the moles of OH^- added.

The addition of H^+ to the $HC_2H_3O_2/C_2H_3O_2^-$ system results in a reaction with $C_2H_3O_2^-$ (and a consumption of H^+) forming $HC_2H_3O_2$; this causes a shift in the buffer system to the *left*, a shift equal to the moles of H^+ added.

$$HC_2H_3O_2(aq) \rightleftharpoons H^+(aq) + C_2H_3O_2^-(aq)$$
$$\uparrow\!\_\ H^+(aq)$$

The change in pH of buffered and unbuffered solutions are compared when a strong acid/base are added in this experiment.

Techniques

- Technique 9a, page 10 — Testing for Odor
- Technique 14, page 18 — Testing with Litmus

Procedure

A. PH MEASUREMENT WITH INDICATORS

1. Clean six 75mm test tubes with soap and water, rinse with distilled (or de-ionized) water and label–#1 through #6.

2. The Data Sheet lists six solutions that are to be tested for pH. Place 2mL of each solution into a test tube and add 5 drops of universal indicator (Figure 27.2).[1] Agitate the mixture until the color is uniform. Compare the color of the solution with the "pH Indicator Chart"; estimate the pH of the solution. Write an equation that shows the presence of free H^+ or OH^- in solution.

---

[1] If universal indicator is unavailable, first test each solution with litmus paper. If the red litmus turns blue (indicating that the solutions is basic), further identify the pH of the solution by testing additional 2mL samples with several indicators, listed in Table 27.1, that change color in the basic range. If the solution tests acidic (blue litmus turns red), similarly test to further identify the exact pH of the solution.

3. Use the same procedure to determine the pH of distilled water, tap water, vinegar, lemon juice, household ammonia, a detergent solution, 409™, and 7–UP™ (or equivalent substitutes).

B. Hydrolysis of Salts

1. Refer to the Data Sheet; transfer 2mL of each solution to a clean 75mm test tube. Add 5 drops of universal indicator and estimate the pH of the solution. Test with litmus paper and other indicators if universal indicator is unavailable. Identify the ion that hydrolyzes and write an equation representing the acidity/basicity of the solution.

2. Ammonium salts readily hydrolyze with the water vapor in air to release NH_3 gas. Write an equation for this hydrolysis. The extent of hydrolysis may be detected merely by detecting the odor of the solid salt. Cautiously smell the solid NH_4Cl, $NH_4C_2H_3O_2$, and $(NH_4)_2CO_3$ salts. The anion having the strongest affinity (the strongest anionic base) for NH_4^+ causes the greatest release of NH_3. List the anionic bases (Cl^-, $C_2H_3O_2^-$, CO_3^{2-}) in order of decreasing strength.

3. Acid salts also hydrolyze in solution. Transfer 2mL of 0.1M $NaHCO_3$, 0.1M $NaHSO_4$, and 0.1M NaH_2PO_4 to 75mm test tubes. Determine the approximate pH of the solutions.

4. Baking powders consist of a combination of baking soda, $NaHCO_3$, and a substance which donates H^+. When the mixture is added to water, CO_2 is produced and the "dough rises".

$$HCO_3^-(aq) + H^+(aq) \rightarrow H_2O + CO_2(g)$$

Some common acids in baking powders are cream of tartar, $KHC_4H_4O_6$, phosphate, $Ca(H_2PO_4)_2$, and alum, $NaAl(SO_4)_2 \bullet 12H_2O$.

Add about 0.1g of each solid acid to separate 75mm test tubes and dissolve with 2mL of water. Test each acid to determine which is most acidic, i.e., the lowest pH.

5. In a 150mm test tube, mix 5mL of the 0.1M $NaHCO_3$ from Part B.3 with 5mL of one of the acid solutions in Part B.4. Observe and record.

C. Buffer Solutions

1. Mix 5mL of 0.10M $HC_2H_3O_2$ with 5mL of 0.10M $NaC_2H_3O_2$ in a 200mm test tube. In a second 200mm test tube, add 10mL of distilled water. Add 5 drops of universal indicator to each sample and estimate the pH.[2]

2. Add 5mL of 0.010M HCl to each test tube, estimate and record the resultant pH, and determine the pH change, ΔpH.[3]

---

[2] If universal indicator is unavailable, add 2 drops of methyl orange indicator.

[3] If methyl orange indicator is used, add, count, and record the drops of 0.010M HCl needed to reach the methyl orange endpoint (when the color change occurs).

3. Again prepare the $HC_2H_3O_2/C_2H_3O_2^-$ test tube and the H_2O test tube as in Part C.1. Add 5 drops of universal indicator.[4] Add 5mL of 0.010M NaOH to each test tube and estimate the pH of the resulting solutions. What is the pH change in each test tube?

[4] If universal indicator is unavailable, first add 2 drops of alizarin yellow R indicator and then add, count, and record the drops of 0.010M NaOH need to reach the endpoint.

PH, HYDROLYSIS, AND BUFFERS-LAB PREVIEW

Date________Name____Lab Sec. _____Desk No._____

1. a. What is an acid-base indicator?

   b. How does it function?

2. a. What color is litmus in an acidic solution?__________

   b. What color is litmus in a basic solution?__________

3. In the equilibrium for hydrocyanic acid, $HCN(aq) + H_2O \rightleftharpoons H_3O^+(aq) + CN^-(aq)$,
   a. state the effect of added H^+.

   b. state the effect of added OH^-.

4. Define and characterize a buffer solution.

5. Predict the ion that hydrolyzes more strongly.

   a. Fe^{3+} or Fe^{2+}. ________ Explain.

   b. Li^{+} or Na^{+}.________ Explain.

6. Write an equation representing the hydrolysis of each base.

   N_2H_4 --

   $C_2H_3O_2^-$ --

7. Write and equation representing the hydrolysis of each acid.

   $(CH_3)_2NH_2^+$____

   Al^{3+}____

8. In preparing an effective buffer solution at a desired pH, it is advisable to select a weak acid–conjugate base pair in which the pK_a of the acid equals the desired pH±1. Over what pH range is the $HC_2H_3O_2/C_2H_3O_2^-$ buffer most effective?

9. Briefly describe the technique for testing the pH of a solution using litmus paper.

PH, HYDROLYSIS, AND BUFFERS-DATA SHEET

Date________ Name____ Lab Sec. _____ Desk No._____

A. PH MEASUREMENT WITH INDICATORS

Solution	Litmus Test (acidic or basic)	Approx. pH	Equation
0.10M HCl	__________	_____	______________________________
0.00010M HCl	__________	_____	______________________________
0.10M $HC_2H_3O_2$	__________	_____	______________________________
0.10M NH_3	__________	_____	______________________________
0.00010M NaOH	__________	_____	______________________________
0.10M NaOH	__________	_____	______________________________
distilled H_2O	__________	_____	
tap water	__________	_____	
vinegar	__________	_____	
lemon juice	__________	_____	
household NH_3	__________	_____	
detergent	__________	_____	
409™	__________	_____	
7-UP™	__________	_____	
other	__________	_____	

B. Hydrolysis of a Salt

1.

Solution	Litmus test (acidic or basic)	Approx pH	Ion hydrolyzed	Balanced Equation
0.1M $NaCl$	----------	-----	--------	------------------
0.1M KNO_3	----------	-----	--------	------------------
0.1M $NaC_2H_3O_2$	----------	-----	--------	------------------
0.1M $NaNO_2$	----------	-----	--------	------------------
0.1M Na_2CO_3	----------	-----	--------	------------------
0.1M Na_3PO_4	----------	-----	--------	------------------
0.1M $FeCl_3$	----------	-----	--------	------------------
0.1M $ZnCl_2$	----------	-----	--------	------------------
0.1M NH_4Cl	----------	-----	--------	------------------
0.1M Na_2SO_4	----------	-----	--------	------------------
0.1M $Al_2(SO_4)_3$	----------	-----	--------	------------------

2. Equation for the hydrolysis of NH_4^+ ______________________________

List the anionic bases in order of decreasing strength according to the decreasing strength of NH_3 smell.

__________ > __________ > __________

Write a balanced equation for the hydrolysis of each anion (if hydrolysis occurs).

Cl^- ____

$C_2H_3O_2^-$ ____

CO_3^{2-} ____

3.

Acid salt solution	Litmus test (acidic or basic)	Approx pH	Ion hydrolyzed	Balanced Equation
0.1M $NaHCO_3$	----------	-----	--------	------------------
0.1M $NaHSO_4$	----------	-----	--------	------------------
0.1M NaH_2PO_4	----------	-----	--------	------------------

4. Dry acid	Litmus test (acidic or basic)	Approx pH	Ion hydrolyzed	Balanced Equation
$KHC_4H_4O_6$	__________	_____	________	__________________
$Ca(H_2PO_4)_2$	__________	_____	________	__________________
$NaAl(SO_4)_2 \bullet 12H_2O$	________	_____	________	__________________

5. Observation of reaction of 0.1M $NaHCO_3$ with an acid from Part B.4

Balanced equation for the reaction ______________________________

C. Buffer Solutions

Observation	$HC_2H_3O_2/C_2H_3O_2$ buffer	Water
initial pH	______________	______
pH after 0.010M HCl	______________	______
ΔpH	______________	______
(drops of 0.010M HCl to methyl orange endpoint)	______________	______
pH after 0.010M NaOH	______________	______
ΔpH	______________	______
(drops 0.010M NaOH to alizarin yellow R endpoint)	______________	______

Comment on the effectiveness of this buffer solution in resisting large changes in pH.

Questions

1. Categorize those ions in Part B.1 that produce an acidic solution and those that produce a basic solution.

   acidic:____

   basic:____

2. a. Which ions in Part B.1 do *not* affect the pH of the solution?________________________

b. Which anion hydrolyzes most extensively?____________

c. Which cation hydrolyzes most extensively?____________

3. At what point does the buffer solution stop resisting a pH change when strong acid is added to the buffer?

4. Predict whether an aqueous solution of each salt produces an solution with a pH > 7, < 7, or = 7.

Salt	pH > 7, < 7, or = 7	Equation justifying your prediction
Na_2SO_3	------	-----------------------------------
KNO_3	------	-----------------------------------
$(CH_3)_2NH_2NO_3$	------	-----------------------------------
$Ca(C_3H_7O)_2$	------	-----------------------------------

5. A student needs 1.00L of buffer solution with a pH = 7.00. She selects the $H_2PO_4^-/HPO_4^{2-}$ buffer system. The $K_a(H_2PO_4^-) = 6.2 \times 10^{-8}$. What must be the $[H_2PO_4^-]/[HPO_4^{2-}]$ ratio for the buffer system?

Experiment 28
Standardized NaOH Solution

Objective

- To determine the molarity of a sodium hydroxide solution.

Principles

The concentration of a solute in solution can be determined quantitatively by adding a second solution of known concentration (a standard solution) until the reaction is complete. The reaction is complete when the mole ratio of the two reacting substances is the same as that which appears in the balanced equation. This is the **stoichiometric point**[1] in a titration. In this experiment the stoichiometric point for an acid-base reaction is detected using a phenolphthalein indicator, colorless in an acidic solution but pink (or red) in a basic solution. The point at which the phenolphthalein indicator changes color is the **endpoint** of the titration. Indicators are selected so that the stoichiometric point and the endpoint occur at essentially the same point in the titration.

In this experiment the molarity of a NaOH solution is determined using potassium hydrogen phthalate as the **primary standard**.[2] Potassium hydrogen phthalate, $KHC_8H_4O_4$, is a white, crystalline, nonhygroscopic, acidic substance with a high degree of purity. To determine the molarity of the NaOH solution, a dried sample of $KHC_8H_4O_4$ is weighed, dissolved in distilled (or de–ionized) water, and titrated with a recorded volume of the NaOH solution until the stoichiometric point is reached—when 1 mol NaOH has been added for each 1 mol $KHC_8H_4O_4$ weighed for the analysis.

$$K^+HC_8H_4O_4^-(aq) + Na^+OH^-(aq) \rightarrow H_2O + K^+Na^+HC_8H_4O_4^{2-}(aq)$$

This is the point at which the phenolphthalein indicator changes from colorless to pink.

Since the mass and formula weight of $KHC_8H_4O_4$ are known, the moles of $KHC_8H_4O_4$ titrated in the analysis are calculated.

$$\text{g } KHC_8H_4O_4 \times \frac{1 \text{ mol } KHC_8H_4O_4}{204.2 \text{ g } KHC_8H_4O_4} = \text{mol } KHC_8H_4O_4$$

At the stoichiometric point equal moles of $KHC_8H_4O_4$ and NaOH have been combined.

$$\text{mol NaOH} = \text{mol } KHC_8H_4O_4$$

---

[1] The stoichiometric point is also called the **equivalence point**, indicating the point at which stoichiometrically equivalent quantities of the reacting substances are combined.

[2] A **primary standard** is a substance that can be obtained in pure form, must be stable in its pure form and in solution, must be dryable and nonhygroscopic, and have a relatively high formula weight.

The molarity of the NaOH solution equals the moles of NaOH per liter of solution. Division of the moles of NaOH by the volume of NaOH solution added to the $KHC_8H_4O_4$ in the titration is the molarity of the NaOH.

$$\text{molarity of NaOH, M} = \frac{\text{mol NaOH}}{\text{liter NaOH solution}}$$

Techniques

- Technique 10, page 12 — Reading a Meniscus
- Technique 11, page 13 — Pipetting a Liquid or Solution
- Technique 12, page 14 & 15 — Titrating a Solution
- Technique 13, page 16 — Using the Laboratory Balance

Procedure

You are to complete at least three trials in the standardization of your NaOH solution. To hasten the analysis, clean, dry, and label three 125mL or 250mL Erlenmeyer flasks and weigh three $KHC_8H_4O_4$ samples while occupying the balance. If all the balances are occupied, proceed to prepare your NaOH solution (Part 2) and buret (Part 4).

In this experiment you are striving for "good" results. The molarity determination for the three trials should be within ±1%. If not, a fourth or fifth trial may be necessary. Part 1 of the experiment may have already been completed for you. Ask your instructor.

1. One week before the scheduled laboratory period, dissolve 10g to 12g of NaOH (pellets or flakes) in 50mL of distilled water in a 125mL rubber–stoppered Erlenmeyer flask. Thoroughly mix and allow the solution to stand for the precipitation of any Na_2CO_3.[3] Dry 4g to 5g of $KHC_8H_4O_4$ at 110°C for several hours in a constant temperature drying oven. Cool the sample in a desiccator (if available).

2. With a graduated cylinder, transfer 10mL[4] of the NaOH solution (**Caution**: *a concentrated NaOH solution causes severe skin burns.*) prepared above into a 500mL polyethylene bottle and dilute to 500mL with previously, boiled, distilled water. The boiled water removes traces of CO_2. Cap the polyethylene bottle, preventing the absorption of CO_2 by the NaOH solution; stir the solution for several minutes and label the bottle. Do not shake the bottle (this increases the probability of CO_2 absorption).

3. Weigh (±0.001g) 0.3g to 0.5g of $KHC_8H_4O_4$ in a clean, dry, 125mL or 250mL Erlenmeyer flask. Add 50mL of distilled water and 2 drops of phenolphthalein.

4. Clean a buret for titration. Rinse the buret with two 5mL portions of your NaOH solution, making certain that the NaOH wets its entire inner surface. Have your instructor approve your buret *before* proceeding to Part 5.

---

[3] $2NaOH(aq) + CO_2(aq) \rightarrow Na_2CO_3(s) + H_2O$. Na_2CO_3 has a low solubility in a concentrated NaOH solution.

[4] If 1L of NaOH solution is to be prepared for this experiment, use 20mL of the concentrated NaOH solution and dilute to 1L.

5. Fill the buret with the NaOH solution and carefully read the meniscus before recording its volume (±0.01mL). Place white paper beneath the Erlenmeyer flask. Slowly add the NaOH solution to the $KHC_8H_4O_4$ solution, swirling the flask after each addition. Consult the instructor (or Technique 12) for proper techniques of titrating with the left hand (if right–handed), rinsing the wall of the flask with water from the wash bottle, and adding half–drops from the buret.

6. As the rate of the phenolphthalein color change (pink, where the NaOH has been added → colorless, after swirling) decreases, decrease the rate of NaOH addition; proceed with drop addition of the NaOH until the phenolphthalein endpoint is reached. This occurs when a single drop finally causes the pink color of the phenolphthalein indicator to persist for 30 seconds. Read the meniscus in the buret (±0.01mL) and record.

7. Repeat the titration at least two more times with varying but accurately known amounts of $KHC_8H_4O_4$ until ±1% reproducibility. **Save** your standardized NaOH solution in the 500mL polyethylene bottle for later experiments (for example, Experiments 29 and 30).

This is a modern titration apparatus. The titrant is automatically added until a desired (programmed) endpoint is reached. The data can be analyzed by pre-programming the computer and can then be printed. The progess of the titration (i.e., a titration curve) can also be printed out.

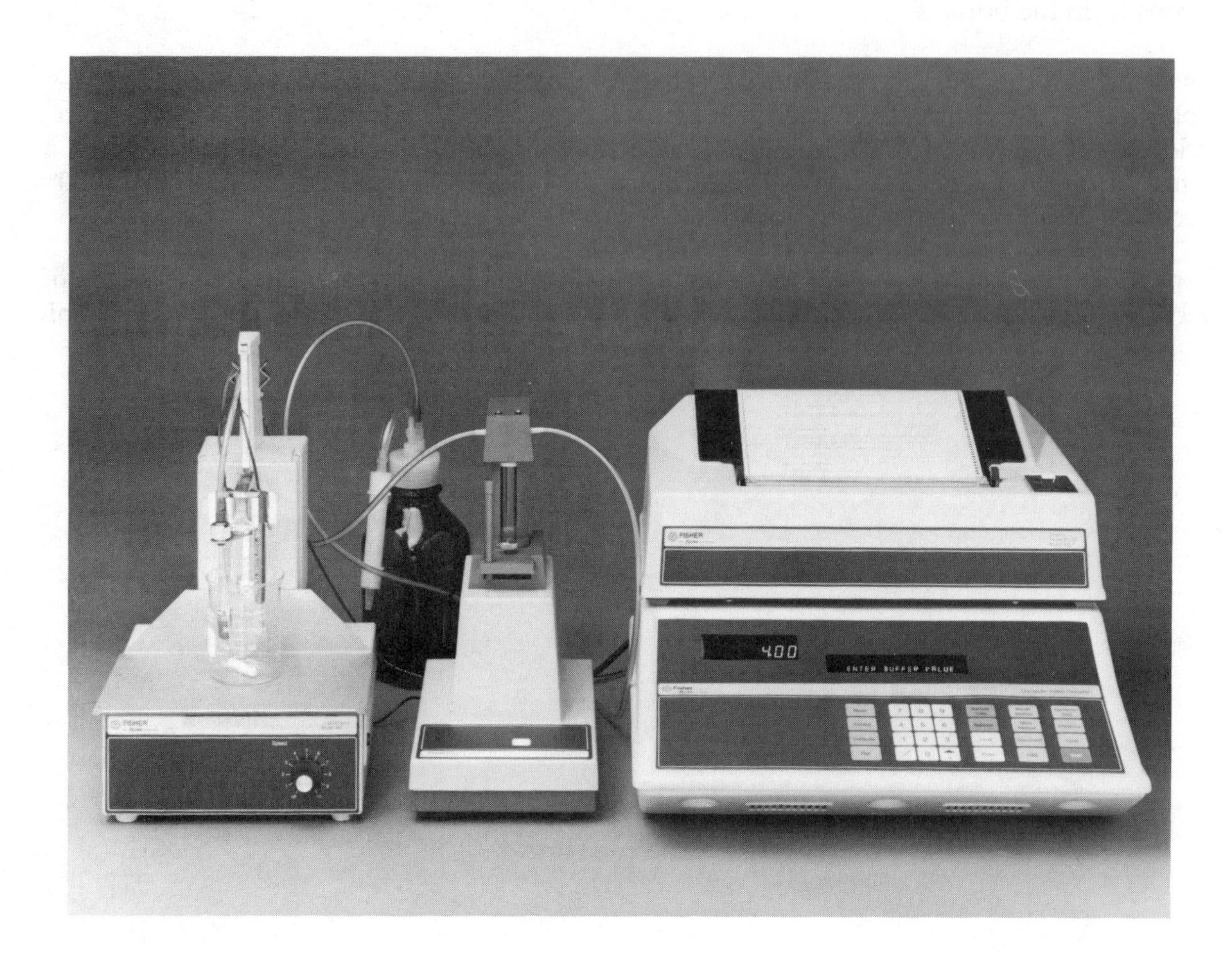

STANDARDIZED NAOH SOLUTION-LAB PREVIEW

Date__________Name____Lab Sec.______Desk No._____

1. Characterize a primary standard.

2. Explain how CO_2 absorbed from the atmosphere affects the molarity of a standardized NaOH solution.

3. What is the purpose for placing white paper beneath the Erlenmeyer flask during today's titration? See Procedure, Part 5.

4. Distinguish between a stoichiometric point and an endpoint.

5. A 0.418g sample of sulfamic acid, NH_2SO_3H, dissolved in 50.00mL of water is neutralized by 27.49mL of NaOH at the phenolphthalein endpoint. What is the molarity of the NaOH solution? The formula weight of NH_2SO_3H is 97.1 g/mol.

$$NH_2SO_3H(aq) + NaOH(aq) \rightarrow NH_2SO_3H^-Na^+(aq) + H_2O$$

6. A 0.377g sample of potassium hydrogen phthalate, $KHC_8H_4O_4$, is dissolved in 100mL of water. If 7.39mL of a NaOH solution are required to reach the stoichiometric point, what is the molarity of the NaOH solution? The formula weight of $KHC_8H_4O_4$ is 204.2 g/mol.

7. a. Where should the meniscus be read when determining the volume of a solution in a buret (or pipet)?

   b. When rinsing a buret after cleaning it with soap and water, should the rinse be dispensed through the buret tip or at the top of the buret? Explain.

   c. What is the criterion for a clean buret?

   d. In preparing a buret for titration, the final rinse (or two) should be with the solution that will subsequently be used in the titration. Why is the solution used instead of distilled water?

   e. How is a half–drop added from a buret into the solution?

   f. How long should the color change from an indicator persist to ensure that its endpoint has been reached?

STANDARDIZED NAOH SOLUTION-DATA SHEET

Date_________Name____Lab Sec. _____Desk No.______

Maintain at least 3 significant figures when recording data and performing calculations.

	Trial 1	Trial 2	Trial 3
1. Mass of flask + $KHC_8H_4O_4$ (g)	__________	__________	__________
2. Mass of flask (g)	__________	__________	__________
3. Mass of $KHC_8H_4O_4$ (g)	__________	__________	__________
4. Moles of $KHC_8H_4O_4$ (mol)	__________	__________	__________
Buret approval by instructor		__________	
5. Buret reading of NaOH, **final** (mL)	__________	__________	__________
6. Buret reading of NaOH, **initial** (mL)	__________	__________	__________
7. Volume of NaOH used (mL)	__________	__________	__________
8. Moles of NaOH neutralized (mol)	__________	__________	__________
9. Molarity of NaOH (mol/L)*	__________	__________	__________
10. Average molarity of NaOH (mol/L)		__________	

*Show calculation for Trial 1

Questions

1. Is it quantitatively acceptable to titrate all $KHC_8H_4O_4$ samples with the NaOH solution to the same *dark red* endpoint, just as long as the red intensity is consistent from one sample to the next?

2. A student titrates the $KHC_8H_4O_4$ to a faint pink endpoint which persists for 30 seconds; the buret reading is recorded and the calculations are completed. When he again looks at the receiving flask, the solution is no longer pink. Explain a probable cause of this change. Assume that the $KHC_8H_4O_4$ was pure, completely dissolved, and the stoichiometric point in the titration had been reached.

3. If the endpoint in the titration of the $KHC_8H_4O_4$ with the NaOH solution is mistakenly surpassed (too pink), what effect does this have on the calculated molarity of the NaOH solution?

4. Oxalic acid, $H_2C_2O_4$, can also be used as a primary standard in this experiment. How many grams of oxalic acid should be weighed to neutralize 25.00mL of your NaOH solution? Oxalic acid is diprotic (Hint: what then in the balanced equation?).

5. If a drop of the NaOH solution adheres to the side of the Erlenmeyer flask during the standardization of the NaOH solution, how does this error affect its reported molarity? Explain.

6. If a drop of the NaOH solution adheres to the side of the buret during the standardization of the NaOH solution, how does this error affect its reported molarity? Explain.

EXPERIMENT 29
ANALYSIS OF ACIDS

Objectives

- To determine the formula weight of an acid
- To determine the percent acid in a commercial product
- To measure the percent acetic acid in vinegar

Principles

FORMULA WEIGHT OF AN ACID

Part A of this experiment determines an acid's formula weight. The standardized NaOH solution prepared in Experiment 28 is used to titrate the acid to the stoichiometric point. The moles of NaOH used in the titration is determined from the volume of NaOH used in the titration and the molarity of the NaOH solution.

$$\textbf{volume NaOH (L) x molarity NaOH (mol/L) = mol NaOH}$$

The moles of unknown acid that the NaOH neutralizes is found from the balanced equation. The unknown acid is either monoprotic, HA, diprotic, H_2A, or triprotic, H_3A.

$$HA(aq) + NaOH(aq) \rightarrow NaA(aq) + H_2O$$
$$H_2A(aq) + 2\,NaOH(aq) \rightarrow Na_2A(aq) + 2\,H_2O$$
$$H_3A(aq) + 3\,NaOH(aq) \rightarrow Na_3A(aq) + 3\,H_2O$$

By knowing the mass of the acid used for the titration, the formula weight of the acid can be determined.

$$\text{formula weight (acid)} = \frac{\text{g acid}}{\text{mol acid}}$$

PERCENT ACID IN VANISH™

Part B determines the percent of acid in a common household acid. Sodium bisulfate, $NaHSO_4$, is a safe, easy–to–use acid that is used to remove rust, calcium deposits, and to acidify home swimming pools. The $NaHSO_4$ solid dissolves in water to produce the weak acid, HSO_4^-. A measure of the moles of NaOH (OH^-) that neutralizes the HSO_4^- in the sample allows us to determine its percent concentration.

$$HSO_4^-(aq) + OH^-(aq) \rightarrow H_2O + SO_4^{2-}(aq)$$
$$\text{mol } OH^- = \text{mol } HSO_4^- = \text{mol } NaHSO_4$$

$$\text{g } NaHSO_4 = \text{mol } NaHSO_4 \times \frac{120.1\text{g } NaHSO_4}{\text{mol } NaHSO_4}$$

$$\%\ NaHSO_4 = \frac{\text{g } NaHSO_4}{\text{g sample}} \times 100$$

PERCENT ACETIC ACID IN VINEGAR

In Part C, the acetic acid, $HC_2H_3O_2$, concentration in various vinegars is determined. Vinegar is a solution that is 4% to 5% (by weight) acetic acid in water (4% is the minimum federal standard). Generally, caramel flavoring and coloring are also added to make the product sell better.

The percent by weight of $HC_2H_3O_2$ in vinegar is determined by titrating a measured weight of vinegar to a phenolphthalein endpoint with a measured volume of standardized base, usually NaOH. The moles of $HC_2H_3O_2$ is calculated from the balanced equation.

$$HC_2H_3O_2(aq) + OH^- \rightarrow H_2O + C_2H_3O_2^-(aq)$$

At the stoichiometric point the moles of OH^- equals the moles of $HC_2H_3O_2$.

$$\text{mol } OH^- = \text{mol } HC_2H_3O_2$$

$$\text{g } HC_2H_3O_2 = \text{mol } HC_2H_3O_2 \times \frac{60.05\text{g } HC_2H_3O_2}{\text{mol } HC_2H_3O_2}$$

$$\% \ HC_2H_3O_2 = \frac{\text{g } HC_2H_3O_2}{\text{g vinegar}} \times 100$$

Techniques

- Technique 10, page 12 — Reading a Meniscus
- Technique 11, page 13 — Pipetting a Liquid or Solution
- Technique 12, page 14 & 15 — Titrating a Solution
- Technique 13, page 16 — Using the Laboratory Balance

Procedure

For completing Parts A, B, and C of this experiment you will need approximately 350mL of the standardized NaOH solution prepared in Experiment 28. You may need to standardize additional NaOH, using potassium hydrogen phthalate as the primary standard. Ask your instructor which parts of the experiment you are to complete.

A. Formula Weight of an Acid

1. Three trials are to be completed in this part of the experiment; successive results should be within ±1%. To hasten the analysis, clean, dry, and weigh (±0.001g) three 125mL or 250mL Erlenmeyer flasks; make all weighings while occupying the same balance. If all balances are occupied, proceed to Part 4 to prepare the buret.

2. Ask the instructor whether the unknown acid is monoprotic, diprotic, or triprotic.

3. In your preweighed Erlenmeyer flask, weigh (±0.001g) 0.3g to 0.4g of your solid acid unknown. Add 50mL of distilled (or de-ionized) water and 2 drops of phenolphthalein.[1]

4. Prepare a buret for titration. Rinse the buret twice with the standardized NaOH solution and drain it through the buret tip. Have your instructor approve the buret before you fill it.

5. Fill the buret with your NaOH solution, remove all air bubbles, and record (±0.01mL) the initial volume. Titrate the sample to the phenolphthalein endpoint. See Experiment 28, Parts 5 and 6 of the Procedure. Read the meniscus after the endpoint has been reached and again record.

6. Reproducibility of successive trials should be <±1%.

B. Percent Acid in Vanish™[2]

1. Read Part A.1 for increasing the efficiency of your time and technique.

2. In a preweighed Erlenmeyer flask, weigh (±0.001g) 0.4g to 0.5g of sample and dissolve in 100mL of distilled (or de-ionized) H_2O. Add 2 drops of phenolphthalein.

3. Prepare the buret as in Part A.4. Obtain your instructor's approval.

4. Titrate the sample to the phenolphthalein endpoint. Follow the procedure in Part A.5.

C. Percent Acetic Acid in Vinegar

1. Read Part A.1 for increasing the efficiency of your time and technique.

---

[1] The acid may be relatively insoluble, but with the addition of the NaOH solution, it gradually dissolves. The addition of 10mL of ethanol may be necessary to hasten the acid's dissolution, especially if the phenolphthalein endpoint is reached before it all dissolves.

[2] Other $NaHSO_4$-containing commercial products may be substituted for the Vanish™.

2. Select a brand of vinegar from the shelf; add 5mL to a previously–weighed Erlenmeyer flask and reweigh (±0.01g). Be sure to use the same balance. Add 2 drops of phenolphthalein to the vinegar and wash down the wall of the flask with 20mL of distilled water.
3. Prepare the buret as in Part A.4. Obtain your instructor's approval.
4. Titrate the sample to the phenolphthalein endpoint. Follow the procedure in Part A.5.
5. Select another brand of vinegar and perform two analyses to determine the percent acetic acid in the vinegar.
6. Compare the percent acetic acid in the two vinegars and then determine which vinegar has the most acetic acid per unit cost.

ANALYSIS OF ACIDS-LAB PREVIEW

Date__________Name____Lab Sec. _____Desk No._____

1. Determine the volume (mL) of 0.0520 M NaOH that neutralizes 0.188g of a diprotic acid having a formula weight of 90.0g/mol.

2. a. How many grams of adipic acid, $C_4H_8(COOH)_2$, will neutralize 28.2mL of 0.188 M NaOH at the phenolphthalein endpoint. Adipic acid is diprotic.

   b. If the mass of the adipic acid sample is 0.485g, what is the percent purity of the sample.

3. If an air bubble is originally trapped in the buret tip but disappears during the titration, how will this affect the reported number of moles in the unknown acid in Part A? Read the Procedure.

4. How can a half-drop of NaOH solution be added from the buret?

5. A 0.793g sample of an unknown monoprotic acid requires 31.90mL of 0.106 M NaOH to reach the phenolphthalein endpoint. What is the formula weight of this acid?

6. a. A 30.84mL volume of 0.128 M NaOH is required to reach the phenolphthalein endpoint in titrating 5.961g of vinegar. Calculate the moles of acetic acid in vinegar.

   b. How many grams of acetic acid are in the vinegar?

   c. What is the percent acetic acid in the vinegar?

ANALYSIS OF ACIDS-DATA SHEET

Date________Name____Lab Sec. ____Desk No.____

Maintain 3 significant figures when recording data and performing calculations.

Data for Part ___ of experiment, entitled ________________________

	Trial 1	Trial 2	Trial 3
1. Mass of Erlenmeyer flask + sample (g)	________	__________	__________
2. Mass of Erlenmeyer flask (g)	__________	__________	__________
3. Mass of sample (g)	__________	__________	__________
4. Instructor's Approval of buret		__________	
5. Buret reading of NaOH, final (mL)	__________	__________	__________
6. Buret reading of NaOH, initial (mL)	__________	__________	__________
7. Volume of NaOH used (mL)	__________	__________	__________
8. Molarity of NaOH (mol/L)		__________	
9. Moles of NaOH used (mol)	__________	__________	__________

A. FORMULA WEIGHT OF AN ACID

1. Unknown number or name of acid _________;

   Molecular form of acid: HA, H_2A, or H_3A __________

2. Balanced equation for the acid's reaction with NaOH:

   ____

3. Moles of unknown acid	__________	__________	__________
4. Formula weight of acid (g/mol)	__________	__________	__________
5. Average formula weight (g/mol)		__________	

B. PERCENT ACID IN VANISH™

Sample number ______	Trial 1	Trial 2	Trial 3
1. Moles of $NaHSO_4$ in Vanish™ (mol)	__________	__________	__________
2. Mass of $NaHSO_4$ in Vanish™ (g)	__________	__________	__________
3. Percent $NaHSO_4$ in Vanish™ (%)	__________	__________	__________
4. Average percent $NaHSO_4$ in Vanish (%)		__________	

C. PERCENT ACETIC ACID IN VINEGAR

Brand of Vinegar	__________		__________	
	Trial 1	Trial 2	Trial 1	Trial 2
1. Moles of $HC_2H_3O_2$ in vinegar sample	______	______	______	______
2. Mass of $HC_2H_3O_2$ in vinegar sample (g)	______	______	______	______
3. Percent $HC_2H_3O_2$ in vinegar by weight (%)	______	______	______	______
4. Average percent $HC_2H_3O_2$ in vinegar by weight (%)	________		________	

5. Comment on the availability of $HC_2H_3O_2$ per unit cost in the two vinegars.

Questions

1. If a drop of NaOH solution adheres to the side of the Erlenmeyer flask during the standardization of the NaOH solution, how will this error affect the reported formula weight of the unknown acid in Part A?

2. If the endpoint is surpassed in determining the percent $NaHSO_4$ in Vanish™, will its reported percent be high or low? Explain.

3. If a drop of standardized NaOH solution adheres to the side of the Erlenmeyer flask and is not washed down into the vinegar with the wash bottle, how does this error affect the reported percent $HC_2H_3O_2$ in vinegar?

4. In determining the percent $HC_2H_3O_2$ in vinegar, the samples were weighed rather than measured volumetrically. Explain.

EXPERIMENT 30
ANTACID ANALYSIS

Objective

• To determine the concentration of antacid in commercial antacids

Principles

Various commercial antacids claim to give the "best relief" for acid indigestion. Pharmaceutical companies issue claims from their laboratories and promote their products through various commercial media.

The stomach generally has a pH range from 1.0 to 2.0; acid indigestion and heartburn occur at a lower pH. Antacids neutralize (or buffer) the excess hydrogen ion in the stomach to relieve this discomfort.

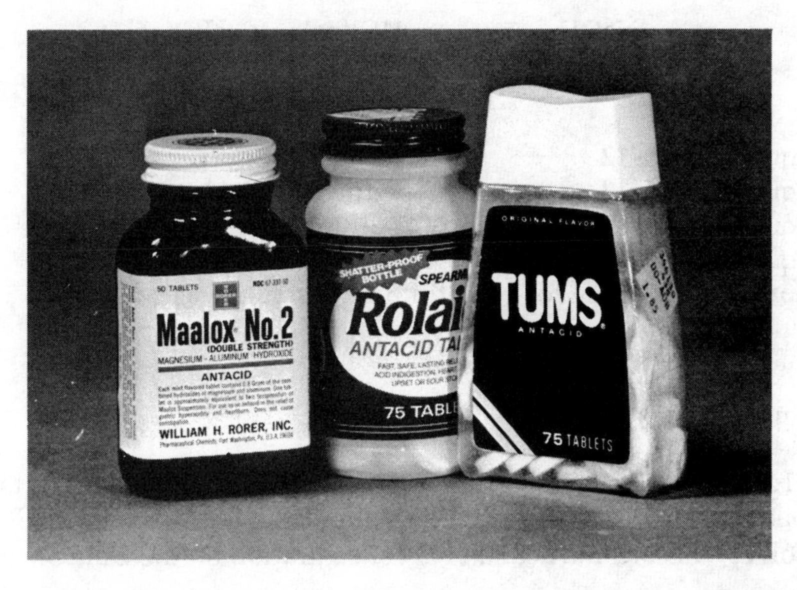

Milk of magnesia, an aqueous suspension of magnesium hydroxide, $Mg(OH)_2$, is an antacid that neutralizes H^+.

$$Mg(OH)_2(s) + 2H^+(aq) \rightarrow Mg^{2+}(aq) + 2H_2O$$

Maalox™, a "double strength" antacid, contains equal masses of $Mg(OH)_2$ and $Al(OH)_3$.

The more common antacids that buffer excess acid in the stomach are those containing calcium carbonate, $CaCO_3$, or sodium bicarbonate, $NaHCO_3$. A HCO_3^-/CO_3^{2-} buffer system is established in the stomach with these antacids.

$$CO_3^{2-}(aq) + 2H^+(aq) \rightleftharpoons HCO_3^-(aq)$$
$$HCO_3^-(aq) + H^+(aq) \rightleftharpoons H_2O + CO_2(aq)$$

Baking soda, which is pure $NaHCO_3$, can be used as an inexpensive antacid. The warmth of the stomach converts the $CO_2(aq)$ to $CO_2(g)$; thus, "gas on the stomach" results.

Rolaids™, containing dihydroxyaluminum sodium carbonate, $NaAl(OH)_2CO_3$, is a combination antacid that also reacts with stomach acid.

$$NaAl(OH)_2CO_3(aq) + 3H^+(aq) \rightarrow Na^+(aq) + Al^{3+}(aq) + 2H_2O + HCO_3^-(aq)$$

This experiment determines the effectiveness of several antacids using a strong acid–strong base titration. To avoid the possibility of a buffer system[1] being established and thus affecting the procedure for analysis, an excess of HCl(aq) is added to the dissolved antacid, driving the $HCO_3^-(aq) + H^+(aq) \rightleftharpoons H_2O + CO_2(aq)$ to the right. The solution is then heated to expel the $CO_2(aq) \rightarrow CO_2(g)$. The excess HCl(aq) is titrated with a standardized NaOH solution.

Since the antacid has the same neutralizing effect on stomach acid as does NaOH, the amount of antacid in a sample is called its $NaOH_{equivalent}$.[2] To determine the $NaOH_{equivalent}$ for an antacid, we subtract the excess moles of HCl(aq) from the total moles of HCl(aq) added to the antacid.

$$NaOH_{equivalent} = HCl(aq)_{total} - HCl(aq)_{excess}$$

Techniques

- Technique 10, page 12 — Reading a Meniscus
- Technique 11, page 13 — Pipetting a Liquid or Solution
- Technique 12, page 14 & 15 — Titrating a Solution
- Technique 13, page 16 — Using the Laboratory Balance

In addition, the technique for back-titrating a solution is used.

Procedure

A. Dissolving the Antacid

1. Weigh (±0.001g) approximately 0.7g of a pulverized commercial antacid tablet in a previously weighed 250mL Erlenmeyer flask. Pipet 50.0mL of standardized 0.1M HCl into the flask and swirl to dissolve.[3] Record the actual HCl concentration on the Data Sheet.

2. Heat the solution to boiling and continue boiling for at least 1 minute to expel the dissolved CO_2. Add 4–8 drops of bromophenol blue indicator.[4] If the solution is blue, add an additional 25.0mL of 0.1M HCl and boil again.

---

[1] A buffer system resists large changes in acidity. In our analysis, we want to "swamp" the system with the excess HCl to remove this buffering effect and then analyze for the HCl that was *not* neutralized by the antacid.

[2] The $NaOH_{equivalent}$ can also be referred to as the number of "equivalents" of antacid in the sample. An **equivalent** of any substance is merely an expression of its amount in the system, just as a **mole** of a substance indicates an amount of that substance in the system.

[3] The inert ingredients, such as the binder used in manufacturing the tablet, may not dissolve.

[4] Bromophenol blue is yellow in an acidic solution and blue in a basic solution.

B. ANALYSIS OF THE ANTACID SAMPLE

1. Prepare a clean 50mL buret. Rinse the clean buret with two 5mL portions of the standardized NaOH solution prepared in Experiment 28. Fill the buret with the NaOH solution; read and record its initial volume (±0.01mL).

2. Titrate the excess HCl to the blue endpoint. Read and record the final volume of NaOH in the buret.

3. Repeat the experiment for a second trial.

4. Select a second antacid for analysis and repeat the procedure. Compare the strengths (amount of antacid per gram of tablet) of the two antacids.

ANTACID ANALYSIS-LAB PREVIEW

Date________Name____Lab Sec. ____Desk No.____

1. a. Write the balanced for the reaction of one mole of the active ingredient in Rolaids™ with *excess* H^+ ion.

   b. Write a balanced equation that represents the antacid effect of sodium citrate, $Na_3C_6H_5O_7$, on an excess of stomach acid.

2. What acid-base indicator is used in this experiment? ________________ Its color in an acidic solution is ____________; in a basic solution it is ____________.

3. a. How much time should be allowed for the titrant to drain from the wall before a reading should be read and recorded?

   b. What color should be the background of the receiving flask in today's titration?

4. A volume of 50.0mL of 0.104M HCl is added to an unknown base. The HCl *not* neutralized by the base (the excess HCl) is titrated to a bromophenol blue endpoint with 26.7mL of 0.0841M NaOH. Calculate the $NaOH_{equivalent}$ for the unknown base.

ANTACID ANALYSIS-DATA SHEET

Date________ Name____ Lab Sec. ____ Desk No.____

Commercial Antacid	________		________	
A. DISSOLVING THE ANTACID	Trial 1	Trial 2	Trial 1	Trial 2
1. Mass of flask + crushed tablet (g)	________	________	________	________
2. Mass of flask (g)	________	________	________	________
3. Mass of crushed tablet (g)	________	________	________	________
4. Volume of HCl added (mL)	________	________	________	________
5. Molarity of HCl solution (mol/L)	________		________	
6. Moles of HCl added (mol)	________	________	________	________
B. ANALYSIS OF THE ANTACID SAMPLE				
7. Buret reading, final (mL)	________	________	________	________
8. Buret reading, initial (mL)	________	________	________	________
9. Volume of NaOH added (mL)	________	________	________	________
10. Molarity of NaOH (Exp't 28) (mol/L)	________		________	
11. Moles of NaOH added (mol)	________	________	________	________
12. Moles of excess HCl (mol)	________	________	________	________
13. $NaOH_{equivalent}$ of antacid in tablet	_____	________	________	________
14. $NaOH_{equivalent}$/g tablet	________	________	________	________
15. Cost of antacid/g tablet (¢/g)	________	________	________	________
Which antacid is the best buy (¢/$NaOH_{equivalent}$)?	______		________	

Questions

1. If the CO_2 is not removed with the boiling after the 0.1M HCl is added, how will this affect the amount of NaOH required to reach the bromophenol blue endpoint? Explain.

2. If the results from Trials 1 and 2 differ by a substantial amount (>5%), what would you do before presenting your results to your laboratory instructor?

3. If the endpoint in the titration is surpassed, will the reported amount of antacid in the sample be too high or too low? Explain.

EXPERIMENT 31
THE ELECTROLYTIC CELL

Objectives

- To identify the reactions occurring at the anode and cathode in an electrolytic cell
- To determine the charge passed through an electrolytic cell

Principles

Nonspontaneous oxidation–reduction reactions occur in an electrolytic cell. Electrical energy is used to direct a flow of electrons through a chemical system causing the reaction to occur. For example, most active metals, such as Na and Al, are prepared by the electrolysis of their molten salts–the directed current causes the reduction of the Na^+ or Al^{3+} at the cathode and the oxidation of its anion at the anode. The products are isolated at the electrodes as they form.

When molten NaCl is electrolyzed using a direct current (D.C.) source, Cl^- is oxidized at the anode and Na^+ is reduced at the cathode.

oxidation, anode: $2Cl^- \rightarrow Cl_2(g) + 2e^-$
reduction, cathode: $Na^+ + e^- \rightarrow Na$

In the electrolysis of an aqueous solution of $CuBr_2$, Cu^{2+} ion is reduced at the cathode and Br^- ion is oxidized at the anode.

reduction, cathode: $Cu^{2+}(aq) + 2e^- \rightarrow Cu$
oxidation, anode: $2Br^-(aq) \rightarrow Br_2(aq) + 2e^-$

Since the electrolysis occurs in an aqueous solution, we must also consider the electrolysis of H_2O at each electrode.

reduction, cathode: $2H_2O + 2e^- \rightarrow H_2(g) + 2OH^-(aq)$
oxidation, anode: $2H_2O \rightarrow O_2(g) + 4H^+(aq) + 4e^-$

If the reduction of H_2O occurs at the cathode, H_2 gas is evolved and the solution near the cathode becomes more basic due to the formation of OH^-. The oxidation of H_2O occurs at the anode, O_2 gas is evolved and the solution near the anode becomes more acidic because of the formation of H^+. The production of H^+ and/or OH^- can be detected using litmus paper.

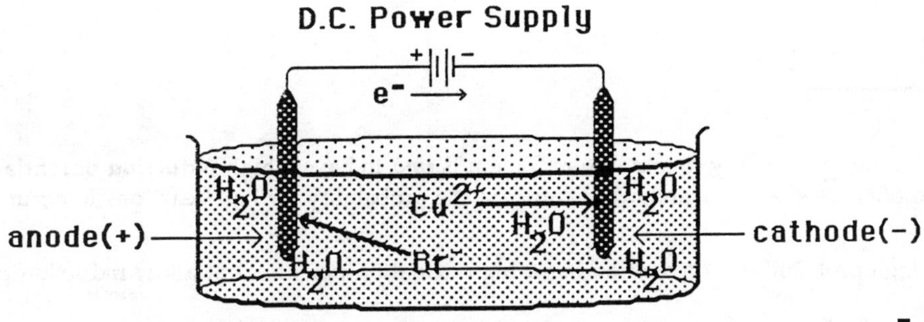

For two competing reactions at the cathode , the reaction that has the highest probability to undergo reduction[1] is the one that occurs. Conversely, for two competing reactions at the *anode,* the reaction having the highest probability to undergo oxidation[2] (or the lowest probability to undergo reduction) is the one that occurs.

MOVEMENT OF CHARGE THROUGH AN ELECTROLYTIC CELL

Electrical current moves by two different modes in an electrolytic cell: ions move and carry charge in the electrolyte and electrons move and carry charge in the external wire that connects the electrodes. The quantity of charge passing through the cell can be measured with either mode.

In the external wire, the quantity of charge (measured in **coulombs, C**) is determined by measuring the current (measured in **amperes,** where an ampere equals a coulomb/second, C/s) for a time period (measured in seconds, s): **C = (C/s) • s.**

In the electrolyte, the quantity of charge is determined by measuring a mass loss at the anode (an oxidation of the metal anode) or a mass gain at the cathode (a deposition of a metal ion from solution onto the cathode). For example, when Zn is the anode and it oxidizes, its mass loss is proportional to the charge that passes through the cell–one mole of Zn is oxidized when two moles of electrons (or two faradays) pass, causing a mass loss of 65.38g Zn from the anode.

$$\mathbf{Zn(anode) \rightarrow Zn^{2+}(aq) + 2e^-}$$

Since one mole of electrons has a charge of 96 500C, the charge passing through the cell for the oxidation of one mole of Zn is 2(96 500C) or 1.93×10^5C.

In this experiment you will electrolyze a number of solutions with an excess of energy (excess voltage). The products formed at each electrode are determined and a half–reaction is written consistent with your tests and observations. In addition, the quantity of charge that passes through a cell is determined by measuring both the electron movement and the mass change of an electrode.

Techniques

- Technique 2, page 2 — Transferring Liquid Reagents
- Technique 13, page 16 — Using the Laboratory Balance
- Technique 14, page 18 — Testing with Litmus

---

[1] A reaction that has a high probability to undergo reduction is said to have a high **reduction potential.** Reduction potentials are often used as the standard to compare the probability for half–reactions to occur.

[2] A reaction with a high probability for oxidation has a high oxidation potential but a low reduction potential.

Procedure

A. Electrolysis of Salt Solutions, Carbon Electrodes

1. Dissolve about 5g of NaCl in 250mL of distilled (or de–ionized) water and transfer it to the U–tube (Figure 31.1). Place 2 carbon electrodes in the solution and connect them to a D. C. energy source.[3] Identify the cathode (the negative terminal) and the anode (the positive terminal).

2. Electrolyze the solution for 5min. During the electrolysis, observe any evidence of a reaction occurring in the anode and cathode chambers.

   • Does the pH of the solution change at the electrodes? Test the solution at each electrode with litmus paper. Compare the color with a similar test on the original salt solution.
   • Is a gas evolved at one of the electrodes? Which one?
   • Does a metal deposit on the cathode? Is there any discoloration of the cathode and/or anode?

3. On the basis of your observations, write balanced half–reactions for the cathode and anode reactions.

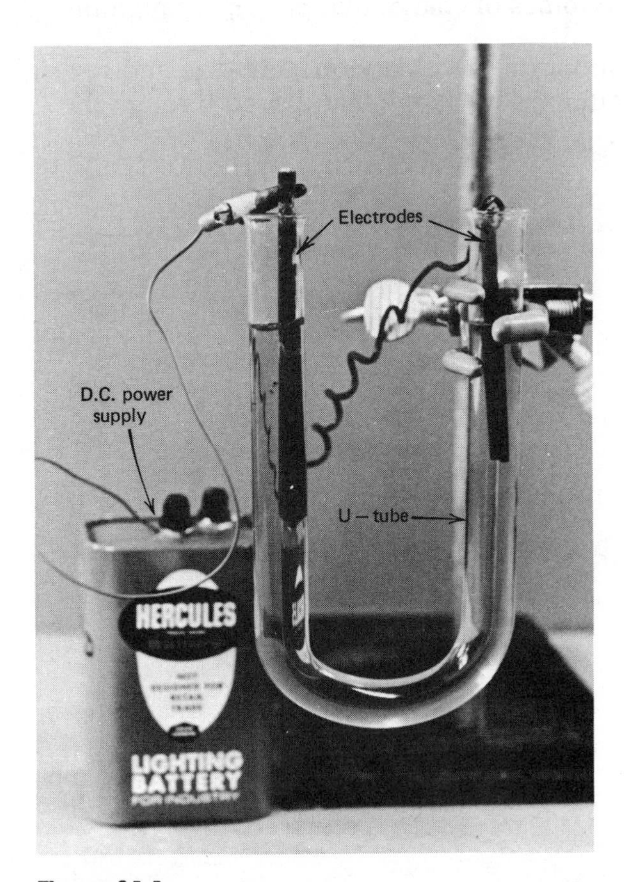

Figure 31.1
An Electrolysis Apparatus

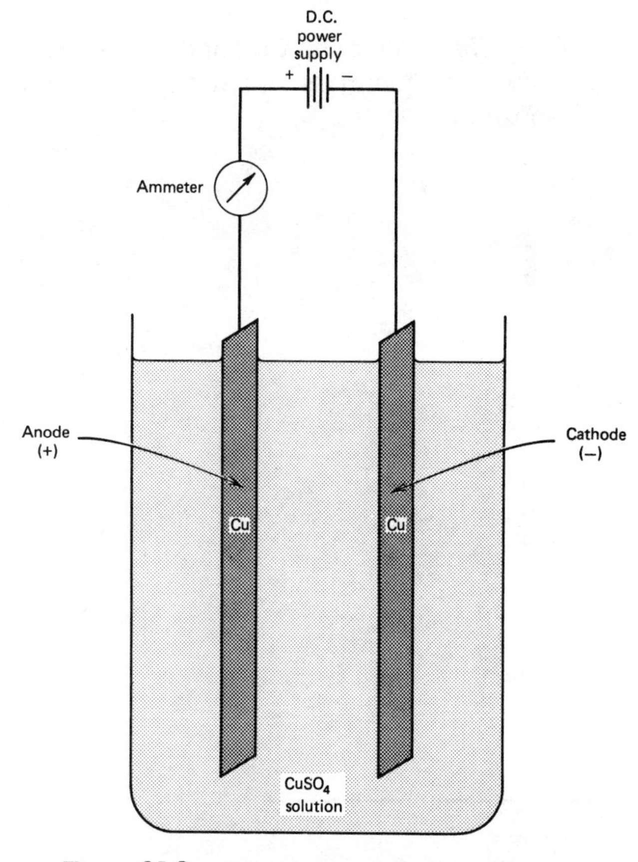

Figure 31.2
Apparatus for the Electrolysis of $CuSO_4$

[3] The most convenient D. C. energy source is a 9V transistor battery.

4. Substitute $CuSO_4$, $Pb(NO_3)_2$, $ZnCl_2$, KI, NaBr, and other salts suggested by your laboratory instructor for NaCl and repeat Part A.2. Be sure to thoroughly clean the U–tube and electrodes after each electrolysis.

B. ELECTROLYSIS OF A $CuSO_4$ SOLUTION, COPPER ELECTRODES

1. Fill the U–tube three–fourths full with 0.1M $CuSO_4$. Insert two copper electrodes, connect them to the D. C. energy source, and electrolyze the solution for 5min. Observe closely the cathode and anode chambers to detect any reactions and test with litmus; write balanced half–reactions that are consistent with your observations.

C. DETERMINATION OF THE QUANTITY OF CHARGE PASSED THROUGH THE CELL

1. Set up the apparatus in Figure 31.2. The D. C. energy source must provide 3–5V[4] and the ammeter should range should be from 0.2 to 1.0 amperes.

2. Polish a strip of Cu metal with steel wool, momentarily dip it into a 6M HNO_3 solution, rinse with distilled (or de–ionized) water and acetone, air–dry, and weigh (±0.0001g).[5] Use this Cu metal for the anode (the positive terminal). Use another Cu electrode as the cathode. Fill the 150mL beaker with 0.1M $CuSO_4$ to a level just below the top of the Cu electrodes. Electrolyze the solution for exactly 15min. Record the average current from the ammeter during the 15min period and calculate the coulombs of charge that passed through the wire.

3. *Carefully* remove the Cu anode from the solution, rinse with acetone, air–dry, and reweigh. Determine the mass loss and calculate the coulombs of charge that passed through the solution.

---

[4] A 3-5V direct current energy source can be two flashlight batteries, a lantern battery, or a transistor battery.

[5] A balance with 0.1mg precision is best; however, if only ±0.001g sensitivity is available, extend the time for electrolysis to 30min.

THE ELECTROLYTIC CELL-LAB PREVIEW

Date________Name____Lab Sec. _____Desk No._____

1. In an electrolytic cell

   a. What is the sign of the anode? ______________________

   b. What is the sign of the cathode? ______________________

   c. Identify the reaction that occurs at the anode. ______________________

   d. Identify the reaction that occurs at the cathode. ______________________

   e. Toward which electrode do cations migrate? ______________________

   f. Toward which electrode do anions migrate? ______________________

   g. Electrons migrate from the _______________ to the _______________ .

2. In the electrolysis of an aqueous solution, water can be oxidized and/or reduced.

   a. Write the equation for the electrolysis of water at the anode.

   b. Write the equation for the electrolysis of water at the cathode.

   c. If water is electrolyzed at the cathode, what will be the color of the litmus paper in a test?

3. A current of 0.5 amperes is passed through an electrolytic cell for 15min.

   a. How many coulombs of charge passed through the wire?

   b. If the this charge deposits zinc metal at the cathode, how many grams of Zn will electroplate? 1 mole of electrons = 96 500C

THE ELECTROLYTIC CELL-DATA SHEET

Date_________Name____Lab Sec. _____Desk No._____

A. ELECTROLYSIS OF SALT SOLUTIONS, CARBON ELECTRODES

Observations, Anode Chamber (+)

Salt	Litmus Test	Evidence of Reaction	Equation for the Reaction
NaCl	-------	------------	--------------------------
$CuSO_4$	-------	------------	--------------------------
$Pb(NO_3)_2$	-------	------------	--------------------------
$ZnCl_2$	-------	------------	--------------------------
KI	-------	------------	--------------------------
NaBr	-------	------------	--------------------------

Observations, Cathode Chamber (-)

Salt	Litmus Test	Evidence of Reaction	Equation for the Reaction
NaCl	-------	------------	--------------------------
$CuSO_4$	-------	------------	--------------------------
$Pb(NO_3)_2$	-------	------------	--------------------------
$ZnCl_2$	-------	------------	--------------------------
KI	-------	------------	--------------------------
NaBr	-------	------------	--------------------------

B. ELECTROLYSIS OF A $CuSO_4$ SOLUTION, COPPER ELECTRODES

1. Anode reaction ____
   Experimental evidence

2. Cathode reaction____
Experimental evidence

Cell reaction____

C. DETERMINATION OF THE QUANTITY OF CHARGE PASSED THROUGH THE CELL

Electrical Measurement

1. Average current (amperes = C/s) ------------------

2. Time for electrolysis (s) ------------------

3. Coulombs passed (C) ------------------

Chemical Measurement ------------------

1. Mass of polished Cu anode before electrolysis(g) ------------------

2. Mass of Cu anode after electrolysis (g) ------------------

3. Moles of Cu oxidized (mol)* ------------------

4. Moles of e^- passed (faradays)* ------------------

5. Coulombs passed (C) ------------------

*Show calculations of mol Cu and mol e^-

Compare the chemical and electrical measurements of charge passed through the electrolysis cell. Comment on any differences.

Questions

1. a. From the chemical measurement in Part C, how many (total) electrons were generated at the anode in 15 minutes?

   b. From the electrical measurement in Part C, and the answer in a, calculate the charge of an electron.

2. a. If nickel electrodes are used instead of carbon electrodes in Part A, could the reaction at the anode change? Explain.

   b. If nickel electrodes are used instead of carbon electrodes in Part A, could the reaction at the cathode change? Explain.

3. a. In the electrolysis of the KI solution, could a high acid concentration affect the product(s) that form at the anode? Explain.

   b. In the electrolysis of the KI solution, could a high acid concentration affect the product(s) that form at the cathode? Explain.

EXPERIMENT 32
THE GALVANIC CELL

Objectives

- To measure the reduction potentials for several redox couples
- To follow the movement of electrons, cations, and anions in a galvanic cell

Principles

A chemical reaction that involves the transfer of electrons from one substance to another is an oxidation–reduction (redox) reaction. The substance that loses electrons is oxidized (its oxidation state increases); the substance that gains electrons is reduced (its oxidation state decreases). In the reaction

$$Zn(s) + 2Ag^{+}(aq) \rightarrow 2Ag(s) + Zn^{2+}(aq)$$

Zn metal is oxidized ($Zn \rightarrow Zn^{2+} + 2e^{-}$) and the Ag^{+} ions are reduced ($2Ag^{+} + 2e^{-} \rightarrow 2Ag$). Because Zn loses electrons, it causes the reduction of Ag^{+}; hence, Zn is called the **reducing agent**. Conversely, because Ag^{+} gains electrons, it causes the Zn metal to oxidize and Ag^{+} is the **oxidizing agent**.

In a redox reaction, oxidation cannot occur unless reduction also occurs–a substance cannot lose electrons unless another substance accepts the electrons. The oxidation and reduction parts of the overall reaction can be divided into two half–reactions, where each half–reaction is called a **redox couple**.

oxidation half-reaction	$Zn \rightarrow Zn^{2+} + 2e^{-}$	Zn^{2+}/Zn **redox couple**
reduction half-reaction	$2Ag^{+} + 2e^{-} \rightarrow 2Ag$	Ag^{+}/Ag **redox couple**

A galvanic cell exists when aqueous solutions of the oxidation and reduction half–reactions are separated such that the electron transfer, from the reducing agent to the oxidizing agent, can only occur through an external wire connecting the two solutions. Each half–reaction is called a **half–cell** in the galvanic cell. The two half–cells are separated by a porous barrier, a barrier that *prevents* the free mixing of the two redox couples but allows for the movement of ions from one half–cell to another. An electrode must be a part of each half–cellμthe electrodes serve to conduct electrons away from the reducing agent (substance being oxidized) through a wire to the oxidizing agent (substance being reduced). The electrode at which oxidation occurs is called the **anode**, the (–) electrode; reduction occurs at the **cathode**, the (+) electrode.

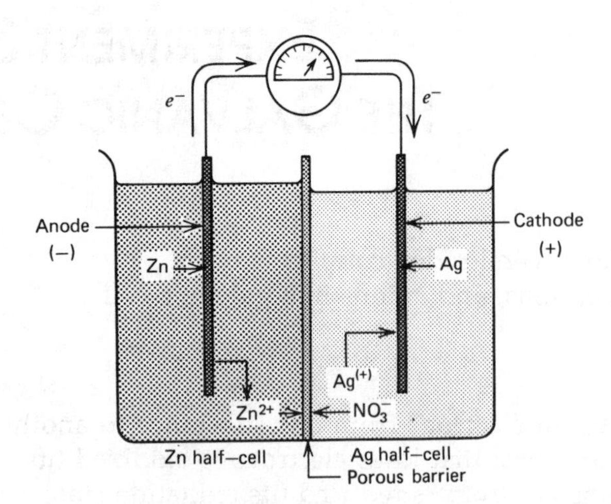

Figure 32.1
Schematic Diagram of a Galvanic Cell

Lets see what happens in a galvanic cell consisting of the Zn^{2+}/Zn and Ag^{+}/Ag redox couples (Figure 32.1): Zn oxidizes to Zn^{2+}; electrons are deposited on the Zn anode and migrate through the wire to the Ag cathode; the Ag^{+} in solution accepts the electrons at the Ag cathode to form Ag metal. In the meantime the Zn^{2+} migrates away from the Zn anode and through the porous barrier–this occurs because the positive charge of the Zn^{2+} is being added to the Zn half–cell but the positive charge of Ag^{+} is being removed from the Ag half–cell. Conversely the anion of the Ag^{+}, such as NO_3^-, can move into the Zn half–cell and out of the Ag half–cell. Both ion movements serve to maintain electrical neutrality in the galvanic cell.

Different metals, such as Zn and Ag, have different tendencies (or potentials) to oxidize; their cations have different tendencies (or potentials) to reduce. These differences are measured in a galvanic cell as a cell potential–the measurement is done with an instrument called a potentiometer (or voltmeter). We can consider the Zn^{2+} as having a potential (call it $\mathcal{E}_{(Zn^{2+}/Zn)}$) to exist in its reduced state, Zn, and Ag^{+} as also having a potential (call it $\mathcal{E}_{(Ag^{+}/Ag)}$) to exist in its reduced state, Ag. The ion having the greatest potential to be in the reduced state will be the one that undergoes reduction when the two redox couples form a galvanic cell; oxidation must then occur in the other redox couple. The difference in these two reduction potentials is measured as the **cell potential,** $\mathcal{E}_{cell}$, for the galvanic cell. Since Ag^{+} has the higher potential for reduction, the positive cell potential is written as

$$\mathcal{E}_{cell} = \mathcal{E}_{(Ag^{+}/Ag)} - \mathcal{E}_{(Zn^{2+}/Zn)}$$

The standard reduction potential (25°C, 1M concentrations of all ions and 1atm pressure for all gases), $\mathcal{E}°$, for the $Zn^{2+}(1M)/Zn$ redox couple is -0.76V and that for the $Ag^{+}(1M)/Ag$ redox couple is +0.80V. A potentiometer connected to the electrodes would show the difference between these potentials.

$$\mathcal{E}°_{cell} = 0.80V - (-0.76V) = 1.56V$$

Deviations from the theoretical potential may be due to surface activity at the electrodes, activity of the ions in solution, and/or current drawn by a voltmeter.

The cell potential, $\mathcal{E}_{cell}$, for a galvanic cell measured at 25°C but at ionic concentrations other than 1M is related to the standard cell potential, $\mathcal{E}°_{cell}$, by the Nernst equation.

$$\mathcal{E}_{cell} = \mathcal{E}^\circ_{cell} - \frac{0.0592}{n} \log Q$$

Q is the mass action expression for the cell reaction and n is the moles of electrons transferred in the cell reaction. For the Zn– Ag galvanic cell (where n = 2) the Nernst equation becomes

$$\mathcal{E}_{cell} = 1.56V - \frac{0.0592}{2} \log \frac{[Zn^{2+}]}{[Ag^+]^2}$$

Notice that the concentrations of Zn and Ag are omitted since they, as solids, maintain a constant concentration. The $\mathcal{E}_{cell}$ of a selected number of redox couples will be measured in the experiment; from the data the redox couples will be listed in order of their relative reduction potentials.

In this experiment the Nernst equation is used to determine an unknown Cu^{2+} concentration using the Zn^{2+}/Zn redox couple as a reference potential. The cell reaction is

$$Cu^{2+}(aq) + Zn(s) \rightarrow Cu(s) + Zn^{2+}(aq)$$

The corresponding Nernst equation is

$$\mathcal{E}_{cell} = 1.10V - \frac{0.0592}{2} \log \frac{[Zn^{2+}]}{[Cu^{2+}]}$$

If 0.1M Zn^{2+} is used for the measurement the Nernst equation becomes,

$$\mathcal{E}_{cell} = 1.10V - \frac{0.0592}{2} \log [0.1] + \frac{0.0592}{2} \log [Cu^{2+}] \text{ or}$$

$$\mathcal{E}_{cell} = 1.13V + \frac{0.0592}{2} \log [Cu^{2+}]$$

Therefore, $\mathcal{E}_{cell}$ is directly proportional to log $[Cu^{2+}]$; cell potentials can therefore be used to determine the concentrations of ions in a solution. This is the basic principle used for pH determinations using a pH meter.

Techniques

- Technique 10, page 12 — Reading a Meniscus
- Technique 11, page 13 — Pipetting a Liquid or Solution

You will develop additional skills in handling and transferring solutions that contain the redox couples used in the collection of the data with the galvanic cell apparatus.

Procedure

A. Reduction Potentials of Selected Half-Reactions

1. Use clean 150mL beakers to obtain 50mL of 0.1M $Zn(NO_3)_2$, 0.1M $Pb(NO_3)_2$, 0.1M $Cu(NO_3)_2$, and 0.1M $FeSO_4$. For each cell in Table 32.1, the anode, the anodic reaction, the cathode, the cathodic reaction, and the cell potential will be determined.

Table 32.1
Measurement of $\mathcal{E}_{cell}$

	Porcelain Cup*		Beaker	
Voltaic Cells	Solution	Electrode	Solution	Electrode
Zn-Cu cell	0.1M $Zn(NO_3)_2$	Zn	0.1M $Cu(NO_3)_2$	Cu
Cu-Pb cell	0.1M $Pb(NO_3)_2$	Pb	0.1M $Cu(NO_3)_2$	Cu
Zn-Pb cell	0.1M $Pb(NO_3)_2$	Pb	0.1M $Zn(NO_3)_2$	Zn
Fe-Pb cell	0.1M $Pb(NO_3)_2$	Pb	0.1M $FeSO_4$	Fe
Zn-Fe cell	0.1M $Zn(NO_3)_2$	Zn	0.1M $FeSO_4$	Fe

*Large diameter glass tubing with dialysis tubing over one end (secured with rubber bands) can be substituted. Also, a KCl salt bridge connecting the two solutions in two 150mL beakers can be used.

2. a. For the Zn–Cu cell, clean the Cu and Zn electrodes. Place the Cu electrode in a 150mL beaker that contains 50mL of 0.1M $Cu(NO_3)_2$. Place the Zn electrode in the porcelain cup that is about $^3/_4$–filled with 0.1M $Zn(NO_3)_2$. Connect one electrode to the positive terminal of the voltmeter and the other electrode to the negative terminal.

   b. Touch the porcelain cup to the surface of the 0.1M $Cu(NO_3)_2$ solution in the beaker. If the needle shows a positive deflection, the connections are correct; if not, reverse the connections to the electrodes.

   c. Immerse the cup in the solution, read, and record the $\mathcal{E}_{cell}$.[1] Which electrode is connected to the positive (+) terminal (cathode)? Does oxidation occur at the Zn or Cu electrode? Write an equation for the reaction at each electrode.

3. Repeat the measurements for the remaining cells in Table 32.1. Take the following precautions in collecting the data.

   a. Polish the electrodes with steel wool and rinse with distilled water before every measurement.

   b. Use distilled (or de–ionized) water to thoroughly rinse the inside *and* outside of the porcelain cup *and* the beaker when changing solutions before each measurement.

4. Assuming the reduction potential of the Zn^{2+}(0.1M)/Zn redox couple is -0.79V, determine the reduction potentials of all other redox couples.[2]

5. Other redox couples. Determine the reduction potential for the following redox couples designated by your instructor. Polish each electrode and again wash the porcelain cup thoroughly after each use.

---

[1]A positive $\mathcal{E}_{cell}$ indicates that the redox couple at the cathode, the one connected to the (+) terminal of the voltmeter, has the higher reduction potential by the voltage read on the voltmeter.

[2]These are not standard reduction potentials because 1M solutions at 25°C were not used for the measurements.

Ag^+/Ag	(0.1M $AgNO_3$ solution)
Ni^{2+}/Ni	(0.1M $Ni(NO_3)_2$ solution)
Mg^{2+}/Mg	(0.1M $MgSO_4$ solution)
Sn^{2+}/Sn	(0.1M $SnCl_2$ solution)
M^{n+}/M (unknown)	(0.1M solution of M^{n+})

B. CONCENTRATION CELL, Cu^{2+}(1M)/CU AND Cu^{2+}(0.001M)/CU REDOX COUPLES

1. Thoroughly wash the cup. Pour 1M $CuSO_4$ into the 150mL beaker and 0.001M $CuSO_4$ into the cup. Immerse a Cu electrode in each solution and connect the electrodes to the voltmeter. Determine the $\mathcal{E}_{cell}$, the anode reaction, and the cathode reaction. Explain *why* a potential is recorded. Write an equation for the reaction at each electrode.

2. Add about 5mL of 6M NH_3 (**Caution:** *Avoid inhalation*) to the 0.001M $CuSO_4$ solution. Observe any changes in $\mathcal{E}_{cell}$. Explain.

C. MEASURING AN UNKNOWN CONCENTRATION

1. Obtain a 1mL pipet or one with 1mL graduations and three 100mL volumetric flasks. To prepare solution **A**, pipet 1mL of 1M $CuSO_4$ solution into a 100mL volumetric flask and dilute to "the mark" with distilled (or de–ionized) water. Prepare solution **B** by pipetting 1mL of **A** into a second 100mL volumetric flask and diluting to "the mark." Similarly, prepare solution **C**. Calculate the $[Cu^{2+}]$ in solutions **A, B,** and **C**. Use the equation, $\mathcal{E}_{cell} = 1.13V + (0.0592/2) \log [Cu^{2+}]$, to calculate the theoretical $\mathcal{E}_{cell}$ for the 1M $CuSO_4$ and solutions A, B, and C.

1M $CuSO_4$	–1mL→	dilute to 100mL **A**	–1mL→	dilute to 100mL **B**	–1mL→	dilute to 100mL **C**

2. Place a Zn electrode in the porcelain cup that is 3/4–filled with 0.1M $Zn(NO_3)_2$. Connect a polished Zn electrode to the negative (-) terminal of the voltmeter and immerse it in the solution. Place about 50mL of solution **C** in the 150mL beaker; connect a polished Cu electrode to the positive (+) terminal and immerse it into solution **C**. Record the $\mathcal{E}_{cell}$.

3. *Thoroughly* rinse the outside of the porcelain cup and the inside of the beaker first with distilled water and then with solution **B**. Discard the rinses. Repeat Part C.2 to record the $\mathcal{E}_{cell}$ for solution **B** relative to the Zn^{2+}(0.1M)/Zn redox couple. Repeat, in order, the measurement of $\mathcal{E}_{cell}$ for solution A and for 1M $CuSO_4$.

4. Obtain a solution with an unknown $[Cu^{2+}]$ from your instructor. As before, determine the $\mathcal{E}_{cell}$. From an analysis of your data in Part C, estimate the $[Cu^{2+}]$ in your unknown.

Activity Series for Hydrogen and Some Typical Metals

Element	Oxidation Product
Gold	Au^{3+}
Mercury	Hg^{2+}
Silver	Ag^{+}
Copper	Cu^{2+}
Hydrogen	H^{+}
Lead	Pb^{2+}
Tin	Sn^{2+}
Cobalt	Co^{2+}
Cadmium	Cd^{2+}
Iron	Fe^{2+}
Chromium	Cr^{3+}
Zinc	Zn^{2+}
Manganese	Mn^{2+}
Aluminum	Al^{3+}
Magnesium	Mg^{2+}
Sodium	Na^{+}
Calcium	Ca^{2+}
Strontium	Sr^{2+}
Barium	Ba^{2+}
Potassium	K^{+}
Rubidium	Rb^{+}
Cesium	Cs^{+}

INCREASING EASE OF OXIDATION

THE GALVANIC CELL-LAB PREVIEW

Date__________ Name____ Lab Sec. ______ Desk No.______

1. State the purpose of the porcelain cup in today's experiment.

2. What is the sign of the anode in a galvanic cell?______________

   What electrochemical process occurs at the anode?____________________

Redox Couples	$\mathcal{E}^\circ$
$Au^{3+} + 3e^- \rightleftharpoons Au$	+1.42 V
$Fe^{3+} + e^- \rightleftharpoons Fe^{2+}$	+0.77 V
$Sn^{4+} + 2e^- \rightleftharpoons Sn^{2+}$	+0.15 V
$Cr^{3+} + 3e^- \rightleftharpoons Cr$	-0.74 V

3. Consider a galvanic cell consisting of the Au^{3+}(1M)/Au and Sn^{4+}(1M)/Sn^{2+}(1M) redox couples.

   a. Write the reduction half–reaction and corresponding $\mathcal{E}^\circ$ for each redox couple.

   b. Which redox couple undergoes reduction?______________

   c. Write an equation for the reaction that occurs at the

   anode____

   cathode____

   d. Write the equation for the cell reaction.

   e. What is the cell potential, $\mathcal{E}^\circ_{cell}$, for the cell?

4. Determine the cell potential, $\mathcal{E}^{\circ}_{\mathbf{cell}}$, of a galvanic cell consisting of the Sn^{4+}(1M)/Sn^{2+}(1M) and Cr^{3+}(1M)/Cr redox couples.

5. Use the Nernst equation to determine the cell potential, $\mathcal{E}_{\mathbf{cell}}$, of the galvanic cell consisting of the following redox couples.

$Sn^{4+}(10^{-3}M) + 2e^{-} \rightleftharpoons Sn^{2+}(10^{-5}M)$
$Fe^{3+}(1M) + e^{-} \rightleftharpoons Fe^{2+}(10^{-5}M)$

THE GALVANIC CELL-DATA SHEET

Date________Name____Lab Sec. ____Desk No.____

A. Reduction Potentials of Selected Half-Reactions

Cell	$\mathcal{E}_{cell}$	anode	anode reaction	cathode	cathode reaction
Zn-Cu	-----	-----	-----------	-----	-----------
Cu-Pb	-----	-----	-----------	-----	-----------
Zn-Pb	-----	-----	-----------	-----	-----------
Fe-Pb	-----	-----	-----------	-----	-----------
Zn-Fe	-----	-----	-----------	-----	-----------
other	-----	-----	-----------	-----	-----------
unknown	-----	-----	-----------	-----	-----------

1. Compare the sum of the Zn-Pb + Cu-Pb cell potentials with the Zn–Cu cell potential. Explain.

2. Compare the sum of the Zn-Fe + Fe-Pb cell potentials with the Zn–Pb cell potential. Explain.

3. Arrange the redox couples in order of *decreasing* reduction potentials, the most positive first. List the reduction potential for each redox couple relative to that of the Zn^{2+}(0.1M)/Zn couple which -0.79 V.

Redox Couple	Reduction Potential
----------------	-----------------
----------------	-----------------
----------------	-----------------
----------------	-----------------
----------------	-----------------
----------------	-----------------

B. Concentration Cell, Cu^{2+}(1M)/Cu and Cu^{2+}(0.001M)/Cu Redox Couples

1. $\mathcal{E}_{cell}$ ____________________

   anode reaction (oxidation)____________________

   cathode reaction (reduction)____________________

   Why is a potential recorded?____________________

2. $\mathcal{E}_{cell}$ ____________________

   Account for the $\Delta\mathcal{E}_{cell}$ from Part B.1

C. Measuring an Unknown Concentration

1.

Solution	Molarity	log $[Cu^{2+}]$	$\mathcal{E}_{cell}$(calc)	$\mathcal{E}_{cell}$(exp'tl)
C	-------	-------	---------	-------
B	-------	-------	---------	-------
A	-------	-------	---------	-------
1M $CuSO_4$	1.0	0	---------	-------

2. Compare $\mathcal{E}_{cell}$ (calc) to $\mathcal{E}_{cell}$ (exp'tl). Are differences consistent? Is the trend of cell potentials as predicted? Explain.

3. $\mathcal{E}_{cell}$ for $Zn^{2+}(0.1M)/Zn–Cu^{2+}(?M)/Cu$ galvanic cell?________________

$[Cu^{2+}]$ in solution of unknown concentration?________________

Questions

1. Identify the oxidizing and reducing agent in each cell in Part A.

Cell	Oxidizing Agent	Reducing Agent
Zn-Cu	--------------	--------------
Cu-Pb	--------------	--------------
Zn-Pb	--------------	--------------
Fe-Pb	--------------	--------------
Zn-Fe	--------------	--------------

2. List two factors accounting for the experimental cell potential not being equal to the theoretical cell potential.

3. If in Part C, the $Zn(NO_3)_2$ solutions were successively diluted instead of the $CuSO_4$ solutions, how would the relative cell potentials have differed?

EXPERIMENT 33
REACTION RATES

Objective

- To identify factors that affect the rate of a chemical reaction

Principles

The rates of chemical reactions can be affected by number of changes in the reaction system. Factors that, as chemists, we can quickly alter include temperature, pressure (if gases), and concentrations (if in solution). In addition the substitution of a more reactive reactant, the subdividing of a reactant (if a solid), or the inclusion of a catalyst can increase a reaction rate. Each of these factors are observed and studied in this experiment.

TEMPERATURE EFFECTS

A temperature increase generally increases reaction rate–a rule of thumb is that a 10°C temperature rise doubles the rate. Because of the temperature increase, the kinetic energy of the reactant molecules (or ions) also increases[1]; when the molecules (or ions) collide, this added kinetic energy becomes a part of the collision system. This added energy is distributed among the atoms, causing an increase in the probability of bonds being broken; subsequent recombination of the fragments can result in the formation of product.

CONCENTRATION EFFECTS

An increase in the concentration of a reactant increases the probability of collision between the reactant molecules (or ions). As a result, most reaction rates increase with an increasing concentration of reactants, although there are systems in which a decrease or no change in reaction rate occurs. Experiment 34 focuses on a *quantitative* investigation of the effect of concentration changes on reaction rate.

NATURE OF REACTANTS

Some substances are naturally more reactive than others, and, therefore undergo more rapid chemical changes. The reaction between water and sodium is very rapid and exothermic, whereas with copper, no reaction occurs. Hydrogen and fluorine react explosively at room temperature; hydrogen and iodine react very slowly, even at elevated temperatures.

---

[1]The kinetic energy–temperature relationship is expressed by the equation

$$K.E. = (1/2)\ mv^2 = (3/2)\ kT$$

k is the called the Boltzman constant and T is in kelvins.

State of Subdivision

The greater surface area of a reactant, the greater its reaction rate. A large chunk of coal burns much more slowly than coal dust. Coal mine explosions and grain elevator explosions are due to the very finely divided (large surface area) dust that is exposed to the oxygen of the air–a spark ignites the very rapid reaction.

Presence of a Catalyst

A catalyst increases the reaction rate without undergoing any *net* change in its properties during the reaction. Catalysts generally reroute the pathway (or mechanism) of the reaction in such a way that the reaction requires less energy and therefore proceeds more rapidly.

Techniques

- Technique 2, page 2 — Transferring Liquid Reagents
- Technique 6d, page 7 — Heating Liquids
- Technique 9b, page 11 — Handling Gases
- Technique 11, page 13 — Pipetting a Liquid or Solution
- Technique 13, page 16 — Using the Laboratory Balance
- Technique 15, page 18 — Graphing Techniques

Procedure

Each portion of the experiment requires that you time the reaction. The time lapse should be recorded in seconds; therefore, before you begin this experiment, make sure you have available a clock or watch that can be read to the nearest second.

A. Temperature Effects

The oxidation–reduction reaction that occurs between HCl and $Na_2S_2O_3$ produces insoluble sulfur as a product. The time required for the cloudiness, due to the sulfur, is a measure of the reaction rate.

$$2HCl(aq) + Na_2S_2O_3\,(aq) \rightarrow S(s) + SO_2(g) + 2NaCl(aq) + H_2O$$

Measure each volume (±0.1mL) with a graduated cylinder. A pair of students can best complete this part of the experiment.

1. Place 5.0mL of 0.1M $Na_2S_2O_3$ into one set of three (clean)150mm test tubes. Into a second set of three 150mm test tubes, measure 5.0mL of 0.1M HCl. Place a $Na_2S_2O_3$ –HCl pair of test tubes in a salt–ice–water bath until thermal equilibrium is established (about 5 or 6 minutes). *O.K., get ready* to time the appearance of the sulfur cloudiness with a watch that reads to the nearest second. As one student pours the solutions together, the other notes the time...pour the two solutions together, immediately stopper, and shake vigorously for several seconds. Return the reaction mixture to the ice bath and record the time for the sulfur to appear. Record the temperature (±0.1°C) of the mixture.

2. Place a second pair of $Na_2S_2O_3$ –HCl test tubes in a hot water bath adjusted to a temperature of about 70°C. Repeat the mixing of the two solutions as in Part A.1 and record the time for the sulfur cloudiness to appear. You can predict the relative time even before the mixing. Record the temperature of the mixture.

3. Combine and time the remaining pair of $Na_2S_2O_3$ –HCl solutions at room temperature.

4. Plot the temperature (°C) *vs* time (seconds) for the appearance of sulfur on linear graph paper. Have your instructor approve your graph.

B. CONCENTRATION EFFECTS

1. Pipet 5.0mL of 0.1M HCl, 1M HCl, 3M HCl, and 6M HCl (**Caution:** HCl *causes a skin irritation*) into a set of four, clean 150mm test tubes. Weigh (±0.01g) four 25mm polished Mg strips.

2. Add a Mg strip (be sure you have recorded its mass) to the 0.1M HCl and start timing (in seconds). When all traces of Mg ribbon disappear, stop timing. Record the elapsed time on the Data Sheet. Repeat the experiment with the 1M HCl, 3M HCl, and 6M HCl solutions.

3. Plot (mol HCl/mol Mg) *vs* time (seconds) on linear graph paper. Have the instructor approve your graph.

C. NATURE OF REACTANTS

1. Into four separate 75mm test tubes containing 3M H_2SO_4, 6M HCl, 6M HNO_3, and 6M H_3PO_4 (**Caution:** *Carefully handle acids*), place a strip of polished Mg ribbon. Note the relative reaction rates and record your observations.

2. Place about 1mL of 6M HCl into each of three 75mm test tubes. Add a polished strip of Zn to the first, one of Pb to the second and another of Cu to the third. Note the relative reaction rates and record your observations.

D. STATE OF SUBDIVISION

1. Set up the apparatus in Figure 33.1. Fill a 50mL graduated cylinder with water and invert it over the collection port. Place about 5g of a piece of chalk, $CaCO_3$, in the generator and cover with 50mL of water. Extend the thistle tube *below the water level* in the generator and add through it 10mL of 6M HCl. Record the time (in seconds) to collect 25mL of CO_2.

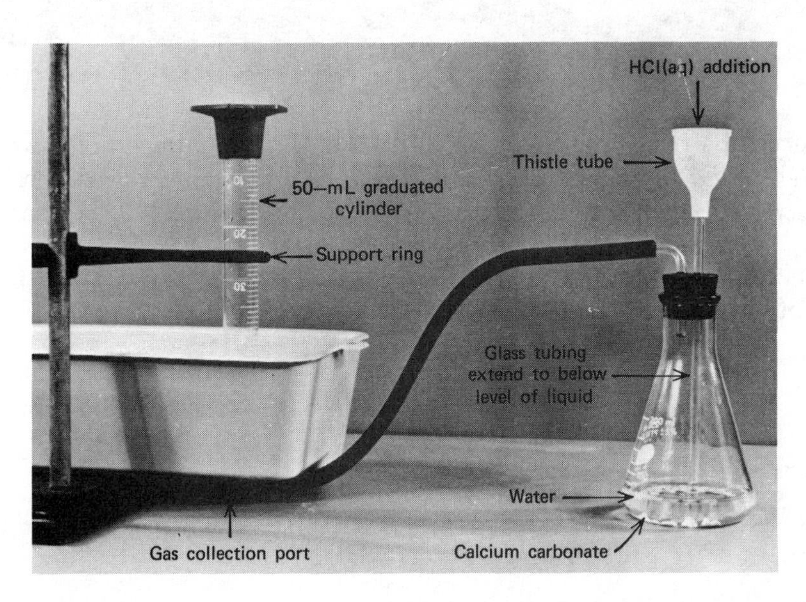

Figure 33.1
Apparatus for the Collection of $CO_2(g)$

2. Clean the generator. Using a mortar and pestle, crush about 5g of chalk (or obtain about 5g of chalk dust) and again cover with 50mL of water. Repeat the addition of 10mL of 6M HCl and record the time to collect 25mL of CO_2.

E. PRESENCE OF A CATALYST

Caution: *This experiment can be dangerous if not performed properly. Before you begin, read the procedure carefully and adhere to the following guidelines:*

- Wear safety glasses
- Use a clean 200mm Pyrex test tube–there should be no evidence of a black residue. **Do not** dry the inside of the test tube with a paper towel.
- Your instructor must approve your apparatus before you begin.

1. Set up the apparatus in Figure 33.2. Into a clean, dry 200mm Pyrex test tube, place about 0.5g of $KClO_3$. Heat the $KClO_3$ and record the time (in seconds) required to collect 10mL of O_2 gas.

$$2KClO_3 (s) \rightarrow 2KCl(s) + 3O_2(g)$$

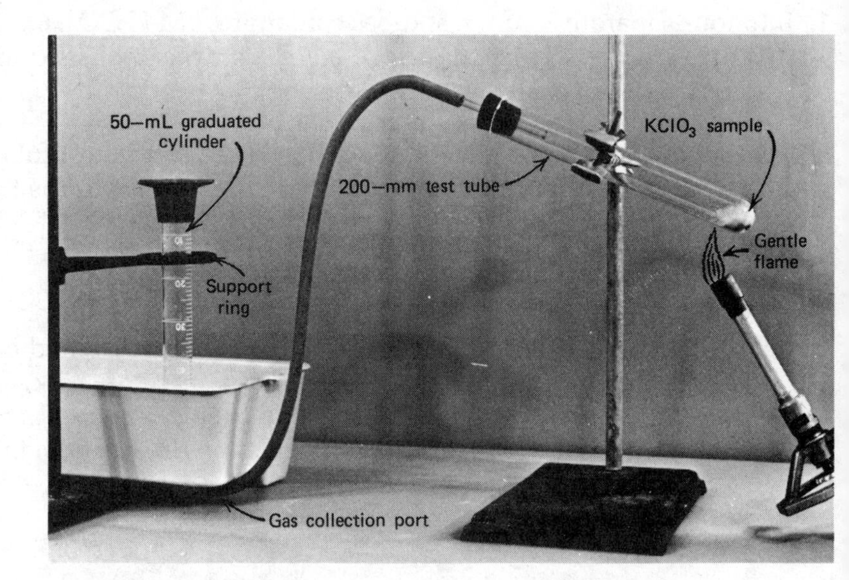

Figure 33.2
Apparatus for the Collection of O_2(g)

2. First, disconnect the gas delivery tube from the test tube. **Caution:** *It is very important that the delivery tube be disconnected while the tube is hot; if it isn't, cool water will be drawn from the trough into the test tube and may cause it to break.* Then, allow the $KClO_3$ to cool.

3. Repeat Parts E.1 and E.2, but in addition, add a pinch of MnO_2 to the 0.5g $KClO_3$. Record the time required to collect 10mL of O_2 gas.

REACTION RATES-LAB PREVIEW

Date________Name______________________________________Lab Sec. _____Desk No._____

1. If the rate of a chemical reaction doubles for every 10°C temperature increase, by what factor will a chemical reaction increase if the temperature is increased over a 30°C range?

2. Explain why coal dust burns more rapidly that larger pieces of coal?

3. Iron rusts, gold does not. Explain.

4. When antiseptic hydrogen peroxide, H_2O_2 , is placed on an open wound, bubbles form because of the reaction

$$2H_2O_2 \rightarrow 2\,H_2O + O_2$$

However, H_2O_2 only slowly decomposes in the bottle. How does the wound accelerate its decomposition?

5. Wood burns more rapidly in a fireplace that has a good draft of air. Explain.

6. List factors that can affect the rate of ammonia production by the Haber process.

$$N_2(g) + 3H_2(g) \rightarrow 2NH_3(g) + 46.1kJ$$

7. What is Technique 6d? How will you need to modify the technique to adapt it to this experiment?

8. Describe how Technique 9b is to be used in this experiment.

REACTION RATES-DATA SHEET

Date__________ Name____ Lab Sec. _____ Desk No._____

A. TEMPERATURE EFFECTS

$Na_2S_2O_3$–HCl Test Pair	Time Elapsed (seconds)	Temperature (°C)
1.	__________	__________
2.	__________	__________
3.	__________	__________

Instructor's Approval of graph. ____________________

Based upon your data, what can you conclude about the effect of temperature on the rate of this reaction?

From your graph of the data and assuming the same set of solutions, estimate the time for the appearance of sulfur when the reaction occurs at 40°C __________; at 95°C __________.

B. CONCENTRATION EFFECTS

Molarity of HCl	mol HCl	Mass Mg (±0.01g)	mol Mg	mol HCl / mol Mg	time (seconds)
0.1	_______	_______	_______	_______	_______
1.0	_______	_______	_______	_______	_______
3.0	_______	_______	_______	_______	_______
6.0	_______	_______	_______	_______	_______

Instructor's Approval of graph ____________________

How does a change in the [HCl] affect the time for a known amount of Mg to react?

From your graph of the data, estimate the time it would take for 1.2g Mg to react in 0.40M HCl.

C. NATURE OF REACTANTS

1. Record the relative reaction rates of the four acids with Mg in order of *increasing* activity.

   __________, __________, __________, __________

   What can you conclude about the relative chemical reactivity of the four acids?

2. List the three metals in order of the increasing reaction rate with 6M HCl.

   __________, __________, __________

   What can you conclude about the chemical reactivity of the three metals?

D. STATE OF SUBDIVISION

1. Time to collect 25mL of CO_2 from a $CaCO_3$ chalk piece. __________________

   Time to collect 25mL of CO_2 from $CaCO_3$ chalk dust. __________________

2. Explain how the physical state of the reactants affects the rate of a chemical reaction.

E. PRESENCE OF A CATALYST

1. Instructor's Approval of apparatus.______________________________

2. Time to collect 10mL of O_2 .__________________

   Time to collect 10mL of O_2 __________________

3. Describe the affect that a catalyst has on the rate of evolution of O_2 gas.

EXPERIMENT 34
RATE LAW DETERMINATION

Objectives

- To determine the rate law for a chemical reaction
- To use a graphical analysis of experimental data

Principles

The rate of a chemical reaction is measured by observing the rate of disappearance of a reactant or the appearance of a product. The choice is dependent on the ease of detection of one of the substances in the reaction. For example, a reactant may lose its color, a product may produce an odor, the reaction may generate heat, a product may be a gas or precipitate, etc.–all of which can be monitored as a function of time. For our purposes, we will monitor a change in concentration as a function of time.

A reaction rate is generally, although not always, proportional to the molar concentration of each reactant raised to some power, called the **order** of the reactant. For the reaction $A_2 + 2B_2 \rightarrow 2AB_2$, the reaction rate is proportional to the molar concentration of A_2 and B_2, raised to their respective order in the reaction, **p** and **q**.

$$\text{rate} \propto [A_2]^p \textit{ and } \text{rate} \propto [B_2]^q \text{ or rate} \propto [A_2]^p[B_2]^q$$

The proportionality sign, $\propto$, is replaced with a proportionality constant, **k**, called the **specific rate constant**.

$$\text{rate} = k[A_2]^p[B_2]^q$$

This equation is the **rate law** for the reaction. The value of **k**, which is determined experimentally, varies with temperature and the presence of a catalyst, but is independent of reactant concentrations.

The orders, **p** and **q**, for the corresponding reactants are also determined experimentally. The effect that a change in a reactant concentration has on a reaction rate is expressed in the order of that reactant. For example, suppose that doubling $[A_2]$ (while holding the $[B_2]$ constant) causes the reaction rate to increase by a factor of four. To maintain the proportionality between rate and $[A_2]$ (rate $\propto$ $[A_2]^p$) **p** must equal 2. By placing the $[B_2]$ concentration in a large excess compared to $[A_2]$, the $[B_2]$ remains nearly unchanged in the reaction–during the reaction the $\Delta[A_2]$ is therefore much larger then the $\Delta[B_2]$ and therefore is more pronounced in affecting reaction rate.

In this experiment, the **k** and **p** for the reaction of the iodate anion, IO_3^-, with the sulfite anion, SO_3^{2-}, in an acidic solution are determined. **p** will be the order of the IO_3^- in the reaction. The reaction occurs in a series of steps.

(step #1)	$IO_3^-(aq) + 3SO_3^{2-}(aq) \rightarrow I^-(aq) + 3SO_4^{2-}(aq)$
(step #2)	$5I^-(aq) + 6H^+(aq) + IO_3^-(aq) \rightarrow 3H_2O + 3I_2$
(step #3)	$3I_2 + 3SO_3^{2-}(aq) + 3H_2O \rightarrow 6I^-(aq) + 3SO_4^{2-}(aq) + 6H^+(aq)$
(Net equation)	$2IO_3^-(aq) + 6SO_3^{2-}(aq) \rightarrow 2I^-(aq) + 6SO_4^{2-}(aq)$

How do we detect the reaction rate? Lets note the function and progress of the SO_3^{2-} anion in the reaction: we see that SO_3^{2-} reacts with not only the IO_3^- anion in step #1, but also with I_2, a product of step #2. What happens when all of the SO_3^{2-} anion has been consumed in the reaction? Now the I_2 is "free" in solution; we detect its appearance by the addition of starch, which forms a deep blue solution when bonded to the I_2.

$$I_2 + \text{starch} \rightarrow I_2 \bullet \text{starch (deep blue color in solution)}$$

This is the visible signal for detecting the consumption of the SO_3^{2-} anion and a method for monitoring the reaction rate. If the concentration of IO_3^- is increased, step #1 proceeds faster and, accordingly, the SO_3^{2-} is consumed more rapidly. Therefore, we have a method for observing the effect that a change in IO_3^- concentration has on the reaction rate.

The rate law for the reaction is

$$\text{rate} = k[IO_3^-]^p[SO_3^{2-}]^q$$

In the experiment, the $[SO_3^{2-}]$ will remain constant in each trial while the $[IO_3^-]$ will change. This modifies the rate law to

$$\text{rate} = k[IO_3^-]^p \bullet \text{constant} \quad \text{or} \quad \text{rate} = k'[IO_3^-]^p$$

The reaction rate is expressed in reciprocal time, the reciprocal of the time it takes to observe the appearance of the deep blue $I_2 \bullet$starch.

$$\text{rate (sec}^{-1}) = \text{appearance of } I_2 \bullet \text{starch/time}$$

DETERMINATION OF THE REACTION ORDER, P, AND RATE CONSTANT, K'

The effect of the $[IO_3^-]$ on the reaction rate is determined by changing its concentration in several reactions while maintaining a constant $[SO_3^{2-}]$ and $[H^+]$. The collected data is graphed to determine **p** and **k'**; the logarithmic form of the rate law, **rate = k'[IO₃⁻]ᵖ** , yields an equation for a straight line.

$$\begin{aligned} \log(\text{rate}) &= \log k' + p \log[IO_3^-] \text{ or} \\ y &= b + m\,x \end{aligned}$$

Therefore, a plot of "log (rate)" *vs* "log $[IO_3^-]$" produces a straight line with a slope of **p**, the order of the $[IO_3^-]$ in the reaction. **q**, the order of $[SO_3^{2-}]$ in the reaction, could be similarly determined.

The reaction rate constant, **k'**, is calculated by substituting the rate, **p**, and $[IO_3^-]$ for a given trial into the rate law. Of course, it can also be determined from the y–intercept of the graphed data— **b = log k'**.

ACTIVATION ENERGY

Reaction rates are temperature dependent. The higher temperatures increase the kinetic energy of the (reactant) molecules, such that when two reacting molecules collide, they do so with a much greater force (more energy is dispersed within the collision system) causing bonds to rupture, atoms to rearrange, and new bonds (products) to form. The sum of these kinetic energies required for a reaction to occur is called the **activation energy** for the reaction.

The relationship between the reaction rate, expressed by **k'**, at the measured temperature, **T(K)**, and the activation energy, E_a, is expressed in the Arrhenius equation.

$$k' = Ae^{-Ea/RT}$$

A is a collision parameter for the reaction and **R** is the gas constant (=8.314J/mol K). When a reaction is performed at two temperatures, **T(1)** and **T(2)**, the ratio of the reaction rates can be determined.

$$\frac{k_1'}{k_2'} = \frac{Ae^{-Ea/RT(1)}}{Ae^{-Ea/RT(2)}}$$

We can simplify the ratio of these two equations by using natural logarithms.

$$\ln\frac{k_1'}{k_2'} = -\frac{E_a}{R}\left[\frac{1}{T(1)} - \frac{1}{T(2)}\right]$$

Therefore, the activation energy for a reaction can be calculated from determinations of rate constants at two temperatures.

In this experiment you will determined the activation energy, E_a, for the $[IO_3^-]/[SO_3^{2-}]$ system by determining the **k'** at room temperature and again at 70°C.

Techniques

- Technique 6c, d, page 7 — Heating Liquids
- Technique 10, page 12 — Reading a Meniscus
- Technique 11, page 13 — Pipetting a Liquid or Solution
- Technique 15, page 18 — Graphing Techniques

In addition, some techniques for timing reactions will be used.

Procedure

Read the entire experimental procedure before beginning. You should work with a partner– one of you will need to determine the time lapse for the reaction and the other will need to handle the solutions.

A. Test Reactions

The test reactions are prepared according to Table 34.1.

Table 34.1
Composition of Test Reactions

	Solution A			Solution B	
Test Reaction	0.1M HIO_3	Starch	Distilled H_2O	0.05M H_2SO_3	Distilled H_2O
1	20mL	5mL	75mL	10mL	90mL
2	15mL	5mL	80mL	10mL	90mL
3	10mL	5mL	85mL	10mL	90mL
4	6mL	5mL	89mL	10mL	90mL
5	5mL	5mL	90mL	10mL	90mL
6	3mL	5mL	92mL	10mL	90mL

1. Test Reaction #1. Prepare solution A for test reaction #1 in a 250mL beaker or Erlenmeyer flask and solution B in a 150mL beaker. The volumes of the HIO_3 and H_2SO_3 solutions should be measured with a 10mL pipet.[1] Stir each solution. Place a white sheet of paper beneath the beaker of solution A so that the appearance of the color is more evident (Figure 34.1).

2. The reaction begins when the H_2SO_3 from solution B is added to the HIO_3 from solution A. Start timing, in **seconds**, the reaction when the two solutions are combined–be as exact as possible in starting and stopping the timing of the reaction.[2] *O.K., get ready* ; quickly pour solution B into solution A ($B \rightarrow A$), **start** timing, and stir. The blue color appears suddenly so watch for its appearance and **stop** the timing. Record the time lapse to the nearest second. Record the temperature of the solution at T_1 on the Data Sheet; T_2 is the temperature for the reactions in Part C.

3. Thoroughly clean each beaker. Prepare solutions A and B for test reaction #2. Repeat the reaction procedure as described in Part A.2.

4. Continue with the data collection for the other test reactions in Table 34.1. If your instructor approves, prepare and collect data for other concentrations of HIO_3 in solution A.

---

[1] Do not intermix pipets, graduated cylinders, or beakers from one trial to the next. Cleanliness is important, especially in this experiment.

[2] The time (seconds) elapsed from the initial mixing of the solutions (A and B) until the deep blue color appears is to be recorded.

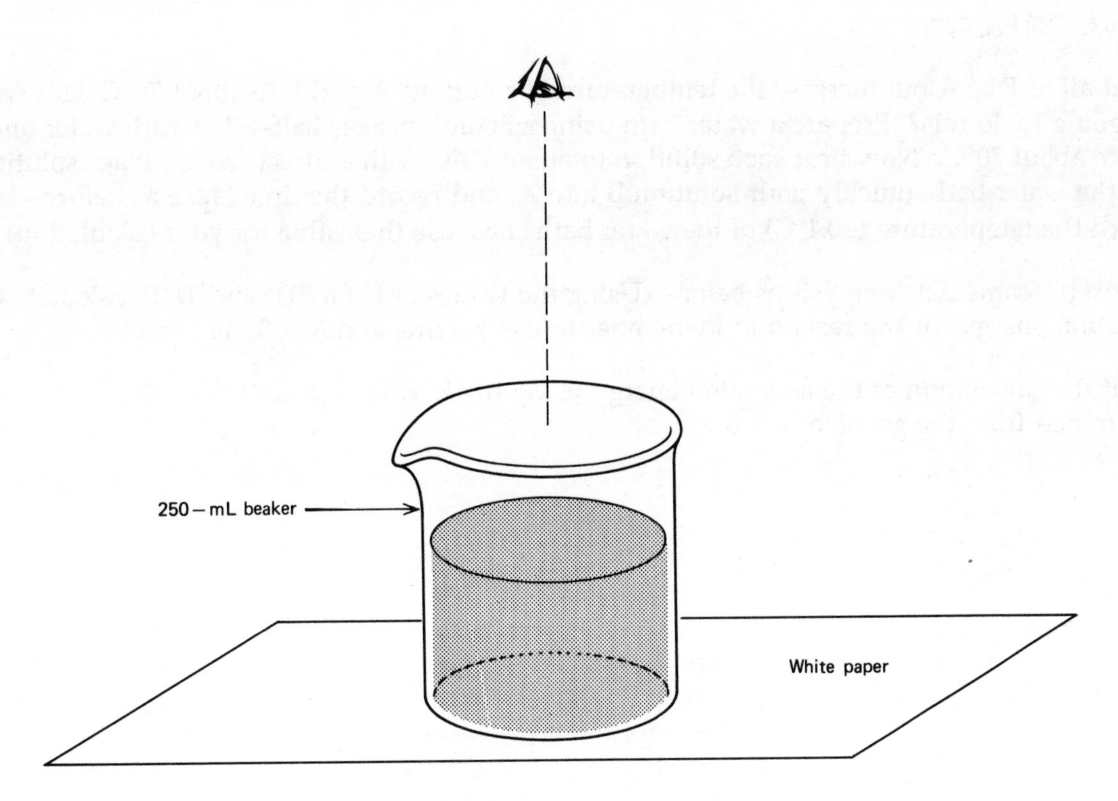

Figure 34.1
Viewing the Appearance of the I_2•starch Complex

B. DATA ANALYSIS AND CALCULATIONS

1. Determine the reciprocal of the time (seconds) elapsed for the reaction and record it as the reaction rate. Calculate the logarithm of this rate and record on the Data Sheet.

2. Calculate and record the initial molarity of the IO_3^-, $[IO_3^-]_i$, and the logarithm of $[IO_3^-]_i$ for each test reaction. Remember the total volume of the test reaction is 200mL.

3. Plot on linear graph paper "log (rate) *vs* log $[IO_3^-]_i$". Draw a straight line that passes as close as possible to the six data points. Calculate the slope of the line; its value should approximate a whole number, but record its actual value on the Report Sheet. This is the order, **p**, of IO_3^- in the reaction. Ask your instructor to approve your graph.

4. Use the values of rate (from Part B.1), $[IO_3^-]$i (from Part B.2), **p** (from Part B.3), and the rate law, **rate = k'$[IO_3^-]^p$**, to determine **k'** for the six test reactions. Calculate the average value of **k'** and express it with its proper units.

5. From the plot, extrapolate the straight line until it intersects the y–axis at x = 0 (or log$[IO_3^-]$ = 0). This y–value equals log k'; calculate **k'**.

C. Activation Energy

1. Repeat all of Part A but increase the temperature of solutions A and B to about 70°C. How are you going to do this? Prepare a water bath using a 600mL beaker half–filled with water and heat to about 70°C. Now heat each solution to about 70°C with a direct flame. Place solution A in the water bath, quickly pour solution B into A, and record the time lapse as before. Record the temperature (±0.1°C) of the water bath[3] and use this value for your calculations.

2. Perform the same data analysis as before. Using the values of k' (at T(1) and T(2)), calculate the activation energy for the reaction. Remember to use kelvins and R = 8.314 J/mol K.

3. Repeat the calculation of the activation energy using the k' values (at T(1) and T(2)) determined from the graph at x = 0.

---

[3] The temperature does not have to be exactly at 70°C, but the *actual* temperature should be recorded to ±0.1°C.

RATE LAW DETERMINATION-LAB PREVIEW

Date__________ Name____ Lab Sec. ______ Desk No. ______

1. a. If 3.5 hours are required to travel 165 miles, what is the *rate* of travel?

b. If 4 minutes and 10 seconds (250s) elapse in completing four laps of a 400m track, what is the running rate, in m/s, of the track star?

c. A total of 27.3 seconds elapses before a reaction is complete. What is the reaction rate?

2. A reaction between the gaseous substances C and D proceeds at a measurable rate to produce a sudden color change. The following set of data was collected for the reaction

$$3C + 2D \rightarrow 2F + G$$

$[C]_i$	$[D]_i$	Time for Color Change (seconds)	Rate (sec^{-1})
1.0×10^{-2}	1.0	30	__________
1.0×10^{-2}	3.0	10	__________
2.0×10^{-2}	3.0	1.3	__________
2.0×10^{-2}	1.0	3.7	__________
3.0×10^{-2}	3.0	0.37	__________

a. What is the order of C in the reaction? ______; of D in the reaction? ______

b. What is the rate law?

c. What is the specific rate constant, k?

d. At any given instant during the reaction,

the rate of appearance of F is ________ times the rate of disappearance of C.

the rate of disappearance of D is ________ times the rate of appearance of G.

the rate of appearance of G is ________ times the rate of appearance of F.

3. A set of rate data is plotted for the reaction of A_2 at several concentrations while the B concentration remains constant for $A_2 + B \rightarrow A_2B$.
   a. From the graph, determine the order of A in the reaction.

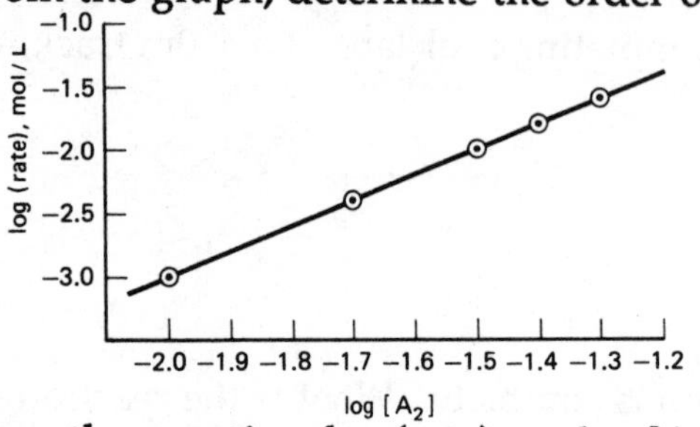

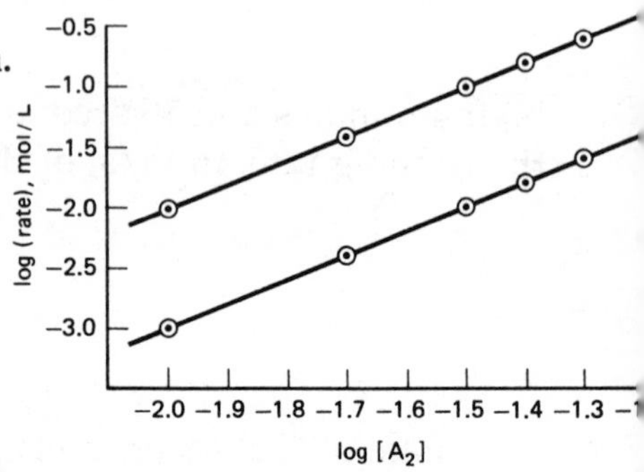

   b. From the equation, log (rate) = p log $[A_2]$ + log k, the graphed data, and the order of A_2 in the reaction, calculate k, the specific rate constant for the reaction.

4. A 1:1 mixture of 0.2M KIO_3 and 0.2M H_2SO_4 is used to prepare the 0.1M HIO_3 solution for today's experiment.
   a. How do you prepare 150mL of a 0.2M KIO_3 solution, starting with solid KIO_3?

   b. How do you prepare 150mL of a 0.2M H_2SO_4 solution, starting with 6.0M H_2SO_4?

RATE LAW DETERMINATION-DATA SHEET

Date________ Name______________________________________ Lab Sec. ____ Desk No.____

A. TEST REACTIONS

Test Reactions		1	2	3	4	5	6
1. Time elapsed, Δt (sec)	at T_1	_____	_____	_____	_____	_____	_____
	at T_2	---	---	---	---	---	---
2. Reaction temperature	T_1	_____	_____	_____	_____	_____	_____
	T_2	---	---	---	---	---	---

B. DATA ANALYSIS AND CALCULATIONS

1. rate (sec^{-1})	at T_1	_____	_____	_____	_____	_____	_____
	at T_2	---	---	---	---	---	---
2. log (rate)	at T_1	_____	_____	_____	_____	_____	_____
	at T_2	---	---	---	---	---	---
3. $[IO_3^-]_i$ (mol/L)	at T_1	_____	_____	_____	_____	_____	_____
	at T_2	---	---	---	---	---	---
4. log $[IO_3^-]_i$	at T_1	_____	_____	_____	_____	_____	_____
	at T_2	---	---	---	---	---	---

5. Plot log (rate) *vs* log $[IO_3^-]_i$

6. Instructor's approval of graph at T_1 ________ at T_2 __________

7. Value of **p** from graph ______________

8. Value of **k'** for each	at T_1	_____	_____	_____	_____	_____	_____
	at T_2	---	---	---	---	---	---

9. Average value of **k'** (calculated) at T_1 _______________

at T_2 ___________

10. Value of **k'** from graph at T_1 ______________

at T_2 ______________

E. ACTIVATION ENERGY

E_a from calculated k' values* at T_1 ______________

E_a from k' values from graph at T_2 ______________

* Show calculation of E_a here.

Questions

1. What would each of these changes in the experimental procedure have on the reaction rate?
   a. a slight increase in the concentration of starch? Assume no volume change. Explain.

   b. a slight increase in the concentration of the HIO_3 in solution A. Explain.

   c. a slight increase in the amount of water. Explain.

2. Two test reactions are the *minimum* needed to obtain the values of **p** and **k'** in this experiment. Explain the advantages that additional test reactions, as were performed in this experiment, have on the data that you analyzed.

3. What was happening in today's reaction between the time solutions A and B were mixed and the appearance of the blue color?

PREFACE TO QUALITATIVE ANALYSIS

A rapid and simple identification of an element's cation in a salt mixture is convenient for determining the components of a sample mixture. It is the characteristic physical and chemical properties of each element's cation that will allow us in the next series of experiments to separate and identify the presence of a particular element. For example, the Ag^+ ion is identified as being present by its precipitation as the chloride, i.e., AgCl(s). While other cations precipitate as the chloride, AgCl is the only one that is soluble in an ammoniacal solution.[1] Similarly, from the solubility rules that you have learned earlier in the course, Ba^{2+} can be separated from a number of other cations with the addition of SO_4^{2-}; $BaSO_4$ forms a white precipitate, while other cations (except Pb^{2+}, Sr^{2+}, and Hg_2^{2+}) generally form soluble sulfates.

Therefore, with enough knowledge of the chemistry of the various cations, a unique separation and identification procedure can be developed. Some procedures are one step, quick tests, while others are more exhausting. The procedure, however, must systematically eliminate all other ions that may interfere with the test. Cations can be classified according to groups that have similar chemical properties; cations within each group can then be further separated and characteristically identified. A procedure that follows this pattern of analysis is called **Qualitative Analysis**, one that we will follow in the next several experiments.

The separation and identification of the cations use many chemical principles, many of which we will cite as we proceed. These principles will include precipitation, ionic equilibrium, acids and base properties, pH, oxidation and reduction, and complex ion formation. To help you to understand these test procedures, each experiment presents some pertinent chemical equations, but you are asked to write equations for other reactions that occur in the separation and identification of the cations. This usually appears in the Lab Preview.

In order to complete these procedures, you will need to practice good laboratory techniques and, in addition, develop several new techniques. The techniques that you should review are

Technique 2, page 2	Transferring a Liquid
Technique 4a, d page 4	Separation of a Solid from a Liquid
Technique 6a, c, d, page 7	Heating Liquids
Technique 7a, b, page 9	Evaporation of Liquids
Technique 9a, c, page 10 & 12	Handling Gases
Technique 14, page 18	Testing with Litmus

In addition, most of the tests will be performed in 3mL (75mm) test tubes and reagents will be added with medicine droppers ($\cong$20 drops/mL). Do not mix the different medicine droppers with the various reagents you will be using. When mixing solutions in a test tube, agitate by tapping the side of the test tube. Break up a precipitate with a stirring rod or cork the test tube and invert, but ***never*** use your thumb! (Figure QA.1).

---

[1]This was one test procedure that prospectors for silver used in the early prospecting days.

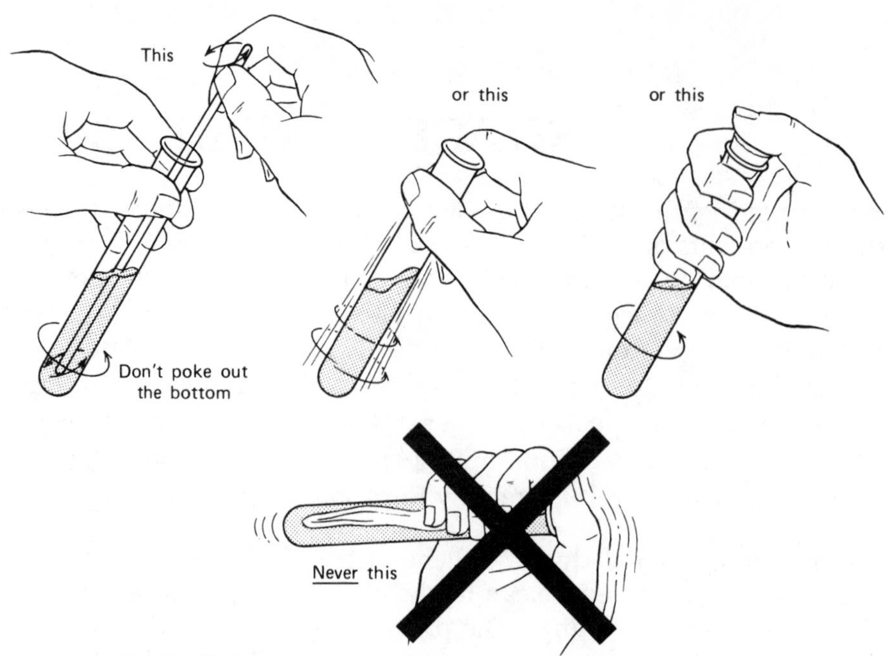

Figure QA.1
Technique for Mixing Solutions in a Test Tube

Washing a precipitate is often necessary to remove occluded impurities in your analysis. Add the water or wash liquid to the precipitate, mix thoroughly with a stirring rod, centrifuge, and decant. Failure to properly wash precipitates often leads to errors in the analysis (and arguments with your laboratory instructor!).

Since you will be frequently using the centrifuge in the next several experiments, *be sure* to read Technique 4d very closely in the Techniques section of this manual.

FLOW DIAGRAMS

To organize the qualitative analysis test procedure for the various cations, a flow diagram is often used. A flow diagram utilizes several standard notations.

- Brackets, [ ], indicate the use of a test reagent, written in molecular form.
- A single horizontal line,___, indicates that separation is made with a centrifuge.
- A double horizontal line, ═══, indicates soluble cations in the "qual scheme".
- Two short vertical lines, ‖, indicate that a precipitate has formed; these lines are drawn to the *left* of the single horizontal line.
- One short vertical line, |, indicates the supernatant and is drawn to the *right* of the single horizontal line.
- Two branching diagonal lines, ⌃, indicates a separation of the solution into two portions.
- A square box, ☐, placed around the compound or test, confirms the presence of the cation.

Flow Diagram for Cation Identification

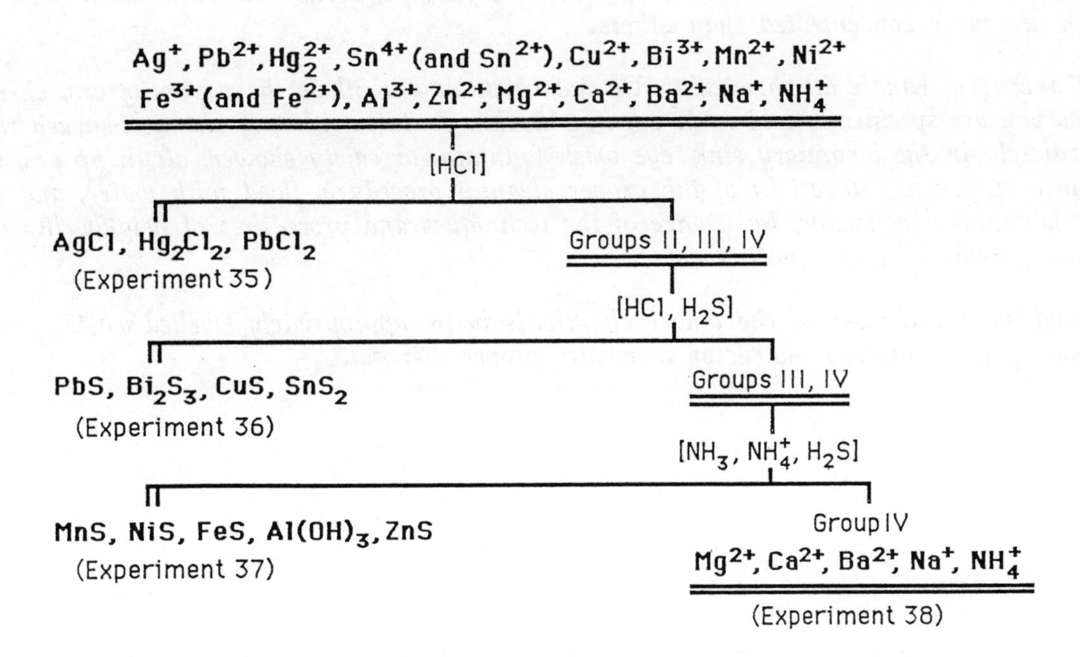

An unknown consisting of any number of the 16 cations analyzed in the next four experiments may be assigned at the conclusion of Experiment 38. Ask your instructor about that assignment. The preceding flow diagram is for these cations. A flow diagram for each group is presented in the corresponding experiment.

The following suggestions are offered while performing these "qual" experiments:

- *Always* read the procedure in detail and with an understanding of the principles before lab. Is extra equipment necessary? Is a hot water bath needed? What cautions are to be taken? Why is this reagent added at this time?
- Mark with a file, 1, 2, and 3mL intervals on a 75mm test tube in order to quickly estimate volumes.
- Keep a wash bottle filled with distilled (or de–ionized) water available at all times.
- Keep a number of *clean* medicine droppers and 75mm test tubes available; always rinse several times with distilled (or de–ionized) water immediately after their use.
- Maintain a "file" of confirmatory tests in the test tubes from the analysis on your *known* solution for each set of ions while conducting your unknown analysis; in that way, comparisons can be quick.
- Closely follow, simultaneously, the principles used in each test, the flow diagram, the Procedure, and the Data Sheet, during the analysis.

Caution: *In the next several experiments you will be handling a large number of chemicals (acids, bases, oxidizing and reducing agents, and, perhaps, even some toxic chemicals), some of which are more concentrated than others.*

Carefully, *handle all chemicals. Do not intentionally inhale the vapors of any chemical unless you are specifically told to do so;* ***avoid*** *skin contact with any chemicals–wash the skin immediately in the laboratory sink, eye wash fountain, or safety shower;* ***clean up*** *any spilled chemical–if you are uncertain of the proper cleanup procedure, flood with water, and consult your laboratory instructor;* ***be aware*** *of the techniques and procedures of neighboring chemists–discuss potential hazards with them.*

And finally, ***dispose of the waste chemicals*** *in the appropriately labelled waste containers. Consult your laboratory instructor to ensure proper disposal.*

PREFACE TO QUALITATIVE ANALYSIS-LAB PREVIEW

Date________Name____Lab Sec. _____Desk No._____

1. a. The approximate volume of a standard 75mm test tube is____________________.

   b. Small volumes of reagents are usually added using ____________________.

   c. A ____________ is commonly used to break up a precipitate.

   d. A ____________ is an instrument used to separate and compact a precipitate in a test tube.

   e. The clear solution above a precipitate is called the ____________________.

   f. The number of medicine drops of water equivalent to 1mL is about ____________.

   g. A solution should be centrifuged for (how long?) ____________________.

   h. On a flow diagram ||, means ____________________________________.

   i. On a flow diagram, a single horizontal line means ____________________.

   j. On a flow diagram, ☐ means ____________________________________.

2. Explain the procedure for balancing a centrifuge.

3. Explain the procedure for washing a precipitate.

EXPERIMENT 35
CATION IDENTIFICATION, I. Ag^+, Pb^{2+}, Hg_2^{2+}

Objective

- To identify the presence of a cation from a mixture of the cations, Ag^+, Pb^{2+}, Hg_2^{2+}

Principles

This experiment is a first in a series in which a set of reagents and a number of separation techniques are used to identify the presence of a particular cation in a group of cations that have similar chemical properties. We will approach this study from that of an experimental chemist–we will make some tests, write down our observations, and then write a balanced equation that agrees with our data.

As an introduction to the chemistry of these cations: the chloride salts of Ag^+, Pb^{2+}, and Hg_2^{2+} are "insoluble" and thus, can be separated from a large number of other cations[1], those that we'll study in subsequent experiments. However, because $PbCl_2$ is more soluble that either $AgCl$ or Hg_2Cl_2, it readily dissolves by heating the solution.[2]

Once the Pb^{2+} is "alone" in solution, its presence can be confirmed with the addition of CrO_4^{2-}; a yellow precipitate will form.

$$Pb^{2+}(aq) + CrO_4^{-}(aq) \rightarrow \underset{\text{(yellow)}}{PbCrO_4(s)}$$

In an ammoniacal solution, $AgCl$ dissolves to form the $Ag(NH_3)_2^+$ ion, but Hg_2Cl_2 undergoes an auto–oxidation–reduction reaction[3] to form Hg metal and mercuric amidochloride, $HgNH_2Cl$, a white insoluble salt. The appearance of grey–black Hg metal masks the white $HgNH_2Cl$ salt.

$$Hg_2Cl_2(s) + 2NH_3(aq) \rightarrow \underset{\text{(black)}}{Hg(l)} + \underset{\text{(white)}}{HgNH_2Cl(s)} + NH_4^+(aq) + Cl^-(aq)$$

---

[1]Review the solubility rules for salts in your text.

[2]If Pb^{2+} is the only cation in a solution, $PbCl_2$ may not precipitate with the addition of Cl^- because of its forming a supersaturated solution, especially if the water is at or above room temperature.

[3]An auto-oxidation-reduction reaction is also called a **disproportionation reaction.**

The $Ag(NH_3)_2^+$ ion is unstable in an acidic solution; the NH_3 forms NH_4^+ and the Ag^+ combines with the Cl^- that is still in solution to reform white insoluble AgCl.

$$Ag(NH_3)_2^+(aq) + Cl^-(aq) + 2H^+(aq) \rightarrow \underset{\text{(white)}}{AgCl(s)} + 2NH_4^+(aq)$$

As a guide to an understanding of the separation and identification of these three cations, use the following flow diagram . Review the Preface to Qualitative Analysis for an understanding of the symbolism.

Flow Diagram for Ag^+, Pb^{2+}, Hg_2^{2+}

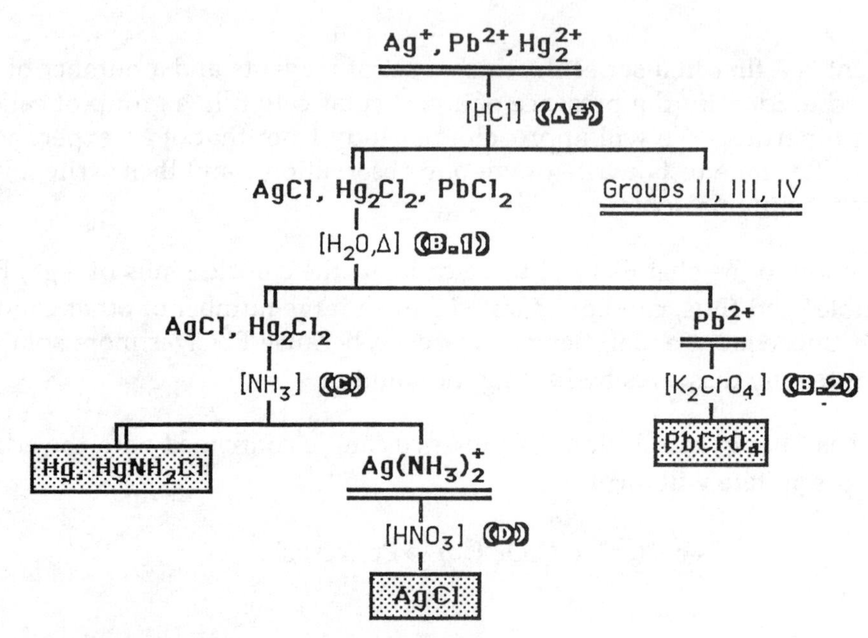

Refers to the Procedure, Part A

Procedure

To become familiar with the identification of these cations, take a sample that contains the three cations and analyze it according to the procedure. At each numbered superscript (example, #1), STOP, and record on the Data Sheet. After the presence of each cation has been confirmed, SAVE the test tube so that its appearance can be compared to that for your unknown sample.

A. Preparation of the Cations for Analysis

1. Place 1mL of the sample solution in a 75mm test tube, add 2 drops of 6M HCl (**Caution:** *Handle acid with care.*) (Figure 35.1) and centrifuge.[#1] Ask the instructor about the safe operation of the centrifuge. Test the supernatant with drops of 6M HCl for the additional formation of precipitate. Again centrifuge, if necessary. Decant the supernatant liquid.[4]

---

[4] If your sample is to be analyzed for additional cations, save this supernatant for that analysis, otherwise discard it.

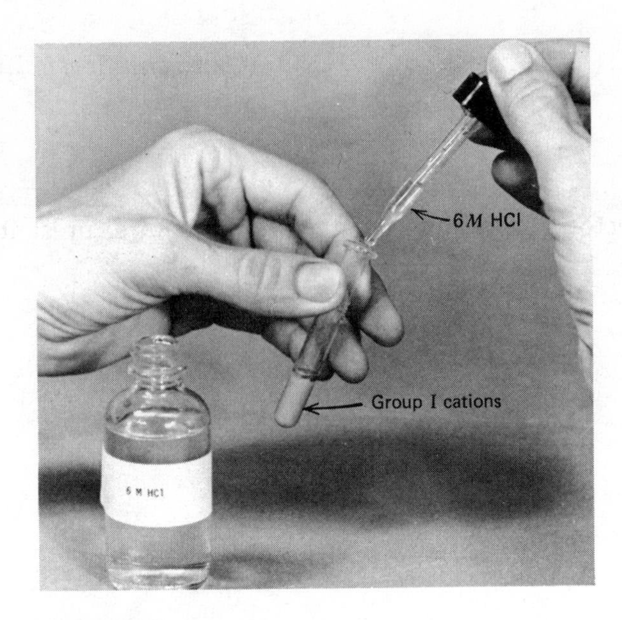

Figure 35.1
Precipitation of Ag^+, Pb^{2+}, Hg_2^{2+}

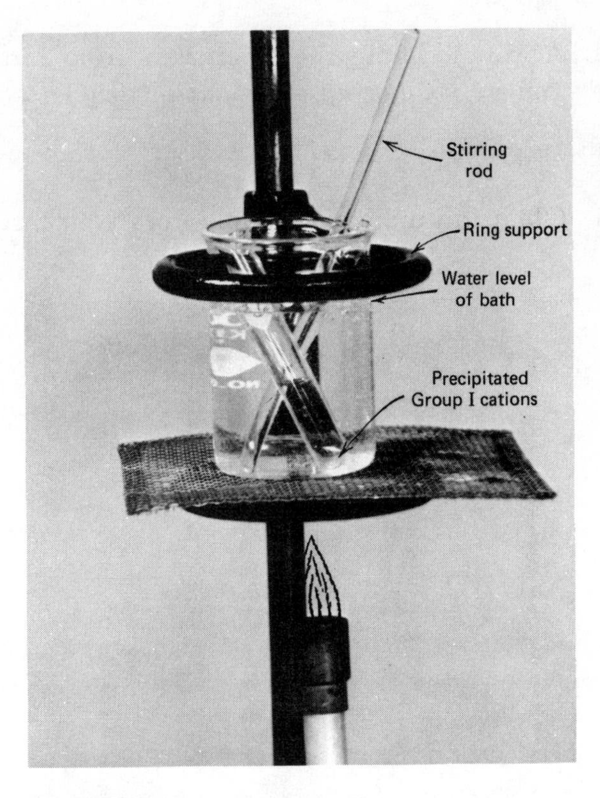

Figure 35.2
The Setup of a Hot Water Bath

B. Lead

1. To any precipitate that has formed (see footnote 2), add 10 drops of distilled (or de–ionized) water and heat nearly to boiling in hot water bath (Figure 35.2). Stir with a stirring rod. Quickly centrifuge. While still warm, decant the supernatant into a 75mm test tube, wash any precipitate with 4 drops of *hot* water, and combine the washing with the supernatant.[5] #2 Save any remaining precipitate for Part C.#3

2. To the supernatant add 2 drops of 1M K_2CrO_4. Centrifuge, if the formation of a precipitate is not evident.#4

C. Mercury

1. To any precipitate that remains after Part B.1, add 5 drops of 6M NH_3 (**Caution**: *Avoid inhalation and skin contact.*) If a grey–black precipitate forms,#5 centrifuge and decant the supernatant.#6

---

[5] Consult your instructor on the technique for washing a precipitate.

D. SILVER

1. Acidify to litmus the supernatant from Part C.1 with drops of 6M HNO_3 (**Caution:** *Avoid skin contact. Clean up spills immediately.*).#7

E. UNKNOWN

1. Obtain an unknown sample from your instructor. Determine which cation(s) is(are) present.

CATION IDENTIFICATION, I. Ag^+, Pb^{2+}, Hg_2^{2+}–LAB PREVIEW

Date________ Name______________________________________ Lab Sec. _____ Desk No._____

1. a. Find the Ksp values of AgCl, $PbCl_2$, and Hg_2Cl_2 in your text book.

   $K_{sp}(AgCl)$ =

   $K_{sp}(PbCl_2)$ =

   $K_{sp}(Hg_2Cl_2)$ =

   b. Calculate the molar solubility of each salt. Which salt is the least soluble? ____________________ Which salt is the most soluble? ____________________

2. Write balanced equations for

   a. the reaction of chloride ion with each of the three cations.

   b. the reaction of NH_3 with Hg_2Cl_2.

   c. the reaction of $Ag(NH_3)_2^+$ with acid and chloride ion.

CATION IDENTIFICATION, I. Ag^{+}, Pb^{2+}, Hg_2^{2+}–DATA SHEET

Date________ Name______________________________________ Lab Sec. _____ Desk No._____

Procedure Number and Ion	Test Reagent or Technique	Observation (Color or General Appearance)		Chemical(s) Responsible for Observation	Check (√) if Observed in Unknown
#1		ppt			
#2 Pb^{2+}		spnt			
#3		ppt			
#4		ppt			
#5 Hg_2^{2+}		ppt			
#6		spnt			
#7 Ag^{+}		ppt			

Cations present in unknown:____

Instructor's Approval:____

Questions

1. Name and give the color for the following.

   a. $AgCl$ ____________________ ____________________

   b. Hg_2Cl_2 ____________________ ____________________

   c. $PbCl_2$ ____________________ ____________________

   d. $HgNH_2Cl$ ____________________ ____________________

   e. Hg metal ____________________ ____________________

   f. $PbCrO_4$ ____________________ ____________________

   g. $Ag(NH_3)_2^{+}$ ____________________ ____________________

2. Why must the precipitate be washed with *hot* water in Part 2 of the Procedure?

3. What happens in Part 1 if 6M HNO_3 is substituted for 6M HCl?

4. What happens in Part 4 if 6M NaOH is substituted for 6M NH_3?

5. Identify a single reagent that separates

   a. AgCl from Hg_2Cl_2. ______________________________

   b. Ag^+ from Cu^{2+}. ______________________________

Experiment 36
Cation Identification, II.
Pb^{2+}, Sn^{4+}(and Sn^{2+}), Cu^{2+}, Bi^{3+}

Objective

- To identify one or more cations from a mixture of the cations, Pb^{2+}, Sn^{4+}(and Sn^{2+}), Cu^{2+}, and Bi^{3+}.

Principles

This is the second grouping of cations that are typically identified in the separation of a more complete listing of cations. While this second group consists of additional cations, namely, As^{3+}, Cd^{2+}, Hg^{2+}, and Sb^{3+}, we will learn of the chemical and physical properties of only those listed. Any qualitative analysis text has a more complete testing procedure for these additional cations.

In a *total* qualitative analysis of a sample, most of the lead(II) cation is removed as the chloride precipitate (Experiment 35), but, because of its relatively high solubility, especially in warm water, some may remain in solution. Therefore, it is important that we address its presence in this second group of cations as well.

The sulfide salts of the cations in this group are insoluble in solutions that are at least 0.3 M H^+ (pH = 0.5). If the pH is greater than 0.5, sulfide precipitation of cations in the third group (Experiment 37) may also precipitate. Therefore, pH control is extremely important in the separation of these two groups–there will be interference with the cation identification in these two groups if we are not careful to control pH.

The sulfide ion is produced *in situ* in the from of $H_2S(aq)$. The H_2S is generated from the thermal hydrolysis of thioacetamide, CH_3CSNH_2, in an acidic or basic solution.

$$CH_3CSNH_2(aq) + 2H_2O \xrightarrow{\Delta} CH_3COO^-(aq) + NH_4^+(aq) + H_2S(aq)$$

The slow generation of H_2S in the solution minimizes the concentration of the foul–smelling, highly toxic gas in the air. When H_2S is produced slowly, more compact sulfide precipitates form, making them easier to separate by centrifugation.

$H_2S(aq)$ is a weak, diprotic acid ionizing slightly to produce the sulfide ion that is necessary for cation precipitation.

$$H_2S(aq) + 2H_2O \rightleftharpoons 2H_3O^+(aq) + S^{2-}(aq)$$

Using the concept of LeChatelier's Principle, a high H^+ concentration (low pH) shifts this equilibrium to the left, leaving a low sulfide concentration; a low H^+ concentration (high pH) shifts the equilibrium to the right, increasing the sulfide concentration. Therefore, if cations precipitate at a low pH, then only small amounts of sulfide ion are needed for precipitation; this indicates that these cations have a low solubility and, also, a small K_{sp} value.

The chemical principles and laboratory techniques that are used to separate and identify the four cations in this experiment require you to be a careful, conscientious chemist. Read carefully the Procedure, review your laboratory techniques, and complete the Lab Preview before beginning the analysis–it will save you time and minimize frustration.

The flow diagram outlines the procedure for the separation and identification of the Pb^{2+}, Sn^{4+}(and Sn^{2+}), Cu^{2+}, and Bi^{3+} cations.

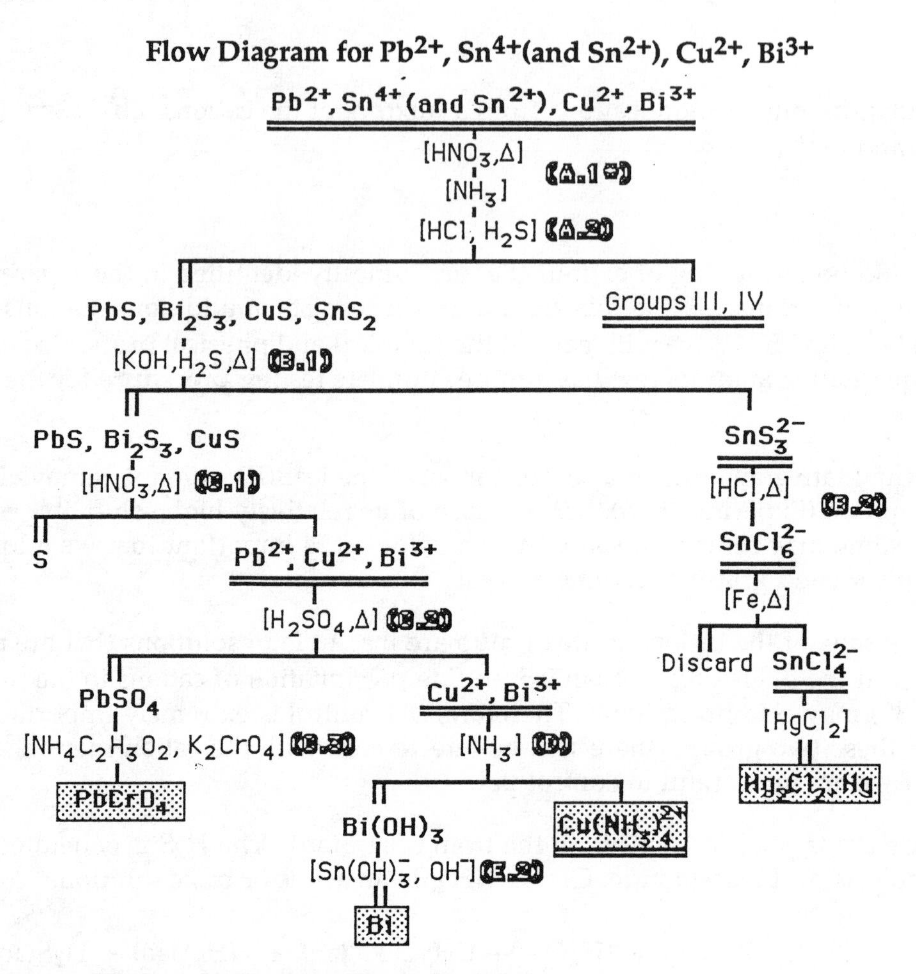

*Refers to the Procedure, Part A.1

PREPARATION OF THE CATIONS FOR ANALYSIS

We begin the analysis by treating our sample with hot HNO_3. Because SnS is a gelatinous precipitate and therefore difficult to separate and identify, any Sn^{2+} ion is oxidized to Sn^{4+} with the hot HNO_3. Thioacetamide is then added, generating the sulfide ion that precipitates the four cations.

TIN

SnS_2, a yellow precipitate, is separated from the other insoluble sulfides by adding additional thioacetamide, but in a *basic* solution. The higher sulfide ion concentration results in the formation of a soluble tin sulfide complex ion, SnS_3^{2-}, leaving PbS, Bi_2S_3, and CuS as precipitates.

Conc HCl, added to the SnS_3^{2-} complex produces a soluble hexachloro complex of tin(IV), $SnCl_6^{2-}$, which when treated with iron metal reduces it to a soluble tetrachloro complex ion of tin(II), $SnCl_4^{2-}$.

The confirmatory test for the presence of tin in the sample is the appearance of a grey–black precipitate when mercuric chloride, $HgCl_2$, is added to the solution containing $SnCl_4^{2-}$ ion.

$$2SnCl_4^{2-}(aq) + 3HgCl_2(aq) \rightarrow \underset{\text{(white)}}{Hg_2Cl_2(s)} + \underset{\text{(black)}}{Hg(s)} + 2SnCl_6^{2-}(aq)$$

Lead

The PbS, Bi_2S_3, and CuS precipitates are dissolved with hot HNO_3. The hot HNO_3 oxidizes the sulfide ion in the precipitate to (free) elemental sulfur, forming NO(g) as its reduction product. The solution, containing the dissolved cations, Pb^{2+}, Bi^{3+}, and Cu^{2+}, is treated with H_2SO_4 and strongly heated to drive off any excess HNO_3. Pb^{2+} then precipitates as $PbSO_4$.

To confirm the presence of Pb^{2+}, the $PbSO_4$ is dissolved with an excess of $C_2H_3O_2^-$; this forms the $Pb(C_2H_3O_2)_4^{2-}$ complex ion, which, when treated with CrO_4^{2-}, forms a yellow $PbCrO_4$ precipitate.

$$Pb(C_2H_3O_2)_4^{2-}(aq) + CrO_4^{2-}(aq) \rightarrow \underset{\text{(yellow)}}{PbCrO_4(s)} + 4C_2H_3O_2^-(aq)$$

Copper

The addition of aqueous NH_3 to the Bi^{3+} and Cu^{2+} cations still remaining solution precipitates Bi^{3+} as a white hydroxide, but complexes the Cu^{2+} as a deep–blue $Cu(NH_3)_4^{2+}$ complex ion, a confirmation of the presence of Cu^{2+}.

$$Cu^{2+}(aq) + 4NH_3(aq) \rightarrow \underset{\text{(blue)}}{Cu(NH_3)_4^{2+}(aq)}$$

Bismuth

When a freshly prepared sodium stannite, $Na_2Sn(OH)_4$, solution is added to the bismuth hydroxide precipitate, the Bi^{3+} ion is reduced to black bismuth metal, confirming in the presence of bismuth in the sample.

$$2Bi(OH)_3(s) + 3Sn(OH)_4^{2-}(aq) \rightarrow \underset{\text{(black)}}{2Bi(s)} + 3Sn(OH)_6^{2-}(aq)$$

Procedure

To become familiar with the identification of these cations, take a sample that contains the four cations and analyze it according to the procedure. At each numbered superscript (example #1), STOP, and record on the Data Sheet. After the presence of each cation has been confirmed, SAVE the test tube so that its appearance can be compared to that for your unknown sample.

A. PREPARATION OF THE CATIONS FOR ANALYSIS

1. To 2mL of test solution in an evaporating dish–either the supernatant from Experiment 35 or a solution containing the four test cations–add 10 to 15 drops of 6M HNO_3 (**Caution!**). Heat until a moist residue remains. Cool and add 1mL of water. Add drops of 6M NH_3 (**Caution!**) until the solution is basic to litmus.[1] Now make the solution acidic to litmus with drops of 6M HCl; then add 2 more drops. This should adjust the pH of the solution to about 0.5. Transfer the solution to a 75mm test tube.

2. Add 10 to 15 drops of 1M CH_3CSNH_2, heat in a hot water bath (≅ 95°C) for several minutes, cool, and centrifuge. Test for complete precipitation by repeating the thioacetamide addition to the supernatant. The precipitate contains the SnS_2, PbS, Bi_2S_3, and CuS salts;[#1] the supernatant contains cations that do not precipitate under these conditions (Experiments 37 and 38).

3. Wash the precipitate twice with 2M NH_4NO_3, stir, and discard each washing.

B. TIN

1. To the washed precipitate in Part A.3, add 2mL of 6M KOH and 3 drops of CH_3CSNH_2 solution. Heat the solution in a hot water bath (≅ 95°C) for at least 5 minutes. While the solution is still warm, centrifuge and then decant the supernatant[#2] into a 75mm test tube. Same the precipitate for Part C.[#3]

2. Reacidify (to litmus) the supernatant with 6M HCl (**Caution**: *Avoid inhalation or skin contact*); then add 12 more drops. Again heat in the hot water bath for 5 minutes. Cool, centrifuge, and decant the supernatant[#4] into a 75mm test tube. Add two *polished* pieces[2] of iron wire to this supernatant and reheat in the hot water bath for 3 to 5 minutes. Remove the excess iron wire and centrifuge. To this supernatant,[#5] add 5 drops of 0.2M $HgCl_2$. A grey–black precipitate confirms the presence of tin in the sample.[#6]

C. LEAD

1. Wash *twice* the precipitate from Part B.1 with 2–3 drops of 2M NH_4NO_3 and discard each washing. Add 15 drops of 6M HNO_3 and heat to boiling with a direct flame (**Caution**: *Read the technique for heating test tubes with a direct flame*) until the precipitates dissolve. Cool and centrifuge. Save the supernatant,[#7] discard any free sulfur.

---

[1]Review this technique for testing acidity and basicity.

[2]Remove the rust from the iron wire with 6M HCl and then rinse thoroughly with distilled water.

2. Transfer the supernatant to a crucible and set up an apparatus in the fume hood to heat the crucible with a direct flame (read your technique section of the manual). Add 3 to 5 drops of conc H_2SO_4 (**Caution**: *conc* H_2SO_4 *causes severe skin burns!*) and heat cautiously until the white dense fumes of SO_3 appear. This is a very important step in the analysis.[3] Cool, add 10 drops of water to the precipitate, which is $PbSO_4$,#8 and stir. Allow the precipitate to settle; decant the supernatant#9 into a 75mm test tube and save it for Part D.

3. Dissolve the $PbSO_4$ precipitate with 3 drops of 1M $NH_4C_2H_3O_2$.#10 Stir thoroughly; transfer the solution to a 75mm test tube and add 1 drop of 1M K_2CrO_4. A yellow precipitate,#11 *not* solution, confirms the presence of lead ion in the sample.

D. COPPER

1. Add drops of conc NH_3 (**Caution**: *Avoid inhalation or skin contact!*) to the solution from Part C.2. The deep–blue solution confirms the presence of Cu^{2+} in the sample.#12 Centrifuge, decant the solution, and save the precipitate#13 for Part E.

E. BISMUTH

1. Prepare a fresh $Na_2Sn(OH)_4$ solution by placing 2 drops of 1M $SnCl_2$ in a 75mm test tube, followed by drops of 6M NaOH until the $Sn(OH)_2$ precipitate just dissolves. Shake or stir the solution.

2. Add several drops of the $Na_2Sn(OH)_4$ solution to the precipitate from Part D. The immediate formation of a black precipitate#14 confirms the presence of bismuth in the sample.

F. UNKNOWN

1. Obtain an unknown sample from your instructor. Determine which cation(s) is(are) present.

---

[3] $PbSO_4$ is soluble in a strongly acidic sulfuric acid solution due to the formation of HSO_4^-. Heating the mixture until SO_3 fumes appear assures the removal of HNO_3 (Part C.1), of water (both have relatively low boiling points), and the decomposition of H_2SO_4 ($H_2SO_4 \rightarrow H_2O + SO_3$ (g)) at its boiling point (338°C). This removes all sources of H^+ and prevents the formation of HSO_4^-.

CATION IDENTIFICATION, II. Pb^{2+}, Sn^{4+} (AND Sn^{2+}), Cu^{2+}, Bi^{3+}–LAB PREVIEW

Date________ Name______________________________________ Lab Sec. _____ Desk No._____

1. What reagent converts Sn^{2+} to Sn^{4+} in preparation for the identification of tin in the sample?

2. What reagent separates Cu^{2+} from Bi^{3+} in solution?

3. a. Explain why HNO_3 is added to the precipitate in Part C.1.

   b. What is the origin of the elemental sulfur?

4. What is the color of

   a. $Cu(NH_3)_4^{2+}$ ____________________

   b. bismuth metal ____________________

   c. a mixture of PbS and SnS_2 ____________________

   d. a mixture of Hg_2Cl_2 and Hg ____________________

5. Write balanced equations for the following reactions cited in this experiment. Read the Principles and Procedure to assist you in writing the equations.

a. The precipitation of Sn^{4+}, Pb^{2+}, Bi^{3+}, and Cu^{2+} cations as the sulfide salts.

b. The reaction of SnS_2 with excess sulfide ion.

c. The reaction of iron metal with $SnCl_6^{2-}$.

d. The reactions of the PbS, Bi_2S_3, and CuS precipitates with hot HNO_3.

e. The reaction of $PbSO_4$ with excess $C_2H_3O_2^-$.

f. The reaction of Bi^{3+} with aqueous NH_3.

Cation Identification, II.
Pb^{2+}, Sn^{4+} (and Sn^{2+}), Cu^{2+}, Bi^{3+}—Data Sheet

Date__________Name____Lab Sec. _____Desk No._____

Procedure Number and Ion	Test Reagent or Technique	Observation (Color or General Appearance)		Chemical(s) Responsible for Observation	Check (√) if Observed in Unknown
#1		ppt			
#2 $Sn^{4+(2+)}$		spnt			
#3		ppt			
#4		spnt			
#5		spnt			
#6		ppt			
#7 Pb^{2+}		spnt			
#8		ppt			
#9		spnt			
#10		spnt			
#11		ppt			
#12 Cu^{2+}		spnt			
#13		ppt			
#14 Bi^{3+}		ppt			

Cations present in unknown:____

Instructor's Approval:______________________________________

Questions

1. What happens in Part B if the iron metal is *not* used in the procedure?

2. What happens in Part D if NaOH is substituted for NH_3?

3. What happens if conc HCl is added to the precipitate in Part D?

4. What happens in Part C.1 if 6M HCl is substituted for 6M HNO_3?

5. Cite a reagent(s) that separates

   a. SnS_2 from CuS ____________________

   b. Cu^{2+} from Bi^{3+} ____________________

   c. Cu^{2+} from Pb^{2+} ____________________

   d. Hg_2^{2+} from Hg^{2+} ____________________ (See Experiment 35)

EXPERIMENT 37
CATION IDENTIFICATION, III. Mn^{2+}, Ni^{2+}, Fe^{3+}(AND Fe^{2+}), Al^{3+}, Zn^{2+}

Objective

- To identify one or more cations from a mixture of the cations, Mn^{2+}, Ni^{2+}, Fe^{3+}(and Fe^{2+}), Al^{3+}, and Zn^{2+}.

Principles

In a systematic separation and identification of cations, the cations studied in this experiment form insoluble salts in a solution containing sulfide ion and having a pH near neutrality. While there are a number of additional cations that belong to this category, such as Cr^{3+} and Co^{2+}, we will study the cations listed in the title. The cations of this group do *not* precipitate as the chloride salts (Experiment 35) or as the sulfide salts in a solution with an approximate pH of 0.5 (Experiment 36). Therefore, a separation of the cations in this experiment from those cations in the previous two experiments is rather easily accomplished.

PREPARATION OF THE CATIONS FOR ANALYSIS

The Mn^{2+}, Ni^{2+}, Fe^{3+}(and Fe^{2+}), and Zn^{2+} cations precipitate as sulfides but Al^{3+} precipitates as the hydrated hydroxide. The sulfide ion is generated in the same manner as was done in Experiment 36, that is, from the hydrolysis of thioacetamide, CH_3CSNH_2; the thioacetamide hydrolyzes to produce H_2S, which subsequently ionizes to produce low concentrations of sulfide ion.

$$H_2S(aq) + 2H_2O \rightleftharpoons 2H_3O^+(aq) + S^{2-}(aq)$$

The addition of base, such as NH_3, causes the equilibrium to shift *right*, increasing the sulfide concentration to a high enough level to where the product of the molar concentrations of the cation and the sulfide ion exceeds the salt's K_{sp} value and the salt precipitates. In this basic solution, the Al^{3+} precipitates as $Al(OH)_3$.

The H_2S also serves as a reducing agent, reducing any Fe^{3+} to Fe^{2+}, forming elemental sulfur as its oxidized product. Therefore, when a sample containing all the test cations in this experiment is treated with thioacetamide in a solution made basic with NH_3, a precipitate of MnS, NiS, FeS, ZnS, and $Al(OH)_3$ forms.

The procedure for the separation and identification of these five cations is summarized in the following flow diagram. Follow it as you read through the Principles and Procedure.

Flow Diagram for Mn^{2+}, Ni^{2+}, Fe^{3+}(and Fe^{2+}), Al^{3+}, Zn^{2+} Identification

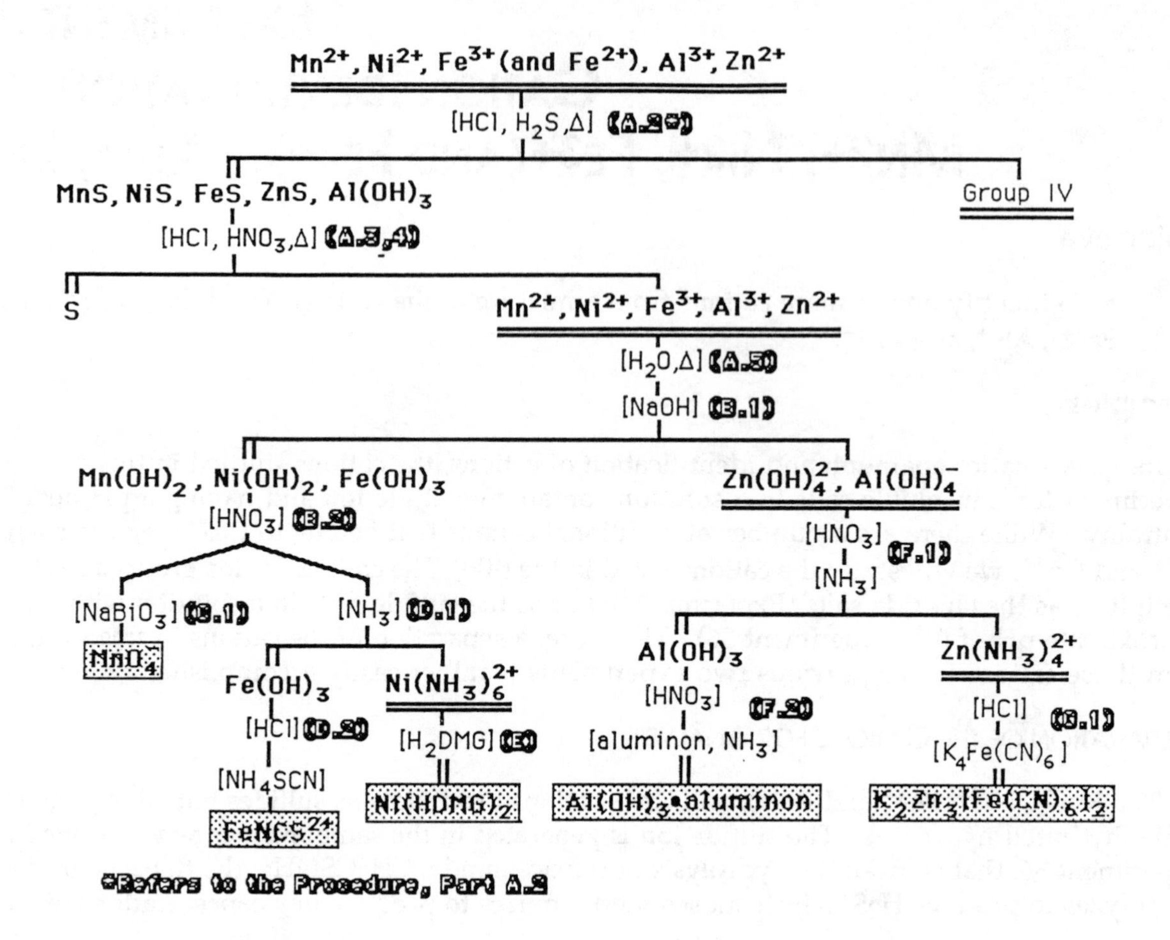

The MnS, FeS, ZnS, and $Al(OH)_3$ precipitates can be dissolved with a strong acid; however hot, conc HNO_3 is necessary to dissolve the NiS. The HNO_3 oxidizes the sulfide ion of the NiS, forming free, elemental sulfur and NO as products; in addition, the HNO_3 oxidizes Fe^{2+} to Fe^{3+} in the solution. Therefore, HNO_3 is used to dissolve the precipitates with the soluble cations, sulfur, and NO gas resulting in products.

SEPARATION OF MN^{2+}, NI^{2+}, AND FE^{3+} FROM ZN^{2+} AND AL^{3+}

Addition of strong base to an aqueous solution of the five cations precipitates Mn^{2+}, Ni^{2+}, and Fe^{3+} as gelatinous hydroxides, but the Zn^{2+} and Al^{3+} hydroxides are *amphoteric* and, thus, redissolve in the strong base, forming the aluminate, $Al(OH)_4^-$, and zincate, $Zn(OH)_4^{2-}$, ions.

The gelatinous hydroxides of Mn^{2+}, Ni^{2+}, and Fe^{3+} dissolve in nitric acid.

MANGANESE

A portion of the solution containing the dissolved hydroxides is treated with sodium bismuthate, $NaBiO_3$, a strong oxidizing agent that oxidizes Mn^{2+} to the characteristic purple permanganate ion, MnO_4^-, confirming the presence of Mn^{2+} in the sample.

$$14H^+(aq) + 2Mn^{2+}(aq) + 5BiO_3^-(aq) \rightarrow \underset{\text{(purple)}}{2MnO_4^-(aq)} + 5Bi^{3+}(aq) + 7H_2O$$

IRON

A second portion of the same solution is treated with an excess of NH_3; this precipitates brown $Fe(OH)_3$ but forms the hexaammine complex of Ni^{2+}, $Ni(NH_3)_6^{2+}$. Acid dissolves the $Fe(OH)_3$ precipitate, whereupon the solution is treated with thiocyanate ion, SCN^-, forming a blood–red complex ion with Fe^{3+}, $Fe(NCS)_6^{3-}$, a confirmation of the presence of Fe^{3+} in the sample.

$$Fe^{3+}(aq) + 6SCN^-(aq) \rightarrow \underset{\text{(blood–red)}}{Fe(NCS)_6^{3-}(aq)}$$

NICKEL

The confirmation of Ni^{2+} is made from the appearance of the bright brick–red precipitate formed with the addition of dimethylglyoxime, H_2DMG,[1] to a solution of the hexaamminenickel(II) complex ion.

$$Ni(NH_3)_6^{2+}(aq) + 2H_2DMG(aq) \rightarrow \underset{\text{(brick–red)}}{Ni(HDMG)_2(s)} + 2NH_4^+(aq) + 4NH_3(aq)$$

ALUMINUM

After acidifying the solution containing the aluminate and zincate ions and then adding NH_3, the Al^{3+} reprecipitates as the gelatinous hydroxide, but the Zn^{2+} forms the soluble tetraammine complex ion, $Zn(NH_3)_4^{2+}$.

The $Al(OH)_3$ precipitate is dissolved with HNO_3, aluminon reagent[2] is added, and the $Al(OH)_3$ is reprecipitated with the addition of NH_3. The aluminon reagent, a red dye, readily adsorbs onto the surface of the gelatinous $Al(OH)_3$ precipitate giving it a pink/red appearance. This confirms the presence of Al^{3+} in the sample.

$$Al^{3+}(aq) + 3NH_3(aq) + 3H_2O + \text{aluminon} \rightarrow \underset{\text{(pink/red)}}{Al(OH)_3 \bullet \text{aluminon}(s)} + 3NH_4^+(aq)$$

---

[1] Dimethylglyoxime, abbreviated as H_2DMG for convenience in this experiment, is an organic complexing agent, specific for the precipitation of the nickel ion.

[2] The aluminon reagent is the ammonium salt of aurin tricarboxylic acid, a red dye.

ZINC

When potassium hexacyanoferrate(II), $K_4Fe(CN)_6$, is added to an acidified solution of the $Zn(NH_3)_4^{2+}$ ion, a light green precipitate of $K_2Zn_3[Fe(CN)_6]_2$ forms, confirming the presence of Zn^{2+} in the sample.

$$3Zn(NH_3)_4^{2+}(aq) + K_4Fe(CN)_6(aq) + 12H^+(aq) \rightarrow \underset{(\textit{light}\ \text{green})}{K_2Zn_3[Fe(CN)_6]_2(s)} + 12NH_4^+(aq)$$

Procedure

To become familiar with the identification of these cations, take a sample that contains the five cations and analyze it according to the procedure. At each numbered superscript (example,[#1]), STOP, and record on the Data Sheet. After the presence of each cation has been confirmed, SAVE the test tube so that its appearance can be compared to that for your unknown sample.

Caution: *A number of concentrated and 6 molar acids and bases are used in the analysis of these cations. Handle each of these solutions with care. Read the Lab Safety section in Experiment 1 for instructions in handling acids and bases.*

A. Preparation of the Cations for Analysis

1. If your sample contains only the cations in this experiment, you will *not* need to precipitate the cations in a sulfide solution at the relatively high pH, but instead you may proceed immediately to Part B.

2. To 2mL of a test solution that may contain cations for analysis in Experiment 38, add 1mL of 2M NH_4Cl. Add 6M NH_3 until the solution is basic to litmus and then add 5 additional drops. Saturate the solution with H_2S by adding 6 drops of 1M CH_3CSNH_2. Heat the solution in a hot water bath ($\cong$ 95°C) for several minutes, cool, and centrifuge. Save the precipitate[#1] for Part A.3 and the supernatant for analysis in Experiment 38.

3. Wash twice the precipitate[3] with 2mL of water and discard the washings. Add 5 drops of 6M HCl and 5 drops of 6M HNO_3 (**Caution:** *Be careful in handling acids*) and again heat in the hot water bath until the precipitates dissolve. Cool and centrifuge; save the supernatant[#2] and discard any free, elemental sulfur.

4. Transfer the supernatant to an evaporating dish and heat until a moist residue remains (*not* to dryness). Add 1–2mL of conc HNO_3 (**Caution:** *conc* HNO_3 *is a strong oxidizing agent–do not allow contact with the skin or clothing!*) and reheat with a 'cool' flame (Figure 37.1) until a moist residue again forms.

---

[3]See the Preface to Qualitative Analysis for the proper technique in washing precipitates.

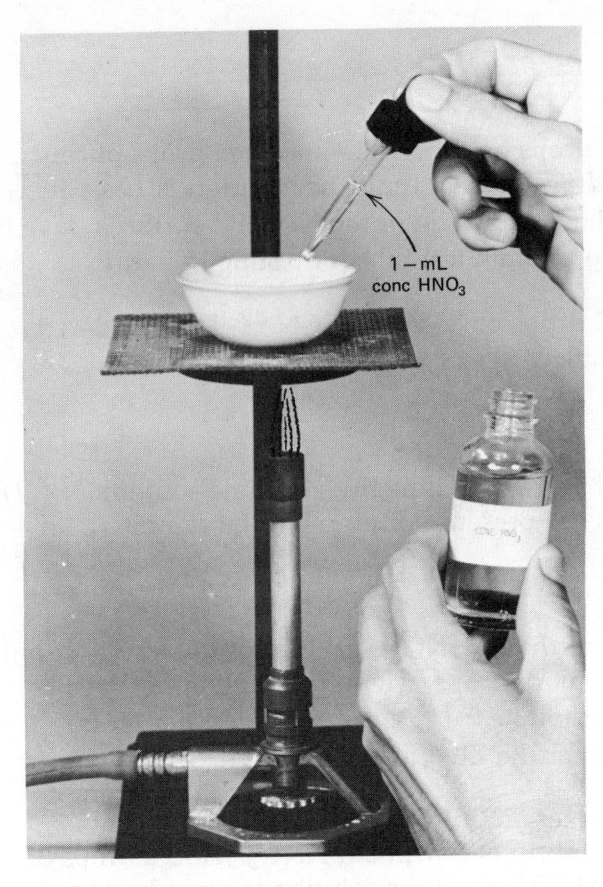

Figure 37.1
Addition of conc HNO_3 (Caution!!) to Moist Residue

5. Dissolve the residue in 1–2mL of water and transfer the solution to a 75mm test tube.

B. Separation of Ni^{2+}, Fe^{3+}, and Mn^{2+} from Zn^{2+} and Al^{3+}

1. To the solution obtained from Part A.5 *or* from a test solution (in a 75mm test tube) containing the five test cations for this experiment only, add 15 drops of 6M NaOH (**Caution:** *Do not permit skin contact*). Centrifuge and save the precipitate.#3 Decant the supernatant#4 into a 75mm test tube for Part F.

2. Dissolve the precipitate#5 with 1–2mL of 15M HNO_3 (**Caution:** *Be careful!!*). Heat the solution in the hot water bath.

C. Manganese

1. Decant about 0.5mL of the solution from Part B.2 into a 75mm test tube. Add a slight excess of $NaBiO_3$ to the solution and centrifuge.#6 The deep–purple MnO_4^- confirms the presence of manganese in the sample.[4]

---

[4]If the deep–purple color forms and then fades, Cl^- is present and is being oxidized by the MnO_4^-; add more $NaBiO_3$.

D. IRON

1. To the resulting solution from Part B.2, add 10 drops of 2M NH_4Cl and then drops of conc NH_3 (**Caution**: *Do not inhale–use a fume hood if available*) until the solution is basic to litmus and then a 1 drop of excess to ensure the complexing of the Ni^{2+}. Centrifuge, save the precipitate,#7 and transfer the supernatant#8 to a 75mm test tube for testing in Part E.

2. Dissolve the precipitate with 6M HCl and add 2 drops of 0.1M NH_4SCN.#9 The blood–red color confirms the presence of iron in the sample.

E. NICKEL

1. To the supernatant solution from Part D.1, add 3 drops of dimethylglyoxime solution.#10 The appearance of a brick–red precipitate confirms the presence of nickel in the sample.

F. ALUMINUM

1. Acidify the supernatant from Part B.1 to litmus with 6M HNO_3. Add drops of 6M NH_3 until the solution is now basic to litmus; then add 5 more drops. Heat the solution in the hot water bath for several minutes to digest the precipitate.#11 Centrifuge and decant the supernatant#12 into a 75mm test tube and save for the Zn^{2+} analysis in Part G.

2. Wash twice the precipitate with 2–3mL of hot water and discard each washing. Centrifugation may be necessary. Dissolve the precipitate with 6M HNO_3. Add 2 drops of the aluminon reagent, stir, and add drops of 6M NH_3 until the solution is again basic to litmus and a precipitate reforms.#13 Centrifuge the solution–if the $Al(OH)_3$ precipitate is now pink or red and the solution is colorless, Al^{3+} is present in the sample.

G. ZINC

1. To the supernatant from Part F.1, add 6M HCl until the solution is acid to litmus; then add 3 drops of 0.2M $K_4Fe(CN)_6$ and stir. A *very light* green precipitate#14 confirms the presence of Zn^{2+} in the sample. Centrifugation may be necessary.

H. UNKNOWN

1. Ask your laboratory instructor for a sample containing one or more of the cations from this experiment and determine those present.

CATION IDENTIFICATION, III
Ni^{2+}, Fe^{3+} (AND Fe^{2+}), Mn^{2+}, Al^{3+}, Zn^{2+}-LAB PREVIEW

Date__________Name____Lab Sec. _____Desk No._____

1. Give the formula of a reagent that precipitates

   a. Pb^{2+} but not Cu^{2+} ____________________

   b. Mn^{2+} but not Zn^{2+} ____________________

   c. Fe^{3+} but not Al^{3+} ____________________

2. Give the formula of a reagent that dissolves

   a. $NiCl_2$ but not NiS ____________________

   b. FeS but not NiS ____________________

   c. $Al(OH)_3$ but not $Mn(OH)_2$ ____________________

3. $Al(OH)_3$ and $Zn(OH)_2$ are amphoteric. What does the word amphoteric mean?

4. What is the color of

   a. $Ni(NH_3)_6^{2+}$ ____________________

   b. $Fe(NCS)_6^{3-}$ ____________________

   c. $Ni(HDMG)_2$ ____________________

   d. MnO_4^- ____________________

   e. the precipitate in Part A.2 ____________________

5. Write balanced equations for the following reactions cited in this experiment. Read the Principles and Procedure to assist you in writing the equations.

   a. The precipitation of the five cations in a sulfide solution with a pH near neutrality.

   b. The reduction of Fe^{3+} with H_2S.

   c. The dissolving of FeS, ZnS, MnS, and $Al(OH)_3$ with strong acid.

   d. The dissolving of NiS with HNO_3.

   e. The reaction of the five cations with an excess of strong base.

   f. The reactions of Fe^{3+} and Ni^{2+} with NH_3.

   g. The reactions of Al^{3+} and Zn^{2+} with NH_3.

   h. The oxidation of S^{2-} with HNO_3–NO is the reduction product of HNO_3.

CATION IDENTIFICATION, III
Mn^{2+}, Ni^{2+}, Fe^{3+} (AND Fe^{2+}), Al^{3+}, Zn^{2+}-DATA SHEET

Date________ Name______________________________________ Lab Sec. ____ Desk No.____

Procedure Number and Ion	Test Reagent or Technique		Observation (Color or General Appearance)	Chemical(s) Responsible for Observation	Check (√) if Observed in Unknown
#1		ppt			
#2		spnt			
#3		ppt			
#4		spnt			
#5 Mn^{2+}		spnt			
#6		spnt			
#7 $Fe^{3+(2+)}$		ppt			
#8		spnt			
#9		spnt			
#10 Ni^{2+}		ppt			
#11 Al^{3+}		ppt			
#12		spnt			
#13		ppt			
#14 Zn^{2+}		ppt			

Cations present in unknown:____

Instructor's Approval:____

Questions

1. How is Fe^{3+} reduced to Fe^{2+} in Part A.2?

2. Nitric acid functions as an oxidizing agent and as an acid in Parts A.3 and A.4.
   a. Identify the two substances that HNO_3 oxidizes.

   b. How does it function only as an acid?

3. What ions precipitate from solution in Part B.1 if NH_3 is substituted for NaOH? Explain.

4. What happens in Part D.1 if NaOH is substituted for NH_3?

5. What happens in Part F.1 if NaOH is substituted for NH_3?

6. Identify a reagent(s) that will

   a. precipitate Cu^{2+} but not Ni^{2+} ____________________

   b. dissolve $Al(OH)_3$ but not $Fe(OH)_3$ ____________________

   c. precipitate Fe^{3+} but not Ni^{2+} ____________________

   d. dissolve ZnS but not NiS ____________________

Experiment 38
Cation Identification, IV. Mg^{2+}, Ca^{2+}, Ba^{2+}, Na^{+}, NH_4^{+}

Objective

- To separate and identify the presence of Mg^{2+}, Ca^{2+}, Ba^{2+}, Na^{+}, and NH_4^{+} ions in a composite sample

Principles

The salts of the cations in this experiment are generally considered soluble, especially those of Na^{+} and NH_4^{+}. The chloride (Experiment 35) and sulfide (Experiments 36 and 37) salts of Mg^{2+}, Ca^{2+}, and Ba^{2+} are also soluble. Since Na^{+} and NH_4^{+} are often used for test reagents, it is always advisable to test for these cations on a fresh (original) sample.

Many cations emit characteristic colors when they are electronically excited with a Bunsen flame. A "flame test" will be made on each of the metallic cations to help in our identification process.

The NH_4^{+} ion is stable in an acidic solution, but when its solution is made basic, the solution equilibrium shifts to evolve NH_3 gas. With gentle heat, the NH_3 gas is driven from the solution and can be detected by its odor or by a litmus test.

$$NH_4^{+}(aq) + OH^{-}(aq) \xrightarrow{\Delta} NH_3(g) + H_2O$$

To separate and identify the alkaline–earth metal ions, Mg^{2+}, Ca^{2+}, and Ba^{2+}, requires you to practice good laboratory technique. We will use the anions, SO_4^{2-}, $C_2O_4^{2-}$, and HPO_4^{2-} to complete the analysis. Some care in pH adjustments will be necessary.

A flow diagram for the analysis is to be completed in the Lab Preview. Be sure to do this prior to entering the laboratory.

Procedure

To become familiar with the identification of these cations, take a sample that contains the five cations and analyze it according to the procedure. At each numbered superscript (example, #1), STOP, and record on the Data Sheet. After the presence of each cation has been confirmed, SAVE the test tube so that its appearance can be compared to that for your unknown sample.

A. Flame Tests

1. Concentrate 5mL of the sample solution to a moist residue in an evaporating dish.

2. Clean a coiled platinum wire, sealed in a non–heat conducting handle, by heating it with the nonluminous flame of a Bunsen burner until the flame is colorless (Figure 38.1). Dip the platinum wire into the moist residue and return it to the flame. Note the color.[#1] Note that if your unknown contains several of these cations, a mixture of colors will result.

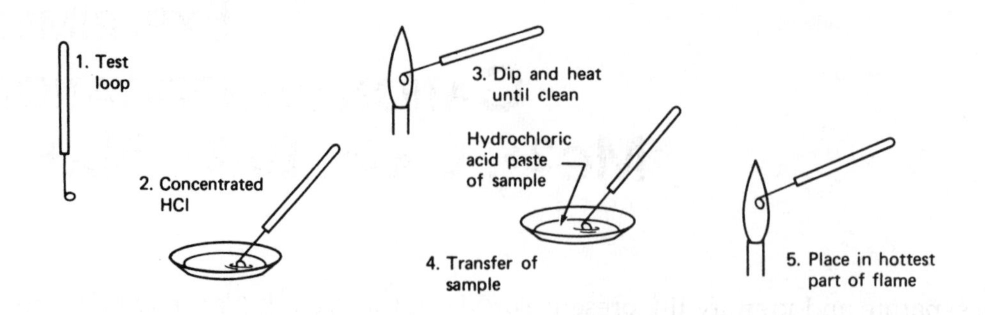

Figure 38.1
Technique for Performing a Flame Test

B. NH_4^+ TEST

1. Place 1mL of the original test solution in a 75mm test tube. Moisten a piece of red litmus paper. Add 5 drops of 6M NaOH (**Caution!!**) (be careful not to let NaOH contact the litmus paper) and, with the litmus paper over the mouth of the test tube, *carefully* warm with a *gentle* flame.#2

2. Using the proper technique, check the odor of the gas evolved.

C. BARIUM

1. To a 1mL of the test sample from Part B, acidify to litmus with 6M HCl and add drops of K_2CrO_4 until any precipitation appears complete.#3 Centrifuge and decant the supernatant.

2. Dissolve any precipitate with 12M HCl (**Caution**: *Do allow skin contact. Avoid inhalation. Clean up any spills.*) and perform a flame test.#4

D. CALCIUM

1. If the supernatant from Part C.1 is acidic, make it slightly basic to litmus with 6M NH_3 (**Caution!**). Add 2-3 drops of 1M $K_2C_2O_4$.#5 Centrifuge and decant the supernatant.

2. Dissolve any precipitate with 6M HCl and perform a flame test.#6

E. MAGNESIUM

1. Add 1-2 drops of 6M NH_3 to the supernatant from Part D.1. Add 2-3 drops of 1M Na_2HPO_4, heat in a hot water (≅90°C) bath, and allow to stand. Any precipitate#7 may be slow in forming; be patient.

F. UNKNOWN

1. Obtain a sample containing an unknown number of cations from this experiment. Analyze it according to this procedure and report your findings to your laboratory instructor.

CATION IDENTIFICATION, IV. MG^{2+}, CA^{2+}, BA^{2+}, NA^{+}, NH_4^{+}–LAB PREVIEW

Date________ Name______________________________________ Lab Sec. ____ Desk No.____

1. Record the K_{sp} values (if they are considered insoluble salts) of the following salts. Use your textbook or other reference book, such as the CRC, *Handbook of Chemistry and Physics*.

K_{sp}	Mg^{2+}	Ca^{2+}	Ba^{2+}
SO_4^{2-}	----------------	----------------	----------------
$C_2O_4^{2-}$	----------------	----------------	----------------
$NH_4PO_4^{2-}$	----------------	----------------	----------------

2. Write balanced equations for the precipitation reactions of

   a. Mg^{2+} and $NH_4PO_4^{2-}$

   b. Ca^{2+} and $C_2O_4^{2-}$

   c. Ba^{2+} and SO_4^{2-}

3. Explain why the tests for Na^{+} and NH_4^{+} are performed on original test solutions.

4. Explain the difficulty that may arise if a flame test were used to identify a single cation in a mixture containing several cations.

5. Construct a flow diagram for the five cations in this experiment. Read the Principles and Procedure to help you in its construction.

Cation Identification, IV. Mg^{2+}, Ca^{2+}, Ba^{2+}, Na^{+}, NH_4^{+}– Data Sheet

Date_________Name____Lab Sec. _____Desk No._____

Procedure Number and Ion	Test Reagent or Technique		Observation (Color or General Appearance)	Chemical(s) Responsible for Observation	Check (√) if Observed in Unknown
#1 Na^{+}	flame	---		-----	
#2 NH_4^{+}		gas			
#3 Ba^{2+}		ppt			
#4	flame	---		-----	
#5 Ca^{2+}		ppt			
#6	flame	---		-----	
#7 Mg^{2+}		ppt			

Cations present in unknown:____

Instructor's Approval:____

Questions

1. What is the color of the flame test for

   a. Na^{+} _______________ c. Mg^{2+} _______________

   b. Ca^{2+} _______________ d. Ba^{2+} _______________

2. What is the color of the salts,

   a. $MgNH_4PO_4$ _______________ b. CaC_2O_4 _______________

   c. $BaSO_4$ _______________

3. Write the formula of a reagent that separates

   a. NH_4^+ from Ba^{2+} ______________

   b. Ba^{2+} from Ca^{2+} ______________

   c. Ba^{2+} from Mg^{2+} ______________

   d. Mg^{2+} from Ca^{2+} ______________

4. What happens if H_2SO_4 is used instead of HCl in Part C.2? Refer to the solubility rules in your textbook.

5. What happens if NaOH is used instead of NH_3 in Part E.1? (Use the solubility rules in your textbook to support your statement.)?

EXPERIMENT 39
COORDINATION CHEMISTRY

Objectives

- To qualitatively prepare several complex ions and compare their stabilities
- To quantitatively synthesize an inorganic coordination compound
- To test for the purity and stability of the prepared compound

Principles

A coordination compound consists of at least one complex ion— a metal ion, usually a transition metal ion, bonds to one or more Lewis bases. These bases, called **ligands**, generally form 2, 4, or 6 bonds to the metal ion; the number of bonds is the **coordination number**. Most ligands have only a single electron–pair donor site, but others have as many as 2, 3, or more rarely, 4 and 6.

In the complex ion, $Ag(NH_3)_2^+$, the two ammonia molecules are the ligands, and the lone electron pair on each nitrogen bonds to the Ag(I) ion. The structure is linear.

$$[H_3N: \rightarrow Ag \leftarrow :NH_3]^+$$

The ferric ion in the complex ion, $[Fe(CN)_6]^{3-}$, has a coordination number of 6; the complex ion forms an octahedral structure, with the electron–pair donor site on the C atom of the $:C{=}N:^-$ ligand.

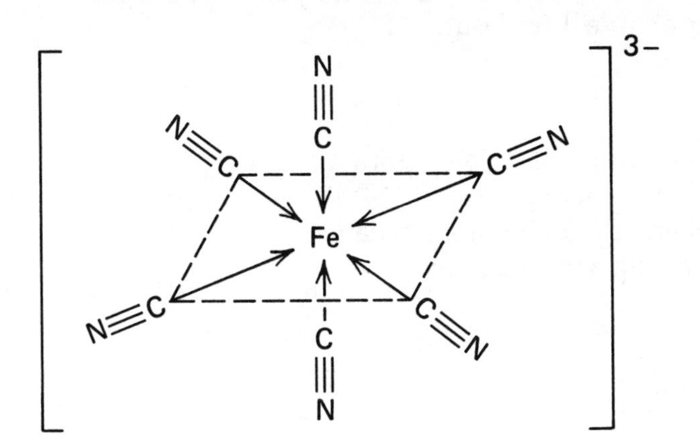

Figure 39.1
The Structure of the $(Fe(CN)_6)^{3-}$ Complex Ion

Since the $[Fe(CN)_6]^{3-}$ complex ion carries a 3^- charge and each CN^- ligand has a 1^- charge, the Fe in the complex ion must be Fe(III).

In this experiment, you will qualitatively prepare a number of Cu(II), Ni(II), and Co(II) complex ions and compare their stability using a number of different ligands. You will then quantitatively prepare the coordination compound $[Co(NH_3)_4CO_3]NO_3$. Four $:NH_3$ molecules and 1 CO_3^{2-} ion form bonds to Co(III). The CO_3^{2-} bonds twice to the Co(III) ion; two oxygen atoms serve as Lewis bases to the Co(III) ion.

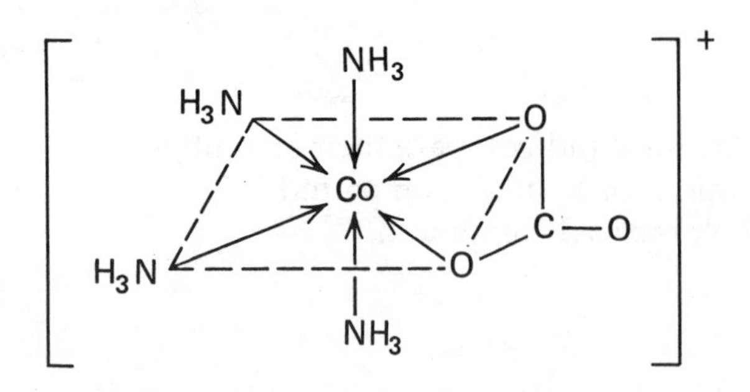

Figure 39.2
The Structure of the $(Co(NH_3)_4CO_3)^+$ Complex Ion

The complex ion has a 1^+ charge and the NO_3^- serves as the neutralizing anion to the complex cation.

$[Co(NH_3)_4CO_3]NO_3$ dissolves in water producing two ions, $[Co(NH_3)_4CO_3]^+$ and NO_3^-; this ionizes in solution just like that of $NaNO_3$.

$$[Co(NH_3)_4CO_3]NO_3 \xrightarrow{H_2O} [Co(NH_3)_4CO_3]^+(aq) + NO_3^-(aq)$$

$$NaNO_3 \xrightarrow{H_2O} Na^+ + NO_3^-$$

Addition of Ca^{2+} to the $[Co(NH_3)_4CO_3]NO_3$ solution yields *no* $CaCO_3$ precipitate because the CO_3^{2-} ion remains bound to the Co(III) ion in the complex ion. How do you suppose the presence of NH_3 affects the pH of the solution? We'll find out.

Techniques

• Technique 4c, d, page 4	Separation of a Solid from a Liquid
• Technique 6d, page 7	Heating Liquids
• Technique 13, page 16	Using the Laboratory Balance
• Technique 14, page 18	Testing with Litmus

Procedure

A. A Look at Some Complex Ions

1. Chloro Complexes. Place 10 drops of 0.1M $CuSO_4$, 0.1M $Ni(NO_3)_2$, and 0.1M $CoCl_2$ in each of three 75mm test tubes. Add 1mL of conc HCl (**Caution:** *Handle carefully, do not allow conc HCl to contact skin or clothing. Flush immediately with water.*) to each. Tap the test tube to stir. Compare the color of the solution with the original solutions. Record your observations on the Data Sheet.

   Slowly add 1–2mL of water to each test tube. Compare the solution colors to 2mL of the original solution. Does the original color return? What does this tell you about the stability of the chloro complex ions?

2. Copper Complex Ions. Place 10 drops of 0.1M $CuSO_4$ in each of five 75mm test tubes. Place 5 drops of conc NH_3 (**Caution**: *Avoid breathing its vapors.*) in the first test tube, 5 drops of ethylenediamine (en) (**Caution**: *Avoid breathing its vapors.*) in the second, 0.2M ammonium tartrate, $(NH_4)_2tar$ in the third, 0.1M KSCN in the fourth. If a precipitate forms in any one of the solutions, add an excess of the ligand–containing solution. Compare the solutions with that in the fifth test tube.

   Add 3 drops of 1M NaOH to each test solution. Explain your observations.

3. Cobalt Complex Ions. Repeat Part A.2, substituting 0.1M $Co(NO_3)_2$ for 0.1M $CuSO_4$.

B. PREPARATION OF $(Co(NH_3)_4CO_3)NO_3$

1. Dissolve about 10g (±0.01g) of $(NH_4)_2CO_3$ in 30mL of distilled (or de–ionized), boiled H_2O. Cautiously add 30mL of conc NH_3[1] (**Caution**: *Avoid skin contact and inhalation.*).

2. Dissolve 8g (±0.01g) of $Co(NO_3)_2 \bullet 6H_2O$ in 15mL of H_2O contained in a 250mL Erlenmeyer flask. While stirring, add the $(NH_4)_2CO_3$. Cool the mixture to 10°C and slowly add 5mL of 30% H_2O_2 (**Caution**: 30% H_2O_2 *causes severe skin burns in the form of white blotches; wash the affected skin immediately with water.*) to this solution.

3. Transfer this mixture to an evaporating dish; using a steam bath (Figure 39.3) reduce the total volume by one–half. During the evaporation periodically add a total of 2.5g (±0.01g) of $(NH_4)_2CO_3$ in small amounts. Vacuum filter the *hot* solution. Cool the filtrate in an ice bath.

4. Red crystals should now be present in the filtrate. Vacuum filter the crystals, wash the crystals twice with 3–5mL of methanol. Air–dry the crystals. Weigh the product. and calculate the percent yield.

C. TESTING THE $(Co(NH_3)_4CO_3)^+$ ION

1. Litmus test and $CaCl_2$ test. Dissolve a small pinch of the coordination compound in H_2O. Test the solution with red and blue litmus paper. Record your observation. Add several drops of 0.1M $CaCl_2$ solution. Record your observation.

2. Na_2CO_3 test. To a second sample dissolved in H_2O, add several drops of 0.1M Na_2CO_3 solution. Centrifuge and record your observations. What is the formula of the precipitate?

---

[1]If a fume hood is available, use it while transferring the conc NH_3.

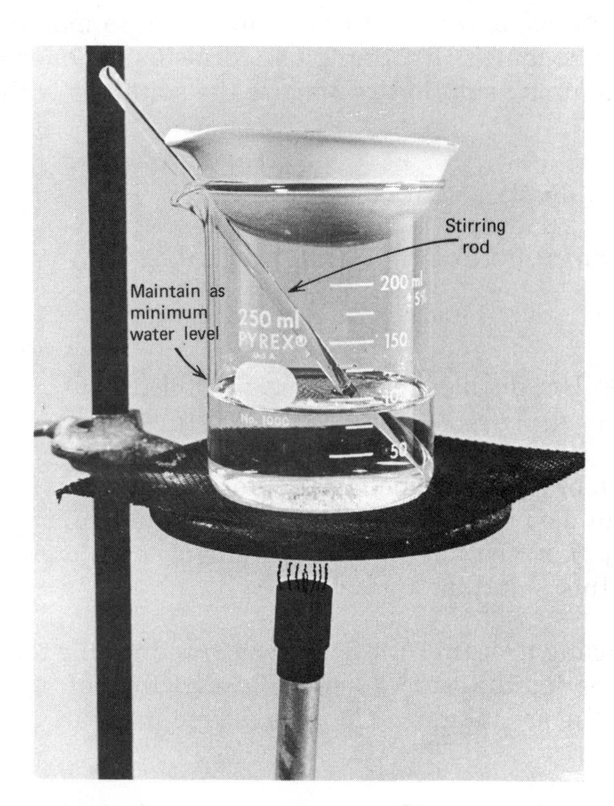

Figure 39.3
The Steam Bath is used for Reducing the Volume of the Reaction Mixture

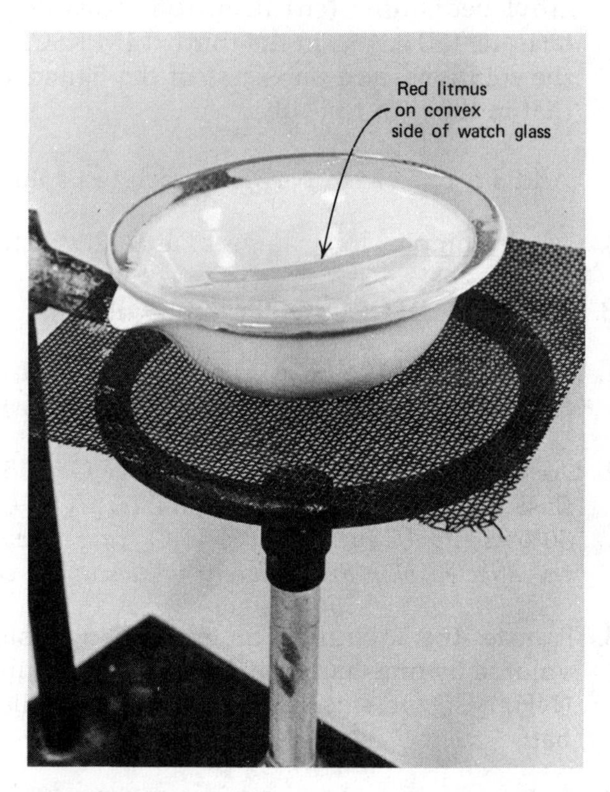

Figure 39.4
The Apparatus for Testing the Stability of the Complex Ion

3. Stability test. Place a pinch of the compound in an evaporating dish and add 5mL of water. Add 20 drops (1mL) of 1M HCl. Adhere a piece of red litmus paper to the convex side of a watchglass (Figure 39.4). Now add 2mL of 0.05M NaOH to the solution and cover with the watchglass, convex side down. Gently heat. What do you observe? Comment on the stability of the $[Co(NH_3)_4CO_3]^+$ ion.

COORDINATION CHEMISTRY–LAB PREVIEW

Date________Name____Lab Sec. _____Desk No._____

1. a. What is a complex ion?

   b. What is a ligand?

   c. What is meant by the statement, "the coordination number of Co(II) is usually six"?

2. Write the formula of the complex ion form between

   a. the ligand, NH_3, and Cu(II), assuming a coordination number of four.

   b. the ligand, CN^-, and Fe(II), assuming a coordination number of six.

   c. the ligand, H_2O, and Ni(II), assuming a coordination number of six.

3. Identify the ligand(s) in the following coordination compounds.

a. $[Pt(NH_3)_2Cl_2]$ ________________________________

b. $[Cu(NH_3)_4]SO_4$ ________________________________

c. $[Cr(NH_3)_4(H_2O)_2]Cl_3$ ________________________________

4. What is the coordination number of the metal in each complex ion in Question 3?

a. __________; b. __________; c. __________

5. The CO_3^{2-} that is a ligand in the $[Co(NH_3)_4CO_3]^+$ complex ion is called a bidentate. Refer to your text and define bidentate.

COORDINATION CHEMISTRY–DATA SHEET

Date________Name____Lab Sec. _____Desk No._____

A. A LOOK AT SOME COMPLEX IONS

Solution	Color/H_2O	Color/HCl	Formula of Complex Ion	Effect of H_2O
0.1M $CuSO_4$	----------	----------	--------------	--------------
0.1M $Ni(NO_3)_2$	----------	----------	--------------	--------------
0.1M $CoCl_2$	----------	----------	--------------	--------------

Ligand with Cu^{2+}	Color	Formula	Effect of OH^-
NH_3	----------	----------	--------------------------
en	----------	----------	--------------------------
tar^{2-}	----------	----------	--------------------------
SCN^-	----------	----------	--------------------------

Ligand with Co^{2+}	Color	Formula	Effect of OH^-
NH_3	----------	----------	--------------------------
en	----------	----------	--------------------------
tar^{2-}	----------	----------	--------------------------
SCN^-	----------	----------	--------------------------

B. PREPARATION OF $(Co(NH_3)_4CO_3)NO_3$

1. Mass of $Co(NO_3)_2 \bullet 6H_2O$ (g) ___________

2. Mass of $[Co(NH_3)_4CO_3]NO_3$ (g) ___________

3. Theoretical yield of $[Co(NH_3)_4CO_3]NO_3$ based on $Co(NO_3)_2 \bullet 6H_2O$ (g)* ___________

4. Percent Yield (%) ___________

*Show sample calculation.

C. TESTING THE $(Co(NH_3)_4CO_3)^+$ ION

Test	Observation	Conclusion
Litmus test	_________________	_________________
$CaCl_2$ test	_________________	_________________
Na_2CO_3 test	_________________	_________________
Stability	_________________	_________________

Questions

1. a. Compare the stability of the chloro complex ions of Cu^{2+}, Ni^{2+}, and Co^{2+}.

   b. Compare the stability of the NH_3 and ethylenediamine complex ions of Co^{2+}.

2. a. What is the purpose of the 30% H_2O_2 in Part B of the experimental procedure?

b. Write an equation for the reaction. Ask your instructor for assistance.

3. In the stability test (Part C), why was it first necessary to add 1M HCl?

4. For the same concentration as $[Co(NH_3)_4CO_3]NO_3$, which of the following salts show the same conductivity? Explain your answer.

a. $Co(NO_3)_3$

b. KCl

c. Na_2CO_3

d. $NaNO_3$

EXPERIMENT 40
ORGANIC CHEMISTRY

Objectives

- To study the chemical behavior of hydrocarbons, alcohols, aldehydes, ketones, acids, bases, and esters

Principles

Carbon has the unusual property of bonding to itself. Although other atoms do this, carbon does it much more extensively. Because of this unique property, over three million carbon containing compounds, called **organic compounds**, have been reported in the literature. As a result, a complete knowledge of the chemical characteristics of all of these organic compounds would indeed by very difficult to acquire. The complexity is lessened somewhat because these compounds form natural classes and because members of each class possess similar chemical properties. We will study five of these classes: the hydrocarbons, alcohols, the aldehydes and ketones, the organic acids and bases, and the esters.

Hydrocarbons are compounds that contain only the elements carbon and hydrogen. According to their chemical properties, the hydrocarbons are subdivided into three subgroups: the saturated hydrocarbons, the unsaturated hydrocarbons, and the aromatic hydrocarbons.

The **saturated hydrocarbons** are commonly referred to alkanes. All C–C and C–H bonds are single bonds and are relatively inert to chemical attack. By far, their most significant reaction is that (for example, propane C_3H_8) with O_2 forming CO_2 and H_2O as products.

$$C_3H_8 + 5O_2 \rightarrow 3CO_2 + 4H_2O$$

The **unsaturated hydrocarbons** have two subgroups: the alkenes and the alkynes. The alkenes have *at least* one C=C bond and the alkynes have at least one C≡C. These double and triple bonds, called unsaturated bonds, are chemically quite reactive. For example, bromine readily reacts across the C=C bond in propene, C_3H_6, to form 1,2–dibromopropane.

$$CH_3\text{–}CH{=}CH_2 + Br_2 \rightarrow CH_3\text{–}CHBr\text{–}CHBr$$

The unsaturated hydrocarbons also react with oxidizing agents, such as $KMnO_4$, to produce alcohols (Baeyer's test). The $KMnO_4$, a purple reagent, is reduced to MnO_2, a brown precipitate, in the reaction. The brown precipitate may appear as a red–brown solution.

$$\underset{\text{(colorless)}}{3\,CH_2{=}CH_2} + \underset{\text{(purple)}}{2MnO_4^-(aq)} + 4H_2O \rightarrow \underset{\text{(colorless)}}{3\,CHOH\text{–}CHOH(aq)} + \underset{\text{(brown)}}{2MnO_2(s)} + 2OH^-(aq)$$

Aromatic hydrocarbons are generally characterized by the presence of the six–membered carbon ring called benzene. Although somewhat resistant to chemical attack, aromatic compounds are usually more reactive that the alkanes. For example, bromine displaces hydrogen from the benzene ring in the presence of a catalyst.

$$C_6H_6 + Br_2 \xrightarrow{FeBr_3} C_6H_5Br + HBr$$

benzene bromobenzene

Many organic compounds contain an atom or group of atoms that substitute for hydrogen or carbon in a hydrocarbon. The atom or group of atoms is commonly referred to as a **functional group**, which imparts characteristic properties to the compound.

An **alcohol** is a hydrocarbon in which an –OH group (called a **hydroxyl** group) replaces a hydrogen atom. An alcohol is also like water in that an alkyl group[1] replaces a hydrogen atom in the water molecule. Therefore, alcohols have physical properties that are intermediate between those of hydrocarbons and water.

Depending upon where the –OH group is attached to the hydrocarbon chain, alcohols are classified as 1°, 2°, or 3°.

$CH_3–CH_2–OH$, a 1° alcohol, called ethanol (grain alcohol)

$CH_3–CHOH–CH_3$, a 2° alcohol, called isopropanol (rubbing alcohol)

$CH_3–COH(CH_3)–CH_3$, a 3° alcohol, called tertiary butanol

Each alcohol reacts differently with a mild oxidizing agent, such $K_2Cr_2O_7$. Those that do react form an aldehyde or ketone as the primary product. A color change from a brilliant orange, due to $Cr_2O_7^{2-}$, to green, due to Cr^{3+}, and/or a change in odor determines if a reaction occurs.

The iodoform test distinguishes between the 1°, 2°, and 3° alcohols. The test is positive when iodine in the presence of a base, such NaOH, oxidizes the alcohol, producing the acid (with one less carbon atom) and iodoform, CHI_3. A very characteristic odor confirms the reaction. This test is also positive for aldehydes and ketones.

Aldehydes and **ketones** are hydrocarbons in which a C=O group (called a **carbonyl** group) has substituted for a $–CH_2–$ group. Testing their solubility in aqueous sodium bisulfite, $NaHSO_3$, generally distinguishes them from other classes of organic compounds. The HSO_3^- ion adds across the C=O bond.

$$R_2C{=}O(l) + Na^+HSO_3^-(aq) \rightarrow R_2COH(SO_3)^-,Na^+(aq)$$

Organic **acids** are hydrocarbons in which a –COOH group (called a **carboxyl** group) substitutes for a $–CH_3$ group. The acids readily react with $NaHCO_3$ releasing CO_2 gas.

---

[1]An alkyl group is simply a hydrocarbon from which a hydrogen atom has been removed; for example, the ethyl group, $C_2H_5–$, results when a hydrogen atom is removed from ethane, C_2H_6.

$$R\text{–}COOH(aq) + Na^+HCO_3^-(aq) \rightarrow R\text{–}COO^-,Na^+(aq) + H_2O + CO_2(g)$$

Organic acids can easily be prepared by an oxidation of an alcohol or aldehyde with a strong oxidizing agent, such as $KMnO_4$.

Organic **bases** are hydrocarbons in which a $-NH_2$ group (called an **amine** group) substitutes for a $-CH_3$ group. The lone electron pair on the nitrogen atom serves as a Lewis base. A pH test of an aqueous solution of many organic compounds shows that most are neutral; however, the acids have a low pH and the bases have a high pH. Organic bases tend to have a "fishy" odor as well.

An organic **ester** is a product from the reaction of an organic acid with an alcohol. Esters tend to be volatile and have a pleasant odor. The natural scents of many flowers and flavors of many fruits are due to the presence of one or more esters. The presence of the R–COO–R bond structure is present in all esters. Table 40.1 list some common esters along with their flavor or aroma.

Important natural esters are fats–butter, lard, and tallow– and oils–linseed, cottonseed, peanut, and olive–used in the synthesis of oleomargarine, peanut butter, and vegetable shortening.

Table 40.1
Chemical Formulas for Natural Flavors and Aromas

Formula	Flavor/Aroma
$C_3H_7\text{–}COO\text{–}C_2H_5$	pineapple
$C_2H_5\text{–}COO\text{–}C_5H_{11}$	apricot
$CH_3\text{–}COO\text{–}C_8H_{15}$	orange
$H_2N\text{–}C_6H_4\text{–}COO\text{–}CH_3$	grape
$CH_3\text{–}COO\text{–}C_5H_{11}$	banana
$H\text{–}COO\text{–}C_2H_5$	rum

Techniques

- Technique 2, page 2 — Transferring Liquid Reagents
- Technique 6a,d, page 7 — Heating Liquids
- Technique 9a, page 10 — Handling Gases
- Technique 13, page 16 — Using the Laboratory Balance
- Technique 14, page 18 — Testing with Litmus

Procedure

Caution: *Many organic compounds are volatile and flammable. Therefore, do not have any open flames in the work area..*

Obtain an unknown compound from your instructor. Perform the following tests on the unknown at the same time you are performing tests on the sample known to have the functional group being studied. Use this procedure to identify the class of organic compounds for your unknown.

A. Chemical Reactivity of Hydrocarbons

A number of suggested organic compounds, listed on the Data Sheet, are to be tested for unsaturation. Unsaturated compounds show a positive test for both Br_2/CCl_4 and $KMnO_4$. On occasion, a test is positive even though the compound has no double or triple bonds. Refer to your text for the structural formulas and note the exceptions on the Data Sheet.

1. Br_2/CCl_4 test[2]. In 150mm test tube, add 4–5 drops of liquid or dissolve 0.1g of solid organic compound in 2mL of CCl_4 (**Caution**: *Avoid breathing* CCl_4 *vapors–use a fume hood if available.*). Add drops of 2% Br_2/CCl_4 (**Caution**: *Avoid contact with the* Br_2/CCl_4. Br_2 *can cause skin burns.*) and agitate after each drop. If the color fades, continue to add drops (5 drops maximum) until the permanent red–brown color persists. Observe and record.

2. $KMnO_4$ test. In a 150mm test tube, add 4–5 drops of liquid or dissolve 0.1g of solid organic compound in 2mL of H_2O or acetone. Add drops of 1% $KMnO_4$ (**Caution**: $KMnO_4$ *is a strong oxidizing agent; avoid skin contact.*) and agitate. If a brown precipitate forms within 3 minutes, an alkene is likely present.

B. Chemical Reactivity of Alcohols

Four alcohols and your unknown are listed on the Data Sheet. Use these while completing the following procedure.

1. Iodoform test. Dissolve 10 drops of each alcohol in 5mL of H_2O in separate 150mm test tubes. Add 5mL of 10% NaOH. While shaking, add drops of 10% KI/I_2 until the definite dark brown of I_2 persists.

   Warm gently in a hot water bath, but do not exceed 60°C. Add more KI/I_2 at the elevated temperature until the dark brown color remains for 2 minutes. Allow the test tube and contents to cool.

   While shaking, add drops of 10% NaOH to expel excess I_2. Nearly fill the test tube with distilled (or de–ionized) water and allow to stand for 10 minutes. Look for CHI_3, a yellow crystalline precipitate. Note any characteristic odor.[3] Summarize the comparative results for the alcohols on the Data Sheet.

---

[2]Since CCl_4 is a suspected carcinogen, dichloromethane, CH_2Cl_2, may be substituted for CCl_4.

[3]Whenever the amount of alcohol is small, iodoform may not separate; however, the characteristic odor establishes its formation.

2. Alcohol Oxidation. In a 150mm test tube, place 2mL of 0.1M $K_2Cr_2O_7$ and *slowly* add 1mL of conc H_2SO_4 (**Caution**: *conc* H_2SO_4 *is a severe skin irritant and causes clothes to disappear; wash the affected area with large amounts of water.*). Swirl to dissolve the $K_2Cr_2O_7$ and cool with tap water. Slowly add 2mL of the test alcohol. Note any color change and odor. Compare its odor with the test alcohol.

C. CHEMICAL REACTIVITY OF ALDEHYDES AND KETONES.

1. In a 150mm test tube, mix 1mL (20 drops) of test solution with 3mL of 40% $NaHSO_3$. Add 1–2 drops of ethanol. Stopper and shake vigorously. The presence of a single phase indicates HSO_3^- addition across the carbonyl group.

D. CHEMICAL REACTIVITY OF ACIDS AND BASES

1. In a 75mm test tube, dissolve 0.1g of solid (or 4–5 drops of liquid) organic acid in 1mL of water. Test the pH of each sample with litmus paper. If the sample is insoluble in water, add drops of ethanol until it dissolves.

   Repeat the litmus test with an organic base.

2. Place 0.1g of solid (or 4–5 drops of liquid) organic acid in a 75mm test tube. Add 1mL of 10% $NaHCO_3$. A distinct "fizzing" sound is detectable even if the visual evolution of CO_2 is questionable. Record.

3. Place 2mL of 0.1M $KMnO_4$ in a 150mm test tube. Slowly add 1mL of ethanol. Watch for a color change. Compare its odor with that of acetic acid. What's your conclusion?

E. PREPARATION OF AN ESTER

1. Banana Oil. Place 2mL of glacial CH_3COOH (**Caution**: *Do not inhale.*) and 3mL of n–$C_5H_{11}OH$ (called n–amyl alcohol) in a 150mm test tube. Slowly and cautiously add 1mL of conc H_2SO_4 (**Caution**: *conc* H_2SO_4 *causes severe skin burns.*). Heat the mixture gently and carefully over a very low flame. If the odor changes, the product is an ester. Have the instructor assist you in writing the balanced equation for the reaction.

2. Oil of Wintergreen. In a 75mm test tube, place a "BB–size" piece of salicylic acid, HO–C_6H_4–COOH. Add 1 drop of 3M H_2SO_4 and 3 drops of water. After one–half minute, add 3–4 drops of methanol, CH_3OH. Place a loose cotton plug into the mouth of the test tube; place the test tube in a hot water bath (about 60°C) for 20–30 minutes. Note the odor. Have the instructor assist you in writing the balanced equation for the reaction.

FUNCTIONAL GROUPS IN ORGANIC COMPOUNDS

Functional Group	Name	Compound Class	Example
$>C=C<$	double bond	alkenes	$H_2C=CH_2$
$-C\equiv C-$	triple bond	alkynes	$H-C\equiv C-H$
$-F, Cl, Br, I$	halogen	halides	CH_3Cl
$-OH$	hydroxyl	alcohols	CH_3OH
$-CHO$	carbonyl	aldehydes	CH_3CHO
$-CO-$	carbonyl	ketones	CH_3COCH_3
$-COOH$	carboxyl	carboxylic acids	CH_3COOH
$-COO-$	–	esters	$C_3H_7-COO-CH_3$
$-NH_2$	amino	amines	CH_3NH_2
$-CONH_2$	amido	amides	CH_3CONH_2
$-NO_2$	nitro	–	2,4,6-trinitrotoluene: CH_3, O_2N, NO_2, NO_2 on benzene ring (TNT)
$-CH(NH_2)COOH$	–	amino acids	$CH_3CH(NH_2)COOH$ (alanine)

ORGANIC CHEMISTRY-LAB PREVIEW

Date________Name____Lab Sec. _____Desk No._____

1. Distinguish between

   a. a saturated and an unsaturated hydrocarbon.

   b. an alkene and an alkyne.

   c. an alcohol and a hydrocarbon.

2. a. Write the chemical equation for the combustion of heptane, C_7H_{16}.

   b. Write the chemical equation for the reaction of Br_2 with CH_3–CH=CH–CH_2Br.

3. How can organic acids be prepared in the laboratory?

4. a. What is the principal product in the oxidation of ethanol with $K_2Cr_2O_7$?

b. What is the principal product in the oxidation of ethanol with $KMnO_4$?

5. How does a chemist quickly distinguish an organic acid from an organic base?

ORGANIC CHEMISTRY-DATA SHEET

Date_________Name____Lab Sec. _____Desk No._____

A. CHEMICAL REACTIVITY OF HYDROCARBONS

Identify the tests as positive (+) or negative (-).

Compound	Structural Formula[4]	Br_2/CCl_4	$KMnO_4$
n–hexane or cyclohexane			
1–pentene or 2–pentene			
cyclohexene			
1–hexyne			
ethanol			
naphthalene			
p–xylene			
methane (see Figure 40.1)			
unknown			

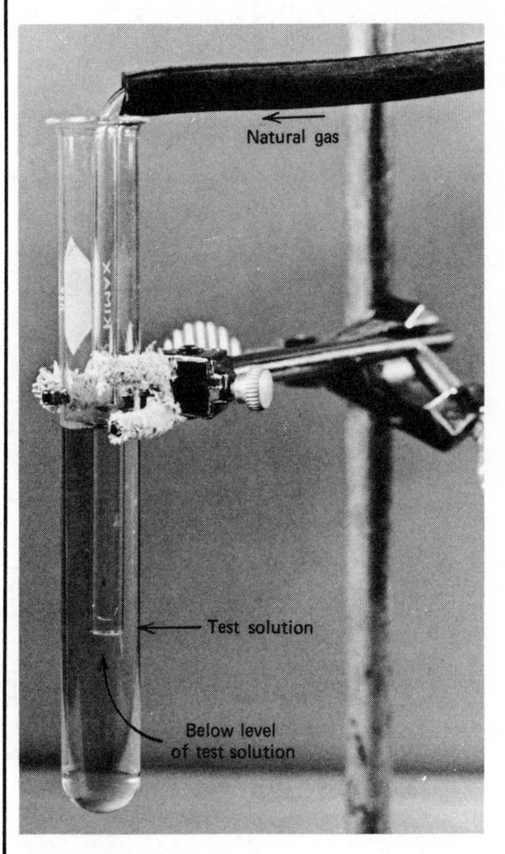

Figure 40.1
Apparatus for Testing Natural Gas for Unsaturation

[4]Use your text to assist you in writing the structural formulas.

B. Chemical Reactivity of Alcohols

1. Iodoform test

Alcohol	Subclass 1°, 2°, 3°	Observation	Oxidation Product
methanol	_______	________________________	_________
ethanol	_______	________________________	_________
i–propanol	_______	________________________	_________
t–butanol	_______	________________________	_________
unknown		________________________	

Conclusion of observations on alcohols.

2. Oxidation of Alcohol

Alcohol	Subclass 1°, 2°, 3°	Observation	Oxidation Product
methanol	_______	________________________	_________
ethanol	_______	________________________	_________
i–propanol	_______	________________________	_________
t–butanol	_______	________________________	_________
unknown		________________________	

C. Chemical Reactivity of Aldehydes and Ketones

Name of Compound	Results of Test	Check (√) for Unknown
____________	___________________________	________
____________	___________________________	________
____________	___________________________	________

D. CHEMICAL REACTIVITY OF ACIDS AND BASES

1.

Name of Acid/Base	Results of Litmus Test	Check (√) for Unknown
____________	__________________________	________
____________	__________________________	________
	__________________________	

2.

Name of Acid	Results of $NaHCO_3$ Test	Check (√) for Unknown
____________	__________________________	________
____________	__________________________	________

3. Oxidation of Ethanol.

Observation ____

Conclusion ____

E. PREPARATION OF AN ESTER

1. Banana Oil. Write an equation for the reaction.

2. Oil of Wintergreen. Write the formula for the compound.

Questions

1. How can you distinguish between the following compounds?

   a. CH_3NH_2 and CH_3OH

   b. C_2H_5OH and tertiary C_4H_9OH, a 3° alcohol

   c. CH_3CHO and CH_3CH_2OH

   d. C_3H_8 and C_3H_7OH

   e. CH_3COOH and CH_3CHO

   f. $CH_3CH_2CH_3$ and $CH_3CH_2CH_2OH$

2. What reactants make the ester characteristic of the following flavors? (See Table 40.1)

   a. pineapple

   b. orange

   c. grape

3. Acetic acid is prepared by the oxidation of ethanol with permanganate ion in an acidic solution. Write a balanced (redox) equation for the reaction. The reduction product of the permanganate ion in an acidic solution is Mn(II).

EXPERIMENT 41
CARBOHYDRATES AND PROTEINS

Objectives

- To identify monosaccharides, disaccharides, and polysaccharides
- To qualitatively test for various carbohydrates and proteins
- To observe some characteristic reactions of proteins

Principles

CARBOHYDRATES

Carbohydrates are a class of organic compounds that consist of only carbon, hydrogen, and oxygen in an approximate 1:2:1 mole ratio. Those having a sweet taste are called **sugars**. Three different stages in the hydrolysis of carbohydrates yield the following classifications.

- **Monosaccharides**, such as glucose, fructose, and galactose, are the simplest sugars. Each molecule consists of a single aldehyde (-CHO) or ketone (-CO-) group with hydroxyl groups on the remaining carbon atoms.

  Glucose is also known as dextrose or blood sugar and fructose is commonly known as levulose or fruit sugar.

- **Disaccharides**, such as sucrose (table sugar), maltose (malt sugar), or lactose (milk sugar), readily hydrolyze in aqueous acidic solutions forming two monosaccharide molecules.

$C_{12}H_{22}O_{11} + H_2O$	$\rightarrow$	$C_6H_{12}O_6$	+	$C_6H_{12}O_6$
sucrose		glucose		fructose
or maltose		glucose		glucose
or lactose		glucose		galactose

Sucrose, maltose, and lactose (as do glucose, fructose, and galactose) have the same molecular formula but different structural formulas.

- **Polysaccharides,** such as starch and cellulose, consist of many monosaccharide units bonded together. The monosaccharide molecules that make up a polysaccharide molecule are generally the same; for example, only glucose molecules form from the complete hydrolysis of starch and cellulose molecules.

  The stages of hydrolysis are

$$\textbf{polysaccharides} \xrightarrow{-H_2O/H^+} \textbf{disaccharides} \xrightarrow{-H_2O/H^+} \textbf{monosaccharides}$$

Sugars oxidized by weak oxidizing agents are **reducing sugars.** Cupric hydroxide, $Cu(OH)_2$, a weak oxidizing agent in the Benedict's reagent, produces a red–orange insoluble salt, Cu_2O, in a reaction with a reducing sugar.

$$\underset{}{\textbf{reducing sugar}} + \underset{\text{(blue)}}{2Cu(OH)_2(aq)} \rightarrow \underset{\text{(red-orange)}}{Cu_2O(s)} + 2H_2O + \textbf{carboxylic acid}$$

Controlled oxidation of monosaccharides in the body is our heat and energy source, about 17kJ per gram of monosaccharide. The final products of its oxidation are CO_2 and H_2O. If the body cannot effectively oxidize the monosaccharides, they are excreted in the urine indicating diabetes (*diabetus mellitus*).

This experiment has a general test for identifying carbohydrates followed with a test that identifies the presence of reducing sugars. Sucrose, starch, and cellulose are hydrolyzed and tested for the presence of reducing sugars.

PROTEINS

Proteins are high molecular weight polymers formed by a chemical combination of 20 to thousands of α-amino acid molecules having formula weights from about 6000 to several million! The peptide linkage, –CO–NH–, connects the α–amino acids; that is, a carboxyl group on one α–amino acid condenses with another's amino group, eliminating water in its formation.

$$\text{H-}\ddot{\text{N}}\text{(H)-CH(R')-C(=O)-}\boxed{\text{OH} + \text{H}}\text{-}\ddot{\text{N}}\text{(H)-CH(R'')-C(=O)-OH} \longrightarrow \text{H-}\ddot{\text{N}}\text{(H)-CH(R')-C(=O)-}\ddot{\text{N}}\text{(H)-CH(R'')-C(=O)-OH} + H_2O$$

R' and R" can be any alkyl group.

In the body, proteins serve many functions: regulate metabolites (hormones, such as insulin), defend against disease (antibodies), catalyze biochemcial reactions (enzymes), and transport oxygen (hemoglobin). Proteins are either *fibrous* (present in hair, wool, muscles, connecting tissue) or *globular* (such as egg or blood albumin, casein in milk, and most enzymes).

Proteins that undergo a change in physical and biochemical properties and, therefore, a change in physiological activity are *denatured*. Heat, alcohol, acids, bases, heavy metal ions, and alkaloids weaken the intermolecular bonding to denature proteins. Coagulation of the protein generally results.

This experiment performs several characteristic tests on egg albumin, a water soluble protein with a formula weight of 34,500. In addition, several color tests are used to detect proteins.

<u>Biuret Test.</u> This is the most general test for proteins. One of the following groups must be present in the protein for a positive test.

$$\text{-C(=O)-NH}_2, \quad \text{-C(=O)-NH-}, \quad \text{-CH}_2\text{-NH}_2, \quad \text{-CHOH-CH}_2\text{-NH}_2$$

Biuret is a compound with the formula, $H_2N\text{-}\overset{\large O}{\overset{\|}{C}}\text{-}NH\text{-}\overset{\large O}{\overset{\|}{C}}\text{-}NH_2$

Xanthoproteic Test. This test detects benzene rings with an attached $-NH_2$ or $-OH$ group.

Million's Test. This test detects the phenolic ring, $-C_6H_4-OH$, in proteins. Since Cl- and NH_4^+ interfere with the test, it is unreliable for a urinanalysis.

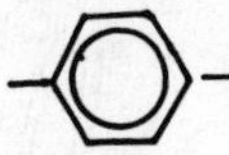

Hopkins–Cole Test. This test is specific for identifying the trytophan system:

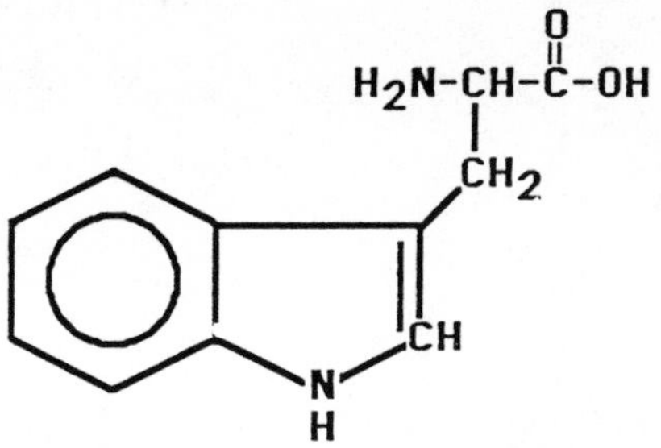

$H_2N\text{-}CH(CH_2\text{-indol-3-yl})\text{-}\overset{\large O}{\overset{\|}{C}}\text{-}OH$

Techniques

- Technique 6a,d, page 7 Heating Liquids

Procedure

A. Carbohydrates. Molisch Test, A General Test for Carbohydrates

1. Pour 5mL of a 1% glucose solution into a 150mm test tube and add 2 drops of Molisch Reagent. Mix the solution.

2. Incline the test tube; *slowly* and *carefully* add 1–2mL of conc H_2SO_4 (**Caution**: *conc* H_2SO_4 *is a severe skin irritant.*) down its side (Figure 41.1). Since conc H_2SO_4 is more dense than the aqueous solution, it forms the lower layer. The purple color at the interface indicates the presence of carbohydrates. Record your observations.

3. Repeat the test with 1% solutions of fructose, sucrose, maltose, lactose, and starch.

B. Carbohydrates. Benedict's Test, A Test for Reducing Sugars

1. Clean six 75mm test tubes and number them in the test tube rack. Fill them as follows.

Test Tube No.	Benedict's Solution	Sugar Solution
1	1mL	3-4 drops 1% glucose
2	1mL	3-4 drops 1% fructose
3	1mL	3-4 drops 1% sucrose
4	1mL	3-4 drops 1% maltose
5	1mL	3-4 drops 1% lactose
6	1mL	3-4 drops 1% starch

2. Place the test tubes in a boiling water bath for 5-7 minutes. Record your observations.

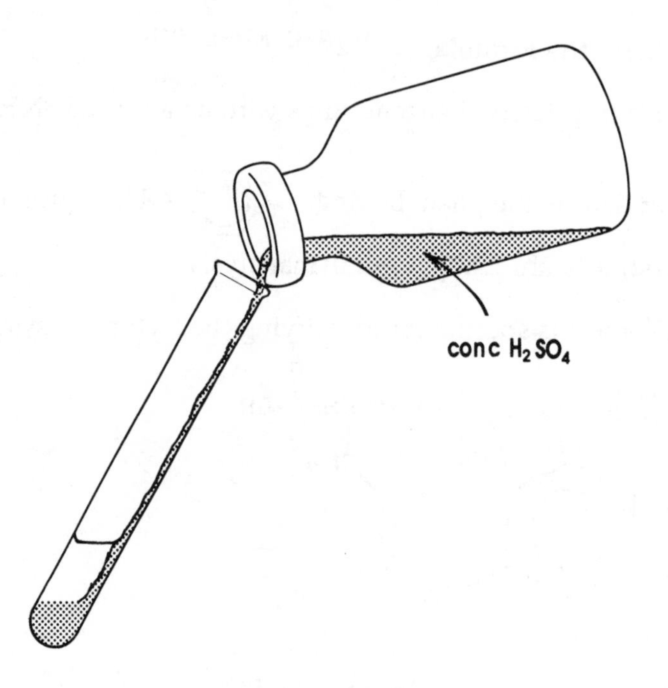

Figure 41.1
Carefully **Pour conc H_2SO_4 Down the Side of the Test Tube**

C. CARBOHYDRATES. HYDROLYSIS OF SUCROSE AND STARCH.

1. Place 10mL of a 1% sucrose solution in one 150mm test tube and 5mL of a 1% starch solution in another; add 5 drops of conc HCl (**Caution**: *Avoid skin contact.*) and place in a boiling water bath for about 10 minutes. Add water to maintain a constant volume in the hot water bath.

2. Cool the solutions and add 2M NaOH until the solutions are just alkaline to litmus.

3. Test for the presence of reducing sugars in each solution, using the Benedict's Solution. Record your observations.

4. A separate test for the presence of starch is to add iodine. A deep–blue I_2•starch complex signifies the presence of starch. To 5mL of a 1% starch solution, add 1–2 drops of a KI/I_2 solution.

5. Repeat Part C.1 and 2 with 5mL of a 1% starch solution and test for the presence of unhydrolyzed starch.

D. PROTEINS. SOLUBILITY AND PH.

1. Place 10 drops of a 2% aqueous albumin solution[1] into 5 separate 75mm test tubes. Add 10 drops of 2M NaOH to the first, 10 drops of 1M HCl to the second, 10 drops of 1M HNO_3 to the third, 10 drops of 0.05M Na_2CO_3 to the fourth, and 10 drops of distilled (or de–ionized) H_2O to the fifth.

2. Record your observations on the Data Sheet.

E. PROTEINS. DENATURATION OF A PROTEIN.

Place the results for each of the following tests in a test tube rack. At the end of Part E, compare the results and record on the Data Sheet as (a) no precipitate, (b) slight precipitate, or (c) heavy precipitate.

1. Place 5mL of the 2% albumin solution in a 150mm test tube and heat the upper portion to boiling. Compare the upper portion to the solution in the lower part of the test tube.

2. Place 3mL of 2% albumin in a 150mm test tube. Hold it at an angle, and very *slowly* and *carefully* pour 3mL of conc HNO_3 (**Caution**: *conc* HNO_3 *is very corrosive to the skin.*) down its side (Figure 41.1). The conc HNO_3 underlays the aqueous layer. The appearance of coagulated protein at the interface confirms the presence of protein. Describe its appearance.

3. Add 5mL of denatured (95%) ethanol to 5mL of a 2% albumin solution.

4. Add 5 drops of a 10% tannic acid solution to 1mL of a 2% albumin solution. Mix.

5. Add 5 drops of 5% $HgCl_2$ to 1mL of 2% albumin. Mix the solution. Repeat with 5% $AgNO_3$, 5% $Pb(C_2H_3O_2)_2$, and 5% NaCl.

6. Compare and record your observations as described above.

F. PROTEINS. COLOR TESTS.

Complete the following tests on 2% solutions of albumin, gelatin, casein, phenol, and glycine.

1. Biuret Test. Mix 2mL of 3M NaOH with 2mL of test solution. Add 1 drop of 0.1% $CuSO_4$. Mix thoroughly and note any color change. If a color does not develop, add more drops (up to 10) of 0.1% $CuSO_4$ and mix.

2. Xanthoproteic Test. Add 1mL of conc HNO_3 (**Caution!!**) to 2mL of test solution. Mix and note any change. Warm carefully; note the color change. Cool and add drops of 3M NaOH. Note the color change.

3. Millon's Test. Add 5 drops of Millon's reagent to 2mL of test solution. Mix; look for a white precipitate. Warm the solution in a boiling water bath until a red color appears.

---

[1]Alternatively, separate an egg white from its yolk. Add the white of the egg to 150mL of water and mix until the solution appears to be homogeneous. A mechanical stirrer works well. Filter the solution and use the filtrate for the various tests.

4. Hopkins–Cole Test. Add 2 mL of Hopkins–Cole reagent to 2mL of test solution and mix. Hold the test tube at an angle. *Slowly* and *carefully* add 2mL of conc H_2SO_4 **(Caution!!)** along its side so that in underlays the aqueous layer (Figure 41.1). Note any color change at the interface. If no change occurs, very gently agitate the tube causing a slight mix at the interface.

CARBOHYDRATES AND PROTEINS-LAB PREVIEW

Date_________Name____Lab Sec. _____Desk No._____

1. Sucrose is a disaccharide of glucose and fructose. Honey is *not* a disaccharide but rather a 1:1 ratio of glucose to fructose. Honey is a quicker energy source than sucrose. Explain.

2. What is a reducing sugar?

3. Candy bars, high in sucrose content, are quicker energy sources than bread even though both foods are high in carbohydrates. Explain.

4. Calculate the energy released in the complete combustion of 1.0g of sucrose.

5. Alcohol is used as a disinfectant to kill bacteria before an injection or surgery. Explain its chemical effect on bacteria.

6. a. To avoid the dangers of metal poisoning, milk or egg white is used as an antidote for patients who have swallowed copper, lead, or mercury salts. Explain why it works.

   b. Why must an emetic follow such treatment?

7. Mercury salts have been used as germicides in the treatment of seed grain (corn, beans, sorghum, wheat, etc.). Explain their chemical effectiveness.

8. Write the formula of the peptide linkage.

CARBOHYDRATES AND PROTEINS-DATA SHEET

Date________ Name____ Lab Sec. _____ Desk No._____

CARBOHYDRATES. MOLISCH TEST AND BENEDICT'S TEST

	Molisch Test-Observation	Benedict's Test-Observation
glucose	______________________	______________________
fructose	______________________	______________________
sucrose	______________________	______________________
maltose	______________________	______________________
lactose	______________________	______________________
starch	______________________	______________________

CARBOHYDRATES. HYDROLYSIS OF SUCROSE AND STARCH.

	Result of Test
hydrolyzed starch	________________________________
hydrolyzed starch	________________________________
starch/I_2	________________________________
hydrolyzed starch/I_2	________________________________

PROTEINS. SOLUBILITY AND PH.

Testing Reagent	Effect on Albumin-Observation
2M NaOH	________________________________
1M HCl	________________________________
1M HNO_3	________________________________
0.05M Na_2CO_3	________________________________
water	________________________________

PROTEINS. DENATURATION OF A PROTEIN

Denaturing Agent	Effect on Albumin-Observation
heat	________________
conc HNO_3	________________
ethanol, 95%	________________
tannic acid (alkaloid reagent[2])	________________
$HgCl_2$, $AgNO_3$, $Pb(C_2H_3O_2)_2$	______, ______, ______

PROTEINS. COLOR TESTS

Substance	Biuret Test	Xanthoproteic Test	Millon's Test	Hopkin's-Cole Test
albumin	______	______	______	______
gelatin	______	______	______	______
casein	______	______	______	______
phenol	______	______	______	______
glycine	______	______	______	______

Questions

1. a. What reducing sugars were identified? ________________

   b. What non–reducing sugars were identified? ________________

2. a. Does a boiled starch solution give a positive Benedict's test? ____________

   b. Does a boiled sugar solution give a positive Benedict's test?____________

   c. Write an equation for the hydrolysis of sucrose, $C_{12}H_{22}O_{11}$.

---

[2]An alkaloid reagent is *not* an alkaloid, but rather a reagent used to extract alkaloids from plants.

3. What effect does a low pH have on the solubility of proteins?

4. Conc HNO_3 causes the skin to turn yellow (Part E.2). Explain.

5. Chemically, what happens to an egg white when it is boiled?

6. A dilute $AgNO_3$ solution is dropped in to the eyes of a newborn infant (Part E.5). Explain.

APPENDIX A
CONVERSION FACTORS

Length

1 meter (m) = 39.37in = 3.281ft = distance light travels in 1/299,792,548 of a second
1 inch (in) = 2.540cm = 0.02540m
1 kilometer (km) = 0.6214 (statute) mile
1 angstrom (Å) = 1×10^{-10}m = 0.1nm
1 micron (μm) = 1×10^{-6}m

Mass (Weight)

1 gram (g) = 0.03527oz = 15.43 grains
1 kilogram (kg) = 2.205lb = 35.27oz
1 metric ton = 1×10^{6}g = 1.102 short ton = 0.9843 long ton
1 pound (lb) = 453.6g = 7000 grains
1 ounce (oz) = 28.35g

Volume

1 liter (L) = 1.057fl qt = 1×10^{3}mL = 1×10^{3}cm^3 = 61.02in^3
1 fluid quart (fl qt) = 946.4mL = 0.250gal
1 fluid ounce (fl oz)= 29.57mL
1 cubic foot (ft^3) = 28.32L = 0.02832m^3

Pressure

1 atmosphere (atm) = 760.0torr = 760.0mm Hg = 29.92in Hg = 14.696lb/in^2
= 1.013bars = 101.325kPa
1 kilopascal (kPa) = 1N/m^2
1 torr = 1mm Hg = 133.3N/m^2

Energy

1 joule (J) = 0.2389cal = 9.48×10^{-4}Btu = 1×10^{7}ergs
1 calorie (cal) = 4.184J = 3.087ft lb
1 British thermal unit (Btu) = 252.0cal = 1054J = 3.93 $\times 10^{-4}$hp hr = 2.93×10^{-4}kw hr
1 liter atmosphere (L atm) = 24.2 cal = 101.3J
1 electron volt (eV) = 1.602×10^{-19}J

Constants and Other Conversion Data

velocity of light (c) = 2.998×10^{8}m/s = 186,272mi/s
gas constant (R) = 0.08205L atm/(mol K) = 8.314J/(mol K) = 1.986cal/(mol K)
= 62.36L torr/(mol K)
Avogadro's number (N_o) = 6.023×10^{23} /mol
°F = 1.8°C + 32
K = °C + 273.2
Planck's constant (h) = 6.625×10^{-34}J sec/photon

APPENDIX B
VAPOR PRESSURE OF WATER

Temperature (°C)	Pressure (torr)
0	4.6
5	6.5
10	9.2
11	9.8
12	10.5
13	11.2
14	12.0
15	12.5
16	13.6
17	14.5
18	15.5
19	16.5
20	17.5
21	18.6
22	19.8
23	21.0
24	22.3
25	23.8
26	25.2
27	26.7
28	28.3
29	30.0
30	31.8
31	33.7
32	35.7
33	37.7
34	39.9
35	42.2
–	–
100	760.0

APPENDIX C
CONCENTRATION OF ACIDS AND BASES

Concentrated Reagent	Approximate Molarity	Approximate Weight (%)	Specific Gravity	mL to dilute to 1L for a 1.0M Solution
acetic acid	17.4	99.7	1.05	57.5
hydrochloric acid	12.1	37.0	1.19	82.6
nitric acid	15.7	70.0	1.41	63.7
phosphoric acid	14.7	85.0	1.69	68.1
sulfuric acid	17.8	95.0	1.84	56.2
ammonia (aq) (ammonium hydroxide)	14.8	29%(NH_3)	0.90	67.6

Caution: *When diluting reagents, add the more concentrated reagent to the more dilute reagent (or solvent).* ***Never*** *add water to a concentrated acid!*

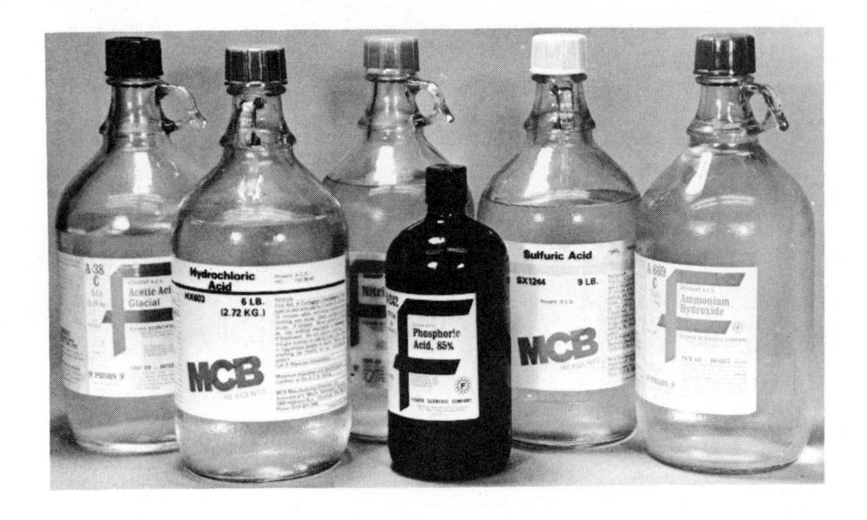

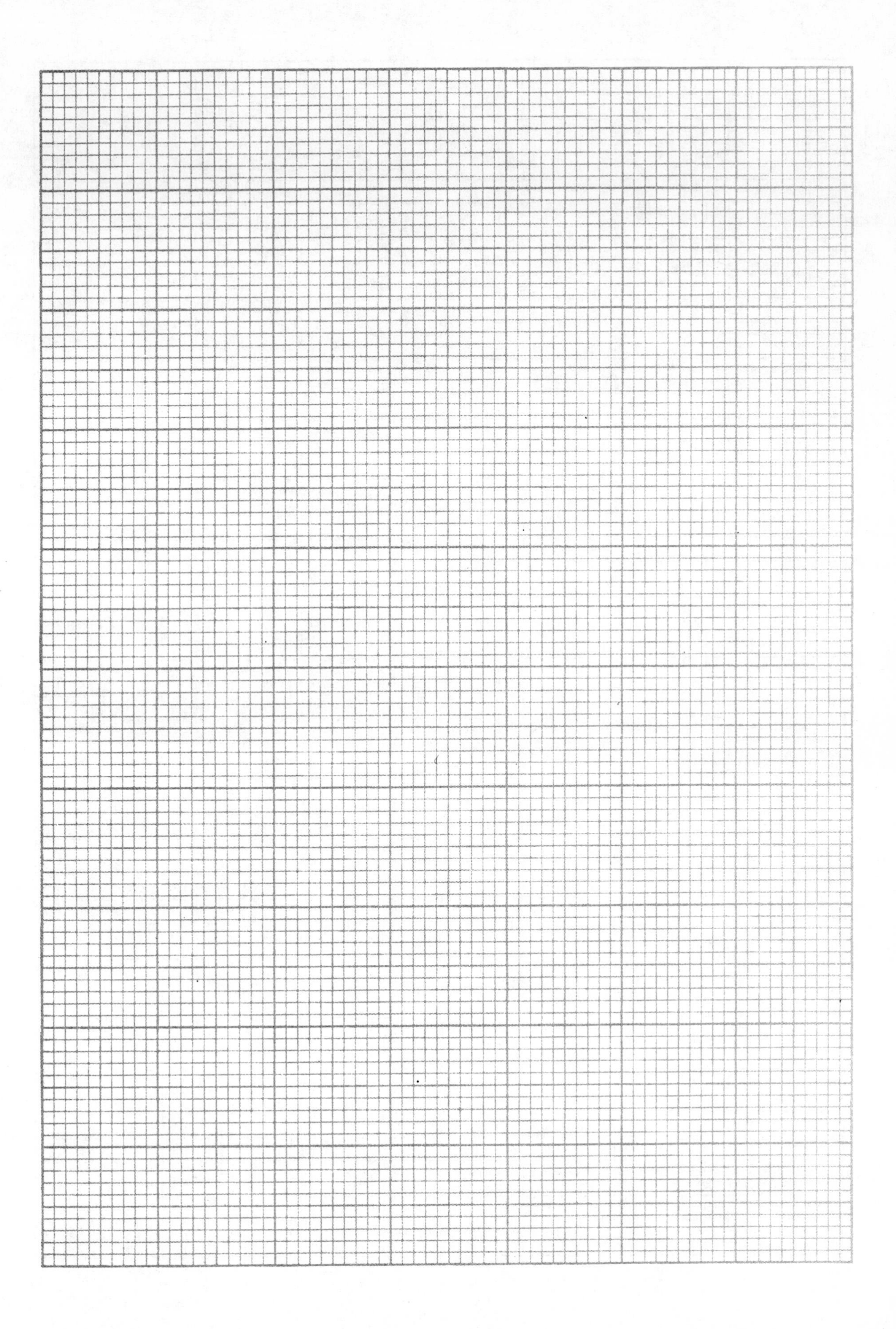

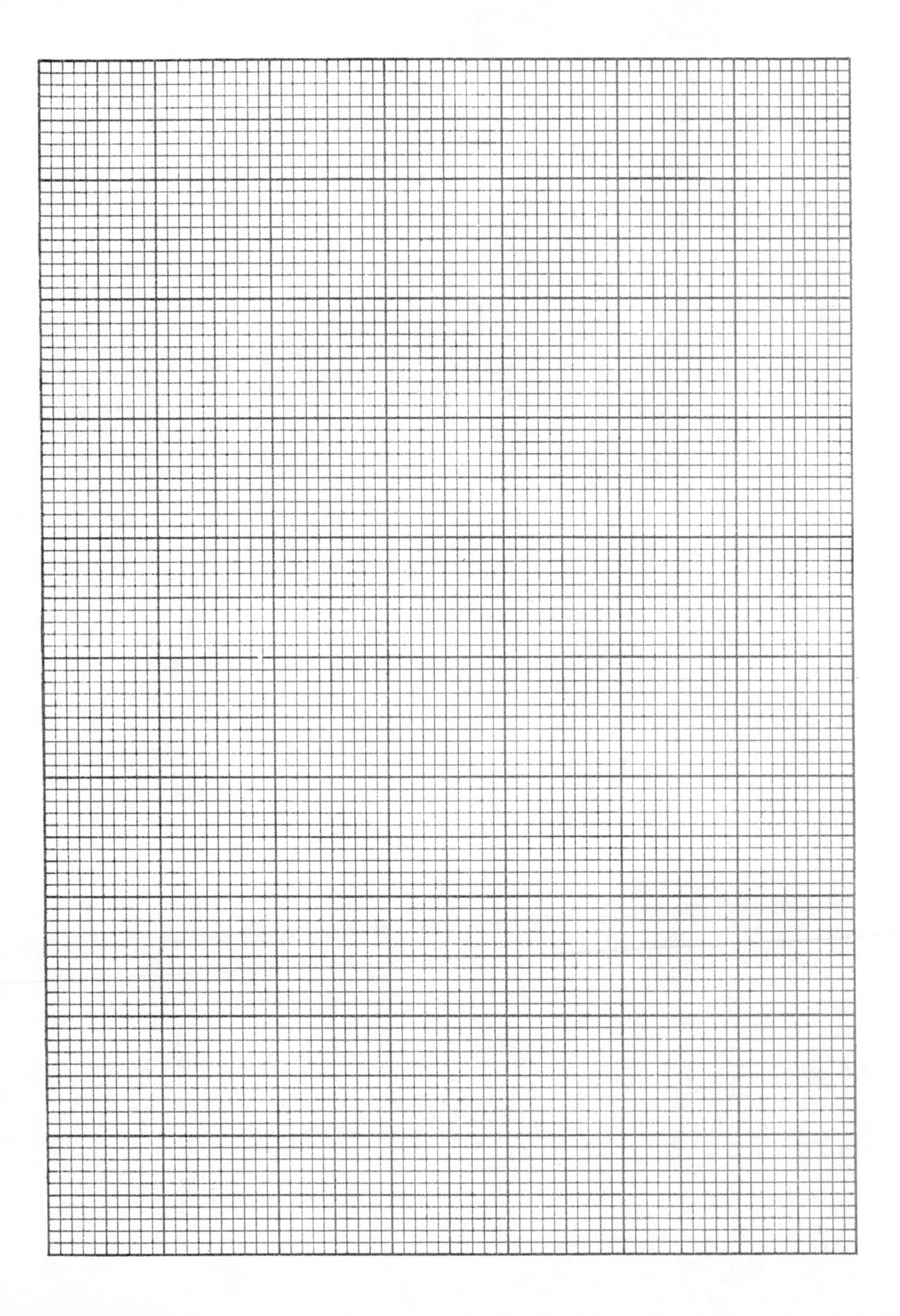

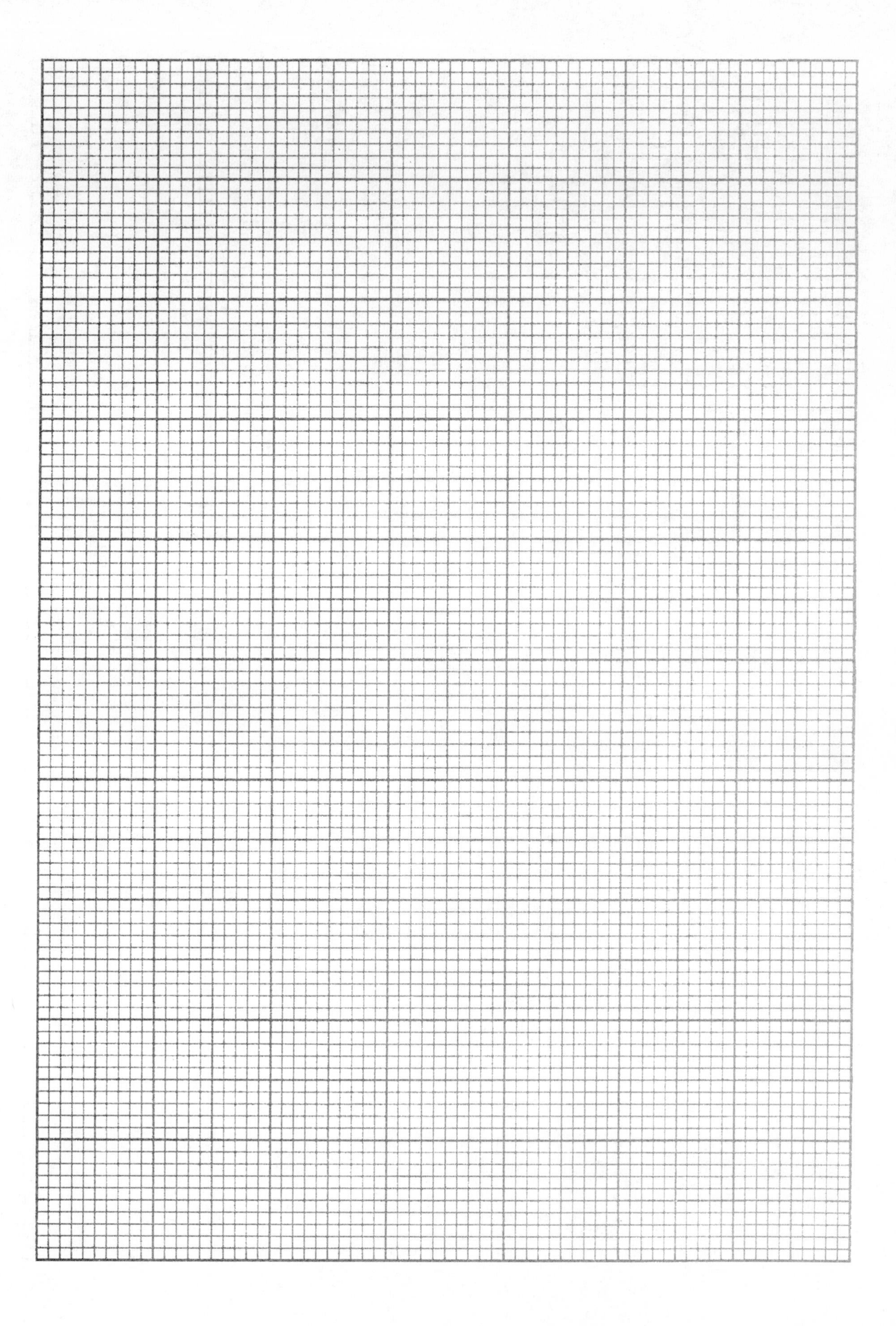

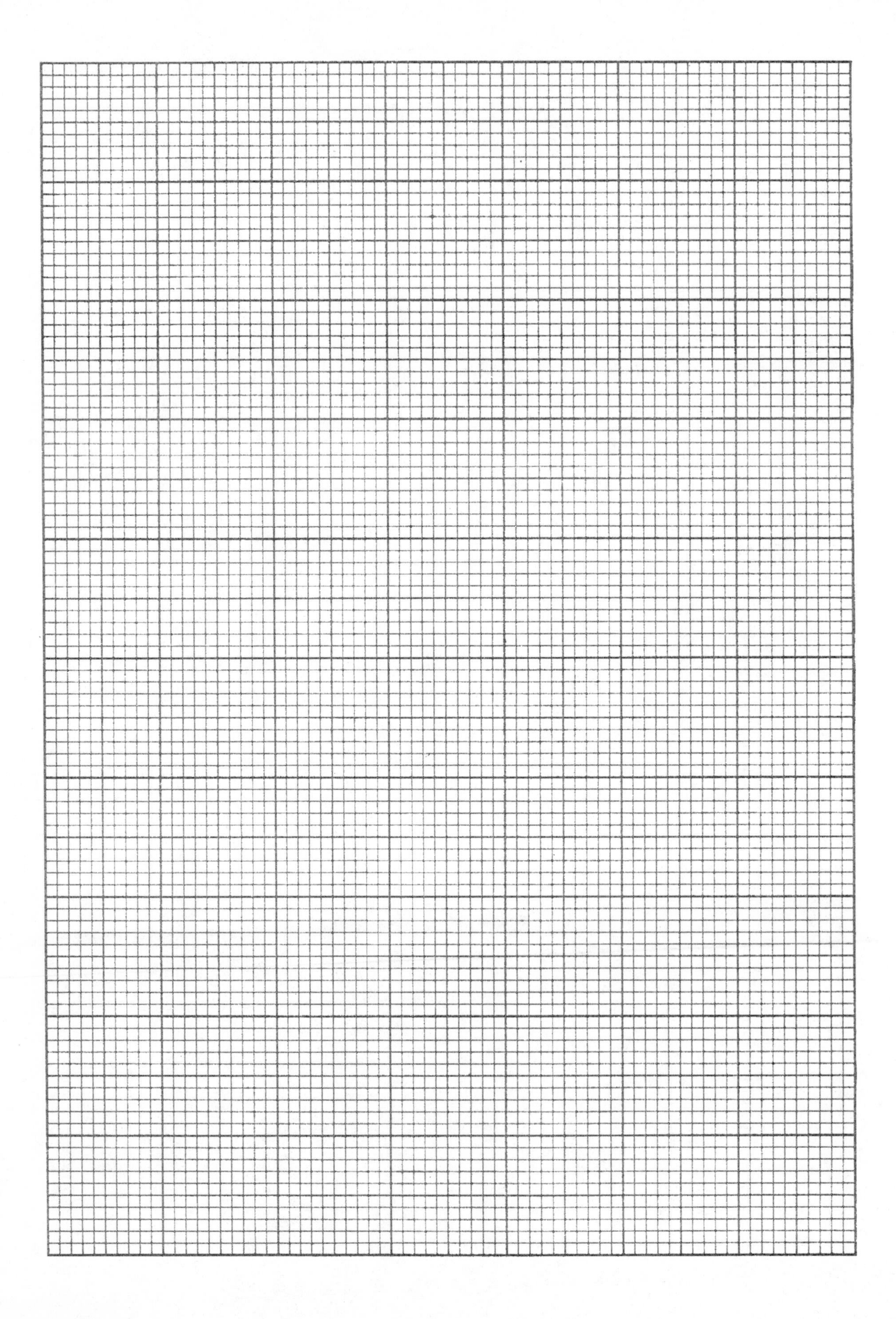

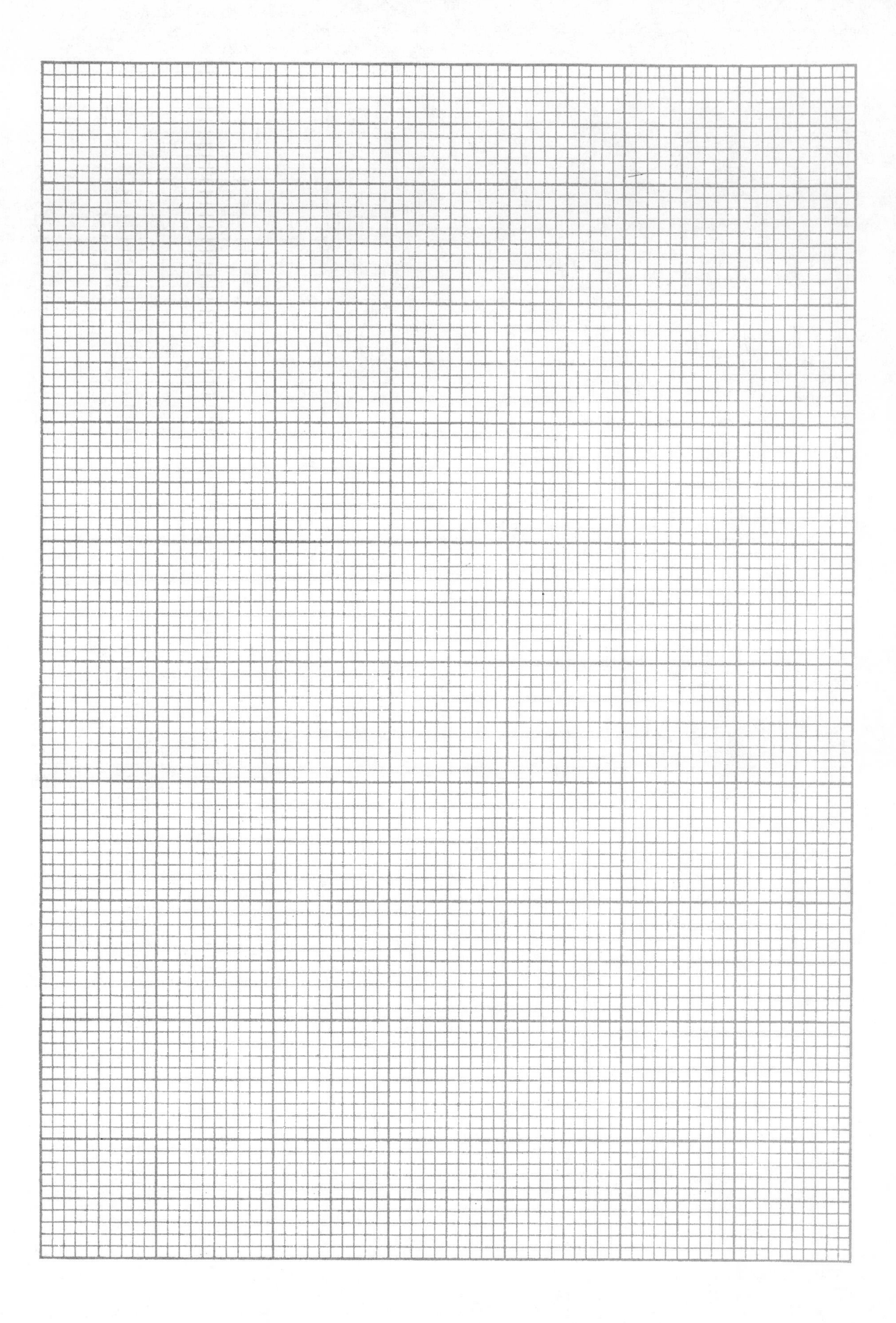

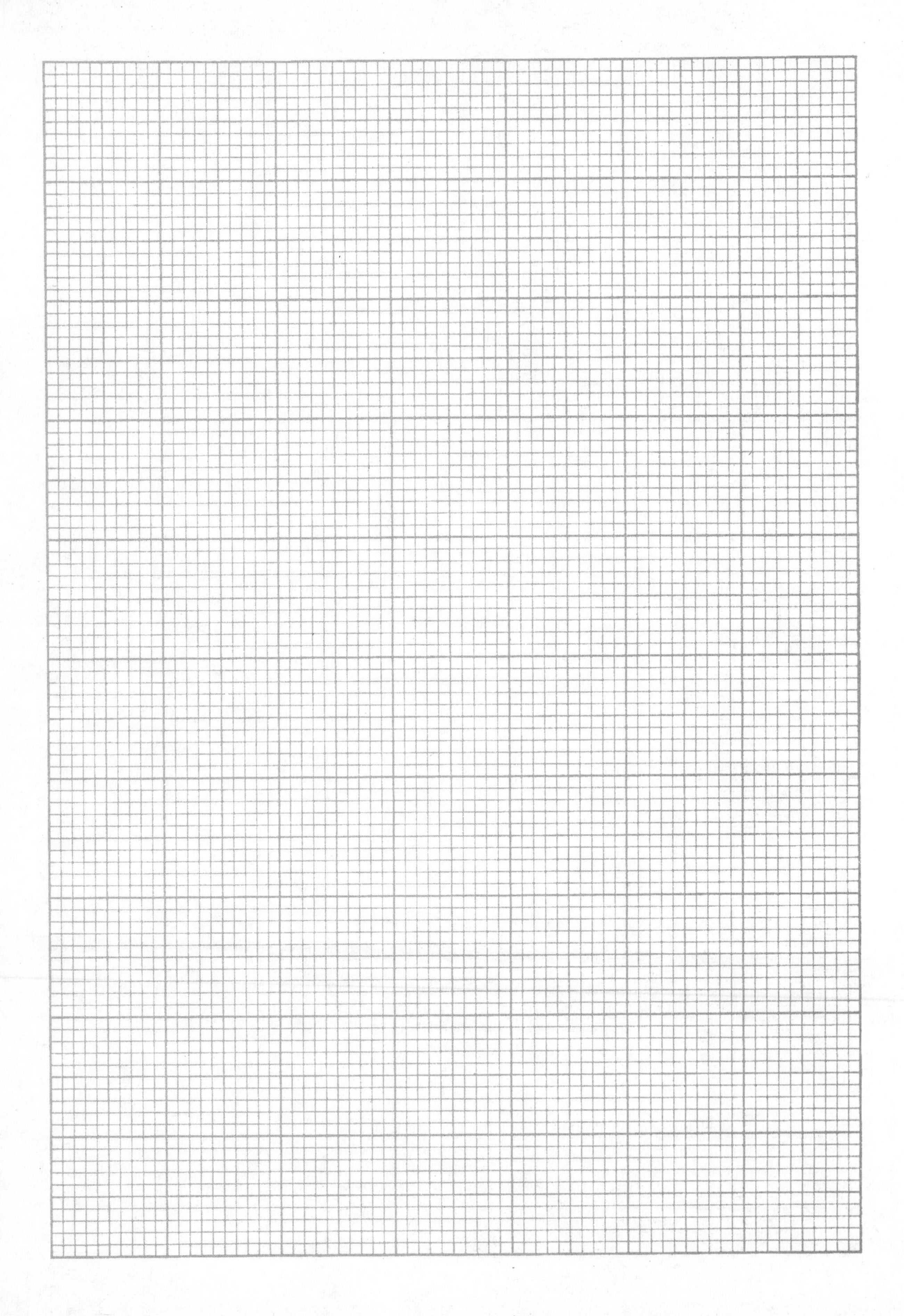